The Channel Tunnel

Conference organized jointly by the Institution of Civil Engineers in London and the Société des Ingénieurs et Scientifiques de France in Paris, 20-22 September 1989.

⅂Ⅰ Thomas Telford, London

Conference organized jointly by the Institution of Civil
Engineers and the Société des Ingénieurs et Scientifiques de
France.

Organizing Committees

The Executive Organizing Committee, as well as the Papers
and Visits Committees, comprise representatives of both the
ICE and ISF, together with Eurotunnel and TML in the UK and
France.

Y. Brooks, Conference Office, ICE, UK
E. P. Courtillot, International Relations, ISF
G. Crighton, Transmanche-Link, UK
M. Frybourg, Président de la Commission 'Transports', ISF
A. Lagier, Vice Président de la commission 'Génie Civil', ISF
A. J. Leggatt, Nachsen Crofts & Leggatt
J. B. Loubriat, Délégué Général, ISF
Y. Missoffe, Transmanche-Link, France
A. C. Paterson, Senior Partner, Bullen and Partners, and
President, ICE
R. Perzo, Président de la commission 'Génie Civil' ISF
J. H. Sargent, Costain Group Plc, and Member of Council, ICE
E. Schwarzcer, Directeur Délégué, Eurotunnel, France
J. Taberner, Eurotunnel, UK
G. Vidal, Eurotunnel, France
R. L. Wilson, Chairman, Travers Morgan Consulting Group,
and Vice President, ICE

ISBN: 0 7277 1546 1

First published 1989

Published for the Institution of Civil Engineers by Thomas Telford Ltd, Thomas
Telford House, 1 Heron Quay, London E14 9XF.
Printed and bound in Great Britain by Billing and Sons Ltd, Worcester.

Contents

Session 8. Transport infrastructure

1. Channel Tunnel — the early stages

Sir Nicholas HENDERSON, Director, Hambros Plc, Chairman,
Channel Tunnel Group 1985—86

SYNOPSIS.   The two Heads of Government announced their
enthusiastic support for a fixed link on 30th November
1984.   In response to an Invitation to Promoters of 2nd
April 1985, various projects were submitted to the two
Governments by 31st October, 1985.   The scheme proposed
jointly by the Channel Tunnel Group and France Manche was
chosen by the Governments and announced on 30th January,
1986.   The two Governments signed the Treaty of Canterbury
on 12th February, 1986.   A concession agreement between the
two Governments and the two companies was signed on 14th
March, 1986.

1. My purpose is to tell you something about how the
Channel Tunnel project came into being - about the aims and
difficulties of those who conceived the idea and how
eventually this particular scheme came to be accepted by the
French and British Governments. I shall be giving you this
background not for the sake of harping on history but
because without some understanding of its origins it is
impossible to comprehend the nature of the present tunnel
enterprise, its importance to both Britain and France - an
importance that's different for each country - and the
conflicting interests and feelings which it serves and
generates. The French have never had any national
inhibitions about it; but the British have had and still
have, as though haunted by the words Shakespeare put into
the mouth of John of Gaunt:

> This precious stone set in the silver sea,
> Which serves it in the office of a wall,
> Or as a moat defensive to a house,
> Against the envy of less happier lands.

2. This fear that a channel link is in conflict with our
deepest national instincts and traditions did and still does
arouse strong feelings in this country. But I believe that,
as with other new enterprises that have run up against

deep-seated objection, it will come to be seen here as part of our heritage, and as adding wealth and diversity to our way of life; and people will wonder why on earth it hadn't happened sooner.

3. What I am going to describe for you will not be the whole picture, but the highlights, as seen at any rate by one observer who was intensely involved at the moment of creation. While my account is therefore a subjective one, I am confident that nothing I say will be misleading - at any rate to an objective listener. Of one shortcoming I am particularly aware, that my account is written from a British viewpoint, and does not adequately cover what was happening on the French side. I can only say in attenuation of this that it simply arises from my ignorance of the detail of what was going on in Paris. I greatly look forward to reading one day an account that will link up with our's just as seamlessly as we expect the two ends of the tunnel to do.

4. For reasons of perspective - and modesty - we must all start by realising that the idea of a fixed link goes back a long time and that what is now being carried out owes much to former ideas and the work of earlier pioneers, particularly engineers. Of course modern technology has been crucial to the present project. But we have drawn a great deal on earlier work and due recognition must be paid to it. To give you one example. In meeting the deadline set by the French and British Governments for submitting proposals for a fixed link we in the Channel Tunnel Group had to put together in 1985 within a space of seven months detailed plans and drawings that filled no less than eleven volumes and required two thousand five hundred pages. We could not have done that had we not had at our disposal a great deal of thorough prior documentation. At Annex 1 I attach a summary of some of the events and personalities connected with a fixed link in the past.

5. By the mid-Fifties it was acknowledged by the British Government that there were no longer the strategic objections to a link that had frustrated the idea for so long on the British side. The last serious attempt was made in the year 1973/4. This scheme, which was for tunnels, was to be financed by way of government-guaranteed loans which were sanctioned by an Act of Parliament. The project foundered not out of military fear but because of the Labour Government's reluctance to spend public money when the Cabinet was badly split over its commitment to Europe and when serious environmental opposition to a new rail link from London to the coast became evident. In December 1974

the British Government announced that the Treaty with France
that had been signed in November 1973 would not be
ratified. However, as if stirred by this governmental
inaction, a rather surprising and stirring experience in
private enterprise then occurred. Immediately after the
abandonment, three British companies* of their own
initiative drove a short length of a service tunnel. This
demonstrated the sophistication of their equipment and the
ideal nature of the lower chalk material below the sea-bed
for high-speed mechnical drilling. This experience and the
tunnel itself came to be turned to good use over a decade
later.

6. Apart from this practical demonstration of its
feasibility the idea of a fixed link was kept alive by a
number of enthusiasts. Three reports were produced in the
early Eighties: one by a Select Committee of the House of
Commons; another by an Anglo-French official study group;
and a third by a group of French and British banks. This
banking group reported that at that stage the financing of
any scheme would be likely to require an element of
Government financial guarantee. Stimulated by these reports
the Channel Tunnel Group (CTG) was put together in 1984, an
amalgam of contractors and banks.** They were convinced
that the only scheme that was technically and financially
feasible in the foreseeable future was a tunnel. This would
take through trains and a shuttle that would carry at
frequent intervals vehicles of all kinds, taking half an
hour or so for the journey. But what was needed at this
stage was some glimmer of a green light from the government.

7. I thereupon led a delegation from the Channel Tunnel
Group to call on Mr. Nicholas Ridley, the Secretary of State
for Transport on the 10th July, 1984. I cannot say that
this meeting was a turning-point but it at least was a
start. It was the first serious discussion that members of
our Group had with Ministers and officials. Not that Mr.
Ridley was encouraging. The line he took with us was that
he was waiting for prospective promoters to come forward.
So far there were two projects in the field, our own and
that of Euroroute for a combined bridge and tube.

* Balfour Beatty Limited, Edmund Nattall Limited and Taylor
Woodrow Construction Limited.

** Balfour Beatty; Costain; Tarmac; Taylor Woodrow; George
Wimpey; National Westminster Bank; Midland Bank.
Merchant bank advisers: Morgan Grenfell and Robert Fleming.

Government prejudice was probably in favour of a
drive-through scheme; but it was up to promoters to persuade
them what was the most financiable and attractive scheme.
He made it absolutely clear that no financial support was to
be expected from the government.

8. None of us came away from that meeting with any sense
of euphoria - of having kicked a goal; but at any rate we
had been encouraged to think that the ball was at our feet.
From talks I held myself with officials I gained the
impression that Mr. Ridley was still neutral about the idea
of a fixed link of any kind. He was neither pro nor anti.
Nevertheless if a credible project emerged then the
government would not stand in its way.

9. I followed up the meeting with a letter on the 23rd
October, 1984 to Ridley explaining that since our meeting
there had been developments. The construction period had
been reduced from six to four and a half years and this
would greatly help the financing. Secondly the financial
experts were now satisfied that a link could be constructed
without government financial guarantee or support.

10. Perhaps partly as a result of this persistence on our
part a certain amount of Ministrial activity took place in
Whitehall that autumn; and in September/October British
Ministers took the decision in principle to back a scheme
provided it was technically and financially viable. The
British and French Ministers of Transport also discussed the
subject. They put their officials to work to draw up
guide-lines to be followed by any would-be promoters.

11. We come now to what I think is a decisive
turning-point in the history of a fixed link. This was the
announcement in support of a fixed link by President
Mitterrand and Mrs. Thatcher when they met on the 30th
November, 1984. Mrs. Thatcher did not simply make a
matter-of-fact statement of support, she enthused saying
that: "It really would be something very exciting .... A
project which can show visibly how the technology of this
age has moved to link the continent and Britain closer
together". The French President was equally enthusiastic;
indeed it was his proposal that the two Leaders should make
the announcement at their joint meeting to stamp the idea
with their authority. However what was surprising was the
strong momentum given to the idea by the British Prime
Minister given the age-long British political hesitancy and
vacillation.

12. I think that the Prime Minister's decision had two main impulses: the desire to encourage some major industrial enterprise such as Britain had not carried out since World War II and one which would provide a good deal of employment; and to make some positive move towards Western Europe with whom we had been wrangling for so long over the budget. Even months later those close to the Prime Minister continued to be surprised by her decision to commit herself so whole-heartedly to the project.

13. Once the two Governments had decided in principle to back a fixed link (provided a project was put forward that was technically and financially viable, without dependence upon governmental resources) they laid down for themselves and for prospective promoters a tight timetable. In order to realise why the project came to fruition when it did it is essential to understand this. The fast pace was set partly to avoid the shoals of elections that were due to be held before long in both countries. But also because the Governments sensed that there was a certain favourable mood in the air and that the groups of possible promoters must be given some definite target if they were to go on spending time and money on working out projects. If the opportunity that seemed to be available between 1984-86 was not seized upon it looked as though the chances of proceeding with a link would be unlikely to recur until the next century. The two deadlines that came to be formulated were firstly the commitment of the Governments to issue guidelines for prospective promoters by the Spring of 1985; and then, as stipulated in the guidelines, the date of the 31st October by which competitors had to submit their schemes for adjudication by the Governments.

14. However the main problem for the Channel Tunnel Group in Britain at the beginning of 1984 and for the first months of 1985 was to find a French partner. There were plenty of bilateral contacts between contractors and banks, and there were many on the French side who agreed with the general ideas of the CTG for a link, but there was nothing there comparable to the British organisation with which to form a joint consortium. Yet that was essential if a scheme was to be acceptable to the French, as to the British Government.

15. I myself became Chairman of the Channel Tunnel Group in early 1985 and I made several visits to Paris in search of a suitable partner. I ran into doubts in French banking circles about the chances of raising money privately in the absence of some public guarantee against the danger of cost over-runs. Many people in Paris were also convinced that Mrs. Thatcher was in favour of a drive-through rather than a

rail link, a preference that was also widely voiced on the
British side. Mrs. Thatcher herself had spoken publicly of
the dream of being able to "get in a car at Dover and drive
all the way through to Calais".

16. Such was the groundswell both in France and the UK in
favour of a drive-through that the Board of CTG agreed in
March 1985 that we had to undertake publicly to re-examine
the feasibility of a drive-through. The scheme that we were
proposing to submit was proclaimed in the middle of March as
being one of twin tunnels, each of 7 metres in diameter, and
a smaller service tunnel. This system would provide a
privately operated, roll-on roll-off shuttle to run between
the British and French coasts, with trains leaving every 5
minutes, giving a capacity for 3,600 vehicles per hour in
each direction. There would also be through rail traffic to
and from all parts of the UK to the Continent. Passenger
services between London and Paris/Brussels would be expected
to take about 4 hours (or 3 hours if a high-speed service
were introduced in France). But we also indicated, as I
have said, that we would consider a road system as well as a
rail one.

17. The guidelines that the two Governments issued on the
2nd April 1985 were entitled "An Invitation to Promoters".
Introducing the guidelines to the House of Commons Mr.
Nicholas Ridley, the Secretary of State for Transport,
emphasised the government's caution: "I cannot say whether
a fixed-link will be built or not"; but, indicating grounds
for hope, he added ".... the private sector now has a unique
opportunity".

18. Apart from defining the requirements with which any
promoter would have to comply, the guidelines contained a
political guarantee that the Governments would not terminate
the link during the concession period.

19. Between the 2nd April and the 31st October, the
dead-line for applications, there was intense activity by
the Channel Tunnel Group (and France Manche, once
co-operation had been established with a French partner)
which is summarised in Annex 2.

20. Although it is not for me to comment on the work of
the officials in Whitehall who were involved with the fixed
link, I cannot refrain from saying how conscious I was of
the heavy burden it imposed and the impressive way in which
the team carried out their tasks without showing particular
favour to any one of the promoters. So far as I could judge
there was also regular contact with the French team; and

this also occurred at Ministerial level. All of it was conducted according to the best traditions of the Entente Cordiale.

21. As I have already mentioned the proposal submitted by CTG/France Manche was voluminous. Every word had to be equally valid in French and English. In the last three weeks ten thousands amendments had to be made. The text ran through fifteen versions. Such was the size of our submission that it required a special van for delivery of the requisite number of copies to the British Government. Yet only a few months had elapsed since the joint team had been put togther and had started working on similiar concepts and configurations.

22. In additional to the eleven volumes of the submission we produced a summary, beautifully illustrated and printed. The preface was designed to give life to the otherwise somewhat terre-à-terre tone of our document. It ran as follows:

> If the peoples of Western Europe are to forge together their separate economies and exploit their varying skills so as to achieve their maximum capabilities by the end of this century, they will have to create fast new means of communication between densely populated regions. Many of these regions are separted by natural barriers - the Alps, the Pyrenees or the Channel. Great improvements have been made in traditional means of transport - motorways and high-speed trains - but new ideas are required to overcome the remaining geographical barriers and increasing transport demands of Western Europe. This is particularly true of the Channel, where traffic is expected to double between now and the end of the century.

23. Having made our submission to the two Governments we were far from being able to take a back seat and await their judgement which they undertook to give by the 20th January 1986. The next two and a half months were to prove as busy as the pre-submission phase - and more intense. We became involved in a dialogue with the Governments which had established assessment teams to examine the different proposals; and there was considerable manoeuvring between the organisations left in the field. On a quite different plane we also had to tune our public relations to reach a climax at the crucial moment of Ministeral decision.

24. We were told that promoters who hoped to gain the mandate would have to provide the Governments with the

details of a hard and fast agreement about rail traffic and rail tariffs. Without information on this it would be impossible for the Governments to assess the financial viability of any scheme. It was with great difficulty that we managed to patch up an understanding on this subject.

25. Some people in government were afraid that our project would lead to trade union disruption and domination. It was suggested that we were going to be dependent upon the co-operation of the British Railway Unions. It was explained in reply that this was not our intention. The shuttle would be run by our organisation. We were going to recruit and employ our own labour force.

26. Meanwhile our competitors were extremely active. The last minute entry into the lists of Mr. James Sherwood's project attracted much interest. The Euroroute scheme also continued to enjoy a great deal of attention and favourable comment. It was evident that the Governments wanted to have a choice, to be sure that various plausible alternatives were put to them and considered seriously. Parliamentary discussion took place in London towards the end of the year. There was a close shave in the Transport Committee of the House of Commons when it needed the casting vote of the Chairman to produce a verdict in favour of our scheme. By contrast our French partners appeared to have to pay little attention to enquiries and pressure from the Chamber of Deputies. This reflected not merely the lesser political interest aroused by the subject in France but the difference in our systems of government.

27. Rabies continued high on the public's indignation hit list. A correspondent wrote in the Yorkshire Evening Press as follows:

> By agreeing to a Channel Tunnel Mrs. Thatcher is leaving this country wide open to a real threat of rabies when the French rats have a free run to our land.

28. CTG and France Manche had a meeting in December with the French and British assessment teams. The purpose was not to cross-examine us on our proposal but to clarify certain things that might arise if we were awarded the mandate, for example the length of the concession period. Similiar discussions about what a concession might contain were also held with other competitors.

29. The French and British Ministers of Transport met on the 7th January 1986. The British Minister referred to

certain important advantages in Mr. Sherwood's Expressway scheme particularly the idea of a drive-through. The French Minister spoke of the advantages of Euroroute which had for long been the favourite of the French Government and had powerful French backers.

30. The French Chairman and I discussed tactics when we heard about this meeting. It seemed to be necessary, as the final hand was being played, to devise some way of ruffing the drive-through ace held by our opponents. It was agreed that I should inform the British Government at the official level that the Channel Tunnel Group and France Manche were prepared to give an undertaking now that, should the traffic grow and the technology permit, we would be ready to consider a drive-through at some stage. This commitment, which, I believe, contributed considerably to the acceptability of our project was conveyed in writing a few days later. The two Governments attached such importance to it that we were asked to incorporate it in a declared decision which we ultimately conveyed after prolonged negotiation less than twenty-four hours before the Heads of Government were due to meet to announce their decision.

31. We were subject to considerable governmental pressure to join up either with Euroroute or Expressway. But neither the British nor French leaders of our project believed that it would be either possible or desirable to modify our scheme to fit in with one of the others. We were confident in the viability of our ideas and in the comprehensiveness of our organisation. We saw no need to bring in additional entrepreneurs. We were anxious to avoid going along with anything in the way of a compromise that might delay the implementation of our project, the timetable of which provided for the tunnels to start operating in the Spring 1993.

32. The British Cabinet came to a decision on the mandate on the 16th January, 1986. Mr. Ridley flew to Paris the same day to see the French Minister, Monsieur Jean Auroux. The two Governments reached agreement that day on which project should be accorded the mandate. However, but nothing was divulged to us or to the public; and, indeed, negotiations with us continued over the drive-through declaration and on the railway agreement.

33. It was only when President Mitterrand and Mrs. Thatcher met at Lille on the morning of Monday, 20th January that the decision was announced in favour of our project.

34. "Today's decision", Mrs. Thatcher said after the issue of the joint communique at Lille, "is a dramatic step in

Anglo French co-operation. The project we have agreed upon
will have immense significance for trade and communications
between our two countries. It is also important for the
enormous opportunity we are giving to the private sector to
demonstrate their abilities and their enterprise in a
project of the utmost public importance".

35. Mr. Ridley wrote to me the same day to let me "know
officially that the Prime Minister and the French President
announced this morning in Lille that the Governments of the
two countries have agreed to take together the necessary
steps to facilitate the development, construction and
operation of a fixed link across the Channel by the Channel
Tunnel Group". He added: "This is a historic decision.
There is much to be done to carry it forward to a successful
outcome, but I am confident that with effort and
co-operation we can succeed".

36. Shortly afterwards, on the 6th February, 1986 the
British Government published a White Paper (Command 9735)
which gave an account of the way in which the Invitation to
Promoters and the decision had come about, the nature of the
different projects and the next steps to be taken. It also
contained details of the Assessment done by the French and
British teams. With great clarity the White Paper also gave
the reasons why the Governments had chosen our project.
These were as follows:

1. it was the soundest financially;
2. it carried the fewest technical risks;
3. it was safest from the traveller's point of view;
4. it presented no maritime problem;
5. it was the least vulnerable to sabotage and
    terrorist action;
6. it had an environmental impact that could be
    contained and limited.

37. The White Paper described the fixed link as the
largest engineering project for many years and the largest
in Europe ever undertaken by the private sector.

38. The White Paper envisaged the setting up of formal
machinery for consultation between the Government and the
local authorities. The Joint Consultative Committee which
was to be chaired by a Minister would be attended by the
Kent County Council, Kent District Councils, British Rail
and Eurotunnel (the organisation into which the Channel
Tunnel Group and France Manche merged), Trans-Manche Link
and British Rail. The Committee starting meeting
immediately after the award of the mandate and has continued
meeting every since.

39. Reflecting on what had most influenced the Governments in reaching their decision I was sure that financial robustness had been a sine as quâ non. Our project certainly showed greater strength in this respect than did the others.

40. Our initial weakness was that our submission made no reference to a drive-through. We were fortunate in the timing of our subsequent commitment on this. Had we given an undertaking earlier that we would promise to build a drive-through, our competitors might well have over-trumped us. As it was, we came forward with the idea at the key moment when Ministers were on the point of reaching a decision. Without this undertaking we might well not have secured the mandate; and we would probably not have won exclusive rights to last until the year 2020, as embodied in the subsequent concession.

50. Our scheme was one that both Governments recognised as corresponding to the needs and hopes, political and economic, of Britain, France and Western Europe generally.

51. The next stage in the story was the signing at Canterbury on the 12th February, 1986 of the Treaty of Canterbury which set out the intentions of and the obligations between the United Kingdom and France. The two Governments had been negotiating this Treaty while the assessment of the rival schemes was going on. The main provisions of the Treaty dealt with the grant of a concession, the political guarantee and compensation between Governments, jurisdiction and legal matters, finance, inter-governmental machinery, and disputes.

52. Before the Treaty could be submitted for legislative authority the two Governments had to conclude a concession agreement with the two companies. This was necessary to establish rights and obligations for the Concessionaires to carry out the development, financing, construction and operation of the tunnel system. The concession agreement, details of which see Annex 3, was signed on the 14th March, 1986. After signing the concession I resigned from the Chairmanship of the Channel Tunnel Group.

53. The Treaty could not be ratified by the British Government until the legislative powers necessary to give its effect had been taken. The Channel Tunnel Bill was introduced into the House of Commons on the 17th April, 1986. It was a Hybrid Bill and had to go before Select Committees in both Houses of Parliament. These stages were in addition to the normal stages of a Public Bill. Individuals and groups affected by the Bill's proposals had

the right to petition the Select Committees. There were
nearly five thousand petitions to the House of Commons
Select Committee and one thousand six hundred in the Lords.
The Commons Select Committee sat for thirty-five sessions,
some of which were held in Folkestone and Dover. In all,
the Bill was before Parliament on more than eighty separate
occasions. The Royal Assent was given to the Channel Tunnel
Bill on 23rd July, 1987.

54. The Parliamentary process in France was much simpler.
The procedure to approve the Treaty and the concession took
barely three weeks. Only two Parliamentary sessions were
involved and approval was unanimous in both the Chamber of
Deputies and the Senate. The Treaty was ratified in Paris
on 29th July 1987.

55. So ended the first chapter of this strange eventful
history. Now that the project is well launched, very much
as proposed in the Joint Submission of the 31st October,
1985, the following features may be worth highlighting in
retrospect.

i)   The resilience of a handful of people in the
        construction and banking industries was responsible
        for keeping the idea alive even when the outlook was
        overcast; and their determination meant that an
        up-to-date project could be put together quickly when
        the climate changed and the two Governments issued an
        Invitation to Promoters.

ii)   It is impossible to exaggerate the importance of the
        declaration by the French President and the British
        Prime Minister on the 30th November, 1984. They not
        only announced their support for a fixed link but they
        both showed an enthusiasm for the idea.

iii)   For various reasons a decision had to be taken within
        a narrow time-scale; but although this involved
        promoters and Governments in intense activity the
        tightness of the timetable probably helped to focus
        people's minds and lead to a clear-cut conclusion.

iv)   The attitudes of public opinion were very different in
        the UK and France. There was practically no
        opposition to the idea in France, merely considerable
        scepticism whether the British who had hesitated so
        often in the past at the decisive moment, were really
        prepared to take the jump now. In Britain there was,
        and still is, both public apathy and hostility to the
        concept of a fixed link. There are also grave

environmental and traffic problems. Initially the
promoters had to cope with this problem of public
opinion as well as competing amongst themselves to
secure the mandate.

v)   It was not a foregone conclusion that the Governments
        would accept any project. They might well have said
        that no promoter had put forward a scheme that
        satisfied the necessary financial and technological
        criteria. As it was our project met these needs and
        we also gave an undertaking about our readiness to put
        forward plans and if necessary and possible construct
        a drive-through in the future.

vi)   The significance of the project in the annals of Anglo
        French industrial collaboration must not be
        overlooked. In putting together the proposal and
        securing the mandate the British and French teams
        worked together constructively and rapidly, the common
        cause benefiting from French flair and British phlegm.

vii)   There were many moments of crisis in the joint
        creation of the scheme, but given the complexity of
        the undertaking it is only reasonable to suppose that
        crisis is the climate in which it will live until the
        first train and shuttle goes through the tunnel.

ANNEX 1

Some of the events and personalities
connected with a fixed link in the past

    1. At some stage a roll call should be drawn up of all
those who have contributed over the years to the idea of a
fixed link that eventually led to the mandate which was
granted at the beginning of 1986 for the implementation of
the present project. But I like to think that when the
tunnel comes to be inaugurated a number of names from the
past should appear above the portals - the names of some of
those who over a period of a century and a half have
believed fervently in and promoted the idea of a link,
whether a tunnel, a tube or a bridge. Thomé de Gamond, for
instance, who spent much of the last century on tunnelling
schemes. He carried out the first systematic geological and
hydrological surveys of the channel in the course of one of
which he plunged naked into thirty-three metres of water
weighed down with stones. After his first dive he
complained that he had been attacked by malevolent fish -
Dover soles perhaps! I tell you these details so that you

13

should realise what a relatively easy and cushy task is that of the present leaders, Bénard, Morton, Ridley or Bertrand.

2.   Then there was Hector Horeau who had the idea of a submerged tube connected at intervals on the surface to a series of gothic pagodas.   On the British side we should not forget Sir John Hawkshaw, a great engineer of the Victorian age who built the Severn Tunnel and who promoted the idea of a single channel tunnel which was adopted by the English Channel Company in 1872.   This led to a tunnel seven foot in diameter being drilled two thousand and twenty-six yards until it was stopped in 1882.   The work was carried out by the world's first compressed air rotary-head tunnel boring machine, the invention of another great Victorian pioneer, Colonel Beaumont.   The tunnel, although never lined, remained in excellent condition until it was flooded at the time of 1973/4 scheme.

3.   The British were responsible for cancelling that eighteen-eighties project which they did on military grounds.   The French not only have had no inhibitions about the idea, as I have already mentioned; but they have been consistently enthusiastic about it.   They have frequently shown themselves to be innovative over transport, e.g. the Suez and Panama Canals.   Not so, I fear the British, at any rate since the pioneering days of sail, railways and steamships.   For the next three-quarters of a century a combination of military fear and the deep-seated instinct of an island people coupled with what I describe as the blue-water-school of British foreign and commercial policy according to which this country's interests were seen as lying mainly overseas, - these have been among the features that have stood in the way of a British decision to build a link.   Many and prominent have been the military leaders from Sir Garnet Wolseley to Field Marshal Montgomery who have fulminated against the dangers of invasion from a tunnel.   Some British leaders at certain moments have been keen on the idea.   Gladstone was a fan.   Queen Victoria gave her blessing initially to a tunnel, partly perhaps because Prince Albert was in favour but also because she saw it as a way of avoiding the sea-sickness from which she had frequently suffered.   "You can tell the French engineer (de Gamond)", she said, "that if he succeeds I will give him my blessing in my personal name and in the name of all the ladies of England".

4.   However she eventually turned against the scheme and over the next hundred years or so there were few people prominent in public life who advocated it consistently and with fervour.   One other name which should appear on any memorial of those who have ignited or kept alive the flame of the tunnel through the ages is that of Leo D'Erlanger a

man who appropriately embodied in himself some of the best virtues of the French and British.

5. A Royal Commission was appointed in 1929 but although it reported in favour of a link which it believed could be financed privately, the idea became submerged following the Wall Street Crash; and the political tension of the Thirties was certainly not conducive to initiating an enterprise of this kind.

ANNEX 2

The following is a check-list of some of the activities of the Channel Tunnel Group and France Manche between the time of the Invitation to Promoters and the submission of our project to the two Governments

1. The search for, and the conclusion of an agreement with, a French partner. I paid several visits to France and saw Ministers, officials, bankers and constructors. My most important meetings were with the President's office and with M. Francis Bougyues upon whose attitude at this stage a great deal depended. Eventually a French team, France Manche, which had similar ideas to the CTG, signed an agreement with CTG on the 1st July. This stated quite simply that "France Manche represented by Monsieur Jean-Paul Parayre, and CTG, represented by Sir Nicholas Henderson, agree to collaborate on an exclusive basis for the purpose of preparing and submitting a proposal and to secure concessions from the French and British Governments ..."

2. Development of the organisation of the CTG in order to produce a project by the 31st October. The Channel Tunnel Group was incorporated on the 26th April. The names of members of the CTG Board and leading executives involved in preparing the submission are set out below.

3. Consultations with Ministers and officials to ensure that the project was compatable with official views, and indeed as attractive to them as possible, particularly in the face of the strong competition. This came initially from Euroroute, a scheme for combination of bridges and tube led by Sir Nigel Broackes which was certainly regarded by both the British and French Governments as a "serious runner". There was also competition from a bridge project led by Lord Layton. At the last minute a proposal for a

Members of the CTG Board and Leading Executives
involved in preparing the Submission of 31st October, 1985

Members	Representative on CGT Board
Balfour Beatty	D. A. Holland
Costain	J. Reeve, succeeded by T. Wyatt
Tarmac	A. Osborne
Taylor Woodrow	F. R. Gibb
George Wimpey	A. Mc Dowall, succeeded by C. J. Chetwood
National Westminster Bank	D. M. Child
Midland Bank	I. Paterson

Merchant Bank Advisers: Morgan Grenfell (represented by J. Franklin) and Robert Fleming (represented by C. Moore)

Managing Director:	Michael Gordon
Financial Director:	Quentin Morris
Other Executives:	Ian Callaghan
	Melville Guest
	Tony Gueterbock
	Martin Hemingway
	Don Hunt
	Bill Shakespeare
	Colin Stannard
	John Taberner

---

drive-through, Expressway, was launched with much brio by Mr. James Sherwood.

4. Ensuring the necessary parliamentary and public understanding and acceptance of the project. This involved not only promoting one's own scheme but trying to counter the apathy or hostility of much opinion towards the idea of any fixed link at all. The favourable prognosis for employment and industry generally was widely canvassed. A good deal of attention was also paid to environmental problems. No less than eighteen different studies were undertaken of the possible environmental impact of the tunnel.

5. Working out the nature and details of our project including, though not necessarily in order of priority, finance, the number and size of the tunnels, methods of loading and unloading, the terminals, rabies control, environmental protection, the formation of the shuttle, relations with the railways, traffic forecasts, the future shape and composition of the joint organisation and the terms of the contract to be agreed between the owners and the constructors. Given the bi-national nature of our project and its many unprecedented features, the ultimate shape and composition of our organisation, including the relationship between owners and those responsible for the construction of the tunnel, the terminals and the shuttle, proved from the outset to be amongst the most difficult issues to be resolved.

ANNEX 3

The Concession Agreement

1. The concession obliged the Concessionaires to construct the fixed link and permitted them to exploit it for the duration of the concession which was fixed for a period of fifty-five years, covering both construction and operation, the Promoters having asked for sixty years. The period began on the 29th July, 1987. The agreement covered matters such as the monitoring of the project by the Inter-Governmental Commission on behalf of the Governments, the freedom of the Concessionaires to operate as a commercial body; frontier, safety and security matters; and exclusivity and the second link.

2. Various matters dealt with by the concession agreement

were, in fact, settled with the Concessionaires before the award of the mandate. One essential matter was that of the guarantee against political interruption or cancellation.

3. The Concessionaires undertook to submit to the Governments by the year 2000 a proposal for a drive-through to be undertaken only when technically feasible and financially viable and provided it would not undermine the return on the original link. They also undertook to expand capacity to meet demand throughout the concession period including, if necessary and if commercially and technologically permissible, the construction of a drive-through.

4. Agreement between the Governments and the Concessionaires also provided for the concept of exclusivity so as not to undermine the financial prospects of the initial project. In return for undertaking to provide an adequate service through the link, the Concessionaires received an undertaking that the Governments would not facilitate any other organisation before the year 2020; nor would they facilitate any second link to be built with public financial support or guarantee throughout the whole concession period.

RESUME

Origines de projet actuel

1. Les projets de construction d'une liaison fixe entre la France et le Royaume-Uni remontent au début de dernier siècle. Les français ont toujours encouragé cette idée; les britanniques se sont montrés réticents, par souci de sécurité ou par crainte, sentiment encore très fréquemment rencontré, qu'une telle liaison ne nuise aux instincts et traditions nationales solidement enracinés dans la population.

2. La technologie moderne s'est révélée cruciale pour le présent projet. Néanmoins les promoteurs mirent largement à profit les travaus précédemment effectués tels que les plans convenus par les gouvernements français et britannique en 1973 qui furent abandonnés par le gouvernement britannique en décembre 1974. La suggestion de l'époque de créer un double tunnel n'en demeura pas moins vivace auprès des ingénieurs et banquiers et leur vue d'avenir fut récompensée quand en novembre 1984 Monsieur Françoise Mitterand, Président de la République française et Madame Margaret Thatcher, Premier Minister britannique, déclarèrent avec enthousiasme qu'ils appuyaient la création d'une liaison fixe - à condition que cela n'entraîne le débours d'aucun fonds gouvernnemental.

3. Au printemps suivant, les deux gouvernements invitèrent les promoteurs à présenter leurs programmes avant le 31 octobre 1985. Plusieurs projects furent donc soumis, et donnèrent lieu a une activité intense de la part des gouvernements et promoteurs hypothétiques. Ceci aboutit à la décision annoncee le 20 janvier 1986 en faveur de l'actuel projet de tunnel qui avait été proposé par un consortium commun du Channel Tunnel Group et de France Manche.

4. Cc programme proposait deux tunnels, chacun de sept mètres de diamètre, et un tunnel de service plus petit. Un service de navettes assuré par une entreprise privée devait assurer la liaison permanente entre les côtes française et britannique, avec des trains partant toutes les cinq minutes environ, donnant une capacité de transport de 3600 véhicules par heure dans chaque sens. Un service ferroviaire devait également utiliser le tunnel pour relier tous les centres britanniques avec le Continent. Les services de passagers entre Londres et Paris/Bruxelles étaient estimés exiger environ quatre heures (ou trois si un service grande-vitesse était implanté en France). Il y était également indiqué qu'en temps voulu les promoteurs envisageraient un système routier en addition au service ferroviaire. Ce dernier point était un caractéristique essentielle de la proposition que remporta le marché; et en fait sans elle il est possible que les deux gouvernements ne se seraient pas décidés en faveur de ce projet. Un engagement fut donné concernant la

soumission de propositions pour un tunnel à l'usage des
voitures. En plus de tunnel ferroviaire, dans la mesure où
la technologie le permettrait et le trafic le justifierait.
Cet engagement fut ensuite porté sur l'accord de concession,
les concessionnaires ayant convenue de soumettre aux
Gouvernements avent l'an 2000 une proposition pour un tunnel
à l'usage des voitures "ne devant etre entrepris que lorsque
cela serait technologiquement faisable et financièrement
viable et à condition que cela ne nuise pas a la rentabilité
de la liaison d'origine".

5. Le 12 février 1986, les gouvernements français et
britannique signèrent le Traité de Canterburuy qui
établissait les intentions et obligations des deux
gouvernements. Le 14 mars 1986, ils conclurent, avec deux
autres sociétiés, un accord de concession qui établissait les
droits et responsabilités pour le développement, la
financement, la construction et l'opération du tunnel.

6. La ratification de Traité de Cantebury eut lieu au
Royaume-Uni le 23 juillet 1987 après d'intenses et longues
discussions publiques et parlementaires. Près de cinq mille
pétitions furent présentées à la Chambre de Communes. Le
projet de loi pour le Tunnel sous la Manche fut présenté au
Parlement à plus de quatre-vingt reprises. En France, la
procédé fut bien plus simple. Le Traité fut ratifié le 29
juillet 1987.

7. Quand au printemps 86 le projet fut enfin lancé, il n'y
avait aucun doute dans l'esprit de ceux qui y avaient pris
part qu'un événement d'une portée sans pareille était entré
dans les annales de la collaboration industrielle franco-
anglaise et que cette liaison fixe allait être d'une profonde
importance pour les affaires et le commerce entre les deux
pays.

2. Développement du projet: l'attitude et l'approche des gouvernements; le rôle de la Commission intergouvernementale

M. LEGRAND, Ingénieur Général des Ponts et Chaussées; Chef de la Délégation française à la Commission intergouvernementale

Si après deux siècles de réflexion et quelques tentatives avortées il nous est aujourd'hui possible d'envisager la mise en service de la LIAISON FIXE à travers la Manche en 1993 c'est tout à la fois parce que les gouvernements français et britannique ont fait preuve en la matière d'une volonté inébranlable et sans faille d'aboutir mais aussi parce qu'ils ont convenu d'une nouvelle approche pour la réalisation de ce projet.

LES LECONS DU PASSE

L'échec de la précédente tentative en 1975 ne les avait pas amené à renoncer à cet ouvrage mais leur en avait fait mieux mesurer les difficultés. Aussi après avoir couvert la moitié des dépenses engagées, ils s'engagèrent, chacun en ce qui le concernait, à prendre les mesures nécessaires à la sauvegarde des travaux déjà réalisés en vue de leur reprise ultérieure éventuelle.

Un temps mort s'imposait à l'évidence avant une telle reprise. Il fut mis à profit pour réaliser de part et d'autre de la Manche une série d'études en vue de tirer toutes les leçons de cette dernière tentative et de son échec.

Dès septembre 1981 il fut décidé au cours d'un sommet franco britannique tenu à Londres de confier à un groupe d'experts des deux pays le soin d'étudier à nouveau les aspects techniques et économiques de la construction éventuelle d'une LIAISON FIXE TRANSMANCHE. Ce groupe devait remettre ses conclusions en avril 1982 et confirmer la faisabilité technique et l'intérêt économique d'une telle liaison.

Dans le même temps la communauté européenne marquait à plusieurs reprises l'intérêt qu'elle portait à un tel projet.

Un groupe franco britannique de banques était constitué en août 1982 et sur la base des projets évoqués dans le rapport technique publié réalisait une étude approfondie des modes de financement envisageables. Cette étude complétée et financée pour partie par la communauté européenne fut communiquée aux gouvernements et publiée en mai 1984.

Elle revêt une importance toute particulière car elle permit aux gouvernements d'avoir confirmation de la part

d'experts indépendants de la rentabilité économique et
financière du projet quand bien même sa réalisation
impliquerait des dérapages importants par rapport aux
prévisions tant en ce qui concerne les coûts que les délais.

Il est vrai que de ce point de vue la situation s'était
considérablement améliorée depuis 1970 et que les recettes
attendues avaient profité d'un accroissement important et
continu des trafics sur le détroit de Calais. Ceux-ci avaient
en effet connu une croissance moyenne annuelle de 8,8 % entre
1971 et 1982 pour les passagers et de 12 % entre 1973 et 1980
pour le fret.

UNE NOUVELLE APPROCHE

Ce qui apparaissait en 1970 difficilement envisageable
semblait devenir possible en 1984 : Nonobstant les
conclusions du rapport des banques qui recommandaient une
concession de 50 ans et une garantie des emprunts par les
Etats la participation directe ou indirecte de ceux-ci
n'apparaissait plus indispensable.

Leur tâche s'en trouvait largement simplifiée et en
particulier vis à vis de leurs Parlements respectifs qui
sollicités de plus en plus par leurs obligations sociales
pouvaient et pourront de moins en moins faire face au
financement de gros investissements.

Dès lors la voie était tracée et le sommet franco-
britannique de RAMBOUILLET réaffirma le 30 novembre 1984 la
volonté commune de réaliser dans les meilleurs délais une
liaison fixe à travers la Manche mais en faisant cette fois
appel à l'initiative privée sans participation ni garantie
financière des Etats. Depuis cette date et sur ce point leur
position commune et ferme n'a pas varié. Et pourtant, si une
telle décision apparaissait tout à la fois comme sage et
simple dans son principe, sa mise en oeuvre exigeait quant à
elle un très gros effort de la part des gouvernements.

L'APPEL D'OFFRES INTERNATIONAL ET SES CONSEQUENCES

En effet et dès lors qu'un appel d'offres international
était envisagé pour la conception, le financement, la
réalisation et l'exploitation d'une liaison fixe à travers la
Manche, il importait d'en fixer les conditions tant au plan
juridique, financier et fiscal que technique. Les Etats se
devaient en outre, de prendre des engagements et donner des
garanties en renonçant en tant que de besoin à leurs
prérogatives habituelles. Ils devaient enfin pour la complète
information des candidats préciser les critères essentiels
auxquels se référerait leur jugement quant à l'évaluation et
la sélection des offres.

Tel a été l'objet des DIRECTIVES rédigées par un groupe
de dix experts franco britanniques et remises aux
gouvernements dès la fin de février 1985.

On sait le rôle prépondérant qu'a joué dans
l'appréciation des offres remises la fiabilité financière des
projets proposés, pour l'examen particulier de laquelle les

gouvernements sollicitèrent l'avis de groupes bancaires indépendants et spécialisés.

De ce fait et dès lors qu'il ne soulevait aucune objection majeure au plan technique le projet qui reprenant, à défaut de la totalité des ouvrages, les dispositions générales du projet de 1970 s'avérait le moins lourd en investissement et donc le moins difficile à financer présentait un avantage incontestable par rapport aux projets concurrents.

Tel fut le choix fait par les gouvernements et annoncé officiellement le 20 janvier 1986 à Lille.

Il leur fallait alors fixer par un traité leurs engagements réciproques et leurs positions communes vis à vis des concessionnaires. Tel fut l'objet du traité signé le 12 février 1986 à CANTERBURY.

Mais il leur fallait aussi matérialiser par un ou des textes de concession les conditions de réalisation de l'ouvrage projeté ainsi que les engagements qu'ils prenaient et les garanties qu'ils donnaient en adaptant au projet retenu et en complétant en tant que de besoin les dispositions énoncées dans les directives ou revendiquées par les concessionnaires dans leur offre.

Si l'élaboration du traité put être étalée sur plusieurs mois la mise au point de l'acte de concession dut être mené à bien en moins de 2 mois, les gouvernements ayant fixé aux experts désignés à cet effet la date limite du 10 mars 1986.

Tous ceux qui ont contribué tant à la rédaction des directives qu'à l'évaluation des offres et à la rédaction de la concession garderont le souvenir d'une extraordinaire tension en raison des délais extrêmement courts impartis et de l'ordre de 45 jours dans chacun de ces cas.

On peut penser qu'il eut été souhaitable de disposer de davantage de temps pour traiter convenablement de problèmes aussi complexes. Mais en faisant peser une telle contrainte sur les négociateurs les gouvernements leur interdisaient en fait ainsi qu'à leurs administrations respectives tout aternoiement et toute mesure dilatoire dont elles sont normalement coutumières.

Ils affirmaient ainsi leur farouche volonté politique d'aboutir en dépit de toutes les difficultés qu'une telle réalisation en commun ne manque pas de soulever.

Cette volonté exprimée d'efficacité et de rapidité se retrouve dans les textes du traité, de la concession et dans leur mise en oeuvre.

LE TRAITE DE CANTERBURY ET LA COMMISSION INTERGOUVERNEMENTALE

Le traité outre les clauses habituelles en la matière comporte d'une façon si ce n'est anormale, du moins inattendue, une série de dispositions concernant la concession qui n'ont d'autre raison d'être que la volonté exprimée par les gouvernements de préciser dans un texte qui régit essentiellement leurs relations, leur position commune vis à vis des concessionnaires. Ces dispositions sont d'ailleurs reprises dans la Concession et en termes identiques pour éviter toute divergence d'interprétation.

Le traité instaure en outre une Commission intergouvernementale pour suivre au nom des deux gouvernements et par délégation de ceux-ci l'ensemble des questions liées à la construction et à l'exploitation de la LIAISON FIXE.

Il met également en place un Comité de sécurité responsable des questions liées à la sécurité de la construction et de l'exploitation de cette LIAISON et qui apporte son aide et ses conseils à la Commission intergouvernementale dans ce domaine.

Ce faisant les deux gouvernements entendaient exercer par un organisme unique binational et paritaire leurs droits et obligations résultant de la concession mais ils témoignaient aussi de leur souci de faciliter la tâche des concessionnaires en leur opposant un interlocuteur unique et responsable.

Certes la Commission intergouvernementale ne saurait se substituer aux autorités nationales dans l'exercice de leurs fonctions habituelles et soustraire le concessionnaire aux législations et règlementations en vigueur mais chacune des deux délégations de la commission aide à l'accomplissement par les concessionnaires de leurs obligations nationales dans le cadre des dispositions de la concession.

Les concessionnaires ayant fait part aux gouvernements de leur intention d'entreprendre les travaux à leurs risques et périls sans attendre la ratification du traité et de la concession, c'est à leur demande que ces derniers ont par échange de lettre et à la signature du traité de CANTERBURY procédé à la mise en place de la Commission intergouvernementale et du Comité de sécurité.

Comptant actuellement respectivement 14 et 10 membres ces deux organismes paritaires ont des présidences alternantes et alternées, chaque chef de délégation assurant pour un an la présidence et les deux présidents à un instant donné étant de nationalité différente. Après plus de trois ans d'existence et de fonctionnement il semble possible d'affirmer que ces organismes ont jusqu'à maintenant répondu aux objectifs qui leur étaient assignés.

Leur tâche est loin d'être terminée pour autant car, comme nous le verrons, la concession leur confie un rôle majeur dans la construction et l'exploitation de la LIAISON.

LA CONCESSION ET SA VALEUR EN DROIT INTERNE

L'élaboration de la concession fut l'occasion de débats de principes importants.

L'exemple des TUNNELS ALPINS aurait conduit à la rédaction de deux actes de concession aussi symétriques que possible mais non identiques signés par chacun des deux états avec la société concessionnaire responsable de la partie de l'ouvrage située sur son territoire et relevant de la juridiction nationale compétente.

Le précédent de 1970 conduisait à la rédaction de deux actes de concession similaires mais non identiques portant sur l'ensemble de l'ouvrage et signés par chacun des deux gouvernements avec chacun des deux concessionnaires, l'un de ces actes relevant de la juridiction britannique l'autre de la juridiction française.

Il fut décidé afin de rendre plus aisée la tâche des concessionnaires et dans un but d'efficacité de recourir à une concession unique signée par les deux gouvernements et les deux concessionnaires.

Cette solution pour logique qu'elle fût n'était pas la plus simple à mettre en oeuvre. Elle exigeait surtout des gouvernements toute une série de renoncements et d'engagements hors du commun et en particulier celui de recourir en cas de litige non à leurs juridictions habituelles mais à des procédures d'arbitrages analogues à celles des contrats privés internationaux entre compagnies et sociétés.

En outre, certaines des garanties données dans le cadre de cette concession étant exorbitantes des législations nationales il fallut donner à ce texte force de loi dans chacun des deux pays.

Pour ce faire un "HYBRID BILL" fut soumis en Grande Bretagne aux Communes et à la Chambre des Lords tandis que la concession était en FRANCE au même titre que le Traité soumis à la procédure de ratification devant l'Assemblée Nationale et le Sénat.

Cette démarche qui exigea plus d'un an aboutit le 29 juillet 1987 à la cérémonie officielle au cours de laquelle à PARIS les deux chefs de gouvernements échangèrent les instruments de ratification. Cette date marque le démarrage officiel de la concession de 55 ans consentie par les deux Etats concédants.

Le texte de la concession reprend, explicite et complète les garanties avancées dans les DIRECTIVES lors de l'appel d'offres lancé le 2 avril 1985.

Tout d'abord une garantie de "bonne fin" (art. 25) par laquelle les gouvernements s'engagent à ne pas interrompre la construction ou l'exploitation de la LIAISON FIXE sauf pour des raisons de DEFENSE NATIONALE ou en cas de carence des concessionnaires.

Une garantie d'exclusivité est donnée à ces derniers jusqu'en 2020 puisque jusqu'à cette date les gouvernements s'interdisent d'autoriser la mise en exploitation de toute autre liaison fixe qui ne pourrait d'ailleurs, en tout état de cause être réalisée que dans les mêmes conditions financières que la première.

Pour assurer la juste rémunération des capitaux investis, les concédants, en dépit de leurs règles nationales propres qui leur permettent d'intervenir habituellement dans la fixation et le contrôle des tarifs publics, accordent aux concessionnnaires une totale liberté dans leur politique commerciale ainsi qu'une totale liberté tarifaire sous la seule réserve du respect en la matière des règles communautaires concernant la discrimination et l'abus de position dominante.

Ils leur garantissent de même et d'une manière extrêmement large l'égalité de traitement avec les modes de transports concurrents.

Enfin et afin d'offrir des garanties indispensables aux prêteurs, les concédants, et en cela le gouvernement français, dérogeait à sa juridiction la plus récente, ont accordé aux investisseurs un droit de substitution en cas de défaillance des concessionnaires.

LES PROGRAMMES D'ACCOMPAGNEMENT

Mais s'il est un domaine dans lequel la volonté des gouvernements s'affirme incontestablement de tout mettre en oeuvre pour que l'ouvrage soit réalisé dans les meilleures conditions c'est celui de leurs engagements quant aux programmes d'accompagnement.

En effet les DIRECTIVES comportaient déjà un important chapitre sur ce sujet et les concédants s'y engageaient en particulier à réaliser un certain nombre d'infrastructures assurant le raccordement des terminaux avec les réseaux autoroutiers et ferrés de part et d'autre de la MANCHE et prenant en compte l'écoulement correct des trafics attendus.

Mais depuis 1985 tout à la fois pour répondre aux aspirations des collectivités locales concernées et pour tirer le meilleur parti de cette liaison au plan EUROPEEN les gouvernements tant français que britannique ont revu leurs programmes d'investissement et les échéanciers de leur réalisation.

C'est ainsi que du côté français outre un programme routier dit d'accompagnement de quelques 5800 MFRS qui doit sensiblement modifier les infrastructures du Nord-Ouest du territoire toute une série de mesures ont fait l'objet de décisions en vue de faciliter à terme les échanges entre l'Angleterre et le Bénélux d'une part et l'Espagne et l'Italie d'autre part.

Plus encore et alors même que lors de l'évaluation des offres les hypothèses économiques retenues prévoyaient la mise en service possible du TGV NORD en 1998 le gouvernement français a décidé en 1987 sa réalisation et sa mise en service entre PARIS et le terminal de FRETHUN en 1993 permettant de mettre ainsi et dès l'ouverture du TUNNEL, LONDRES à moins de 3 heures de PARIS.

Une démarche analogue a conduit le gouvernement britannique et les B.R. à envisager entre LONDRES et CHERITON la réalisation d'une voie nouvelle non prévue initialement mais seule susceptible de garantir un écoulement satisfaisant des trafics prévisibles sans nuire aux besoins croissants des trafics de banlieue. Une telle réalisation devrait permettre d'abaisser encore de quelques dizaines de minutes les temps de parcours entre LONDRES, PARIS et BRUXELLES.

LE ROLE DE LA COMMISSION INTERGOUVERNEMENTALE ET DU COMITE DE SECURITE

Mais si de telles démarches témoignent, s'il en était encore besoin, de la volonté inébranlable et sans faille des deux gouvernements de mener à bien pour ce qui les concerne une telle réalisation, elles ne sauraient être interprêtées comme l'assurance qu'ils viendraient nonobstant les termes du traité ou de la concession financièrement au secours des concessionnaires défaillants.

Bien au contraire non seulement les textes mais leur démarche constante depuis cinq ans attestent de leur volonté de n'entraver en rien les libertés des concessionnaires dans le cadre des juridictions nationales ou internationales et de n'encourir aucune responsabilité dans leurs difficultés éventuelles.

Le rôle dévolu à la Commission intergouvernementale est significatif en la matière.

Certes elle se doit de veiller au nom des gouvernements au respect des dispositions de la concession et des règles nationales et communautaires qui s'appliquent normalement à l'ouvrage dans sa construction et son exploitation

Les deux délégations française et britannique qui la composent ont largement contribué à l'aboutissement des procédures préalables indispensables de part et d'autre du CHANNEL et à l'instauration et au maintien des concertations permanentes indispensables avec les autorités locales et régionales.

Elles n'ont cessé et ne cessent en fait dans les domaines les plus divers d'apporter aux concessionnaires tout ce qui est en leur pouvoir pour les aider dans leur information et leurs démarches mais sans jamais se substituer à eux.

S'agissant du projet lui-même et dès lors que l'ouvrage sera habilité à recevoir du public, les états ne pouvaient renoncer à leurs obligations et prérogatives en la matière.

Mais là encore l'initiative et la responsabilité sont laissées aux concessionnaires tant en ce qui concerne les dispositions techniques retenues que les règlements de sécurité et les procédures d'exploitation à mettre en oeuvre.

S'ils doivent communiquer leurs intentions à la CIG dès que possible et en particulier au niveau des avants projets c'est pour permettre à celle-ci, avec l'aide du Comité de sécurité non seulement de vérifier la conformité d'ensemble des ouvrages avec les propositions initiales retenues par les gouvernements mais aussi d'éviter dès l'origine toute disposition inadéquate et qui serait de nature à compromettre l'agrément de l'ouvrage au moment de sa mise en service (art. 10).

Mais là encore les pouvoirs de la CIG sont très strictement limités dans leur délai et dans leur expression.

En effet et en vertu de l'article 7 de la concession, celle-ci ne dispose que d'un délai de 15 jours francs pour faire connaître son opposition ou ses observations sur les documents qui lui sont officiellement soumis.

Elle se doit en outre de motiver ses avis et n'émettre d'objection que pour des raisons de sécurité, sûreté, environnement et défense.

On imagine sans peine le poids de telles contraintes pour des administrations dont les lenteurs sont souvent critiquées et pour des experts de toute compétence davantage habitués à être sollicités sur des aspects économiques et financiers.

Il est en fait apparu très vite que la mise au point de ces avants projets nécessitait en elle-même des délais et des approches préalables que l'on pouvait utilement mettre à profit pour une meilleure information et une meilleure collaboration des deux parties en présence.

C'est ainsi que chacun des quelques 25 avants projets répertoriés qui ont suivis ou suivront la procédure prévue a été découpé en un nombre variable d'éléments constitutifs dénommés "pièce intégrable".

Celles-ci au fur et à mesure de leur élaboration ont été communiquées au Comité de sécurité et à la Commission intergouvernementale qui les a examinés du point de vue sûreté, défense et environnement en liaison étroite avec les autorités nationales concernées.

Cet important travail préalable a permis tout à la fois de dégager les principaux problèmes qui nécessitaient des études voire des expérimentations complémentaires dont la plupart ont pu être menés à bien durant ces deux dernières années.

Ainsi l'approbation des avants-projets s'en est-elle trouvée facilitée et n'a-t-elle suscitée aucun problème nouveau dont la résolution eût pu entraver la marche des travaux.

Fort de cette expérience les concessionnaires ont souhaité et la commission a admis que des démarches analogues soient utilisées tant en ce qui concerne le suivi et la réalisation des travaux qu'en ce qui concerne l'élaboration d'ores et déjà entreprise des règlements d'exploitation et de sécurité.

Ainsi les concessionnaires assurent-ils dans les meilleures conditions leur maîtrise nécessaire des délais et des coûts sans que la Commission intergouvernementale soit amenée en ce qui la concerne à renoncer en quoi que ce soit aux contrôles et obligations qui lui incombent et que les gouvernements lui ont confié le soin de mener à bien sans faiblesse mais sans retard.

CONCLUSION

En décidant en novembre 1984 de recourir à l'initiative privée sans participation ni garantie financière des Etats les gouvernements français et britannique ont témoigné une nouvelle fois de leur volonté commune de réaliser une LIAISON FIXE à travers la MANCHE.

Mais leur mérite fondamental a certainement été de tirer de cette décision toutes les conséquences qu'elle impliquait de leur part et de se plier aux règles et devoirs qu'elle leur imposait.

Ils ont ainsi donné aux concessionnaires et aux investisseurs non seulement des garanties qu'ils étaient en droit d'attendre mais aussi la meilleure preuve de leur volonté commune de mener à bien quant à eux l'oeuvre entreprise dans les conditions qu'ils avaient fixées.

SUMMARY

Despite the failure in 1975 of the last attempt to build a Fixed Link across the Channel, the French and British governments did not abandon their project, and continued to pursue the technical and financial studies.

The latter, which took into account the very important increases in traffic between 1970 and 1982 in the Straits of Dover, convinced them of the possibility of building the link by resorting to private initiative, without Government financing or financial security.

This new approach led to the reopening of the project in 1984 and on 29 July 1987 resulted in the start of a Concession granted for 55 years by the two governments for the design, financing, realisation and operation of the Fixed Link.

Having displayed such an unshakable and flawless common intention, neither government has since deviated at any moment from the line of action they agreed.

The drawing up of instructions for the invitation to tender, the criteria retained for the evaluation of these proposals, the text of the Franco-British Treaty, as well as that of the Concession Agreement, all testify to this continuity.

In their concern to facilitate the realisation of the work, they have not hesitated in granting exceptional guarantees to the Concessionaires and in surpassing their initial agreements for the accompanying works.

In no way, however, will they be able to substitute the specific responsibilities of the Concessionaires.

To facilitate the task of the latter, they have established an Intergovernmental Commission, assisted by a Safety Authority.

If the Commission monitors on behalf of the two governments, and by delegation of the government, all questions linked to the construction and operation of the Fixed Link, its action, duly specified in texts, is carried out to strict deadlines and in a very restrictive framework.

A procedure of permanent dialogue has been established with the Concessionaires, thus enabling the Intergovernmental Commission to fully exert the tasks entrusted to it, without introducing any delay to the works.

3. The private sector financing of the Channel Tunnel

A. BENARD and A. MORTON, Co-Chairmen, Eurotunnel

INTRODUCTION. In May 1985 the Governments of France and the United Kingdom revived a 200 year old dream by launching an invitation to bid for a 55 year Concession to build and operate Fixed Links, tunnel or bridges or both, across the Channel. In January 1986 the Concession was awarded to Eurotunnel, in July 1987 legislative consent in the UK followed earlier French approval and, in November, the underwriting of Eurotunnel's simultaneous offering of units and warrants in the UK and France in November 1987 was the successful culmination of the campaign. That underwriting confirmed the £6 billion financing package needed to ensure that the Channel Tunnel could at last proceed. The 2½ year campaign had its setbacks, but they were overcome. It demonstrated that the private sector could finance major infrastructure projects, in this case with an innovative simultaneous listing of shares on two national stock exchanges, backed by the largest syndicated project loan financing in history.

DEVELOPMENT OF THE FINANCING PLAN
1.   The campaign started in May 1985 when the French and British governments issued their Invitation to Promoters. There were eventually four competing, and different, offering groups. The Eurotunnel proposal was put forward jointly by a British group of five contractors and two banks, and a French group of five contractors and three banks - a total of 15 internationally known groups.

2.   The core group of banks was Banque Indosuez, Banque Nationale de Paris, Crédit Lyonnais, Midland Bank and National Westminster Bank. They had already worked together as the authors of Finance for a Fixed Channel Link, published in May 1984. They now developed the financing scheme that was included in the submission to the two governments in October 1985 alongside the "technical" proposal. They were assisted by two UK merchant banks, Robert Fleming and Morgan Grenfell. Their scheme envisaged up to £1 billion of equity finance and up to £5 billion of debt finance, the amounts that were ultimately raised. The arranging banks already had

commitments from about thirty banks to join an underwriting group for a satisfactory structure of debt finance and there were positive indications of support from investors in the UK and France on the equity finance.

3.   Not unnaturally, all of these commitments or expressions of support were subject to the satisfaction of various conditions, such as the terms of the Anglo-French Treaty, the Concession to Eurotunnel, the Construction Contract and the Railway Usage Contract with British Rail and the SNCF. The equity investors in particular stressed the need for a board with a clear majority of directors independent of the contractors and banks. The key personnel also had to be acceptable to investors and independent of the banks and the contractors.

4.   Following the award of the Concession in early 1986 to the two concessionaires, France Manche and Channel Tunnel Group, work started to structure the two public companies needed for raising the equity. It was agreed that there should be a French and a UK public company, each the owner of the French and UK concessionaire partners respectively. The shares of the public holding companies were to be twined into units that could only be transferred together.

EQUITY 1 AND 2
5.   The public companies, Eurotunnel SA and Eurotunnel PLC, were formally incorporated in September 1986 at the end of an eight month preparatory period following the award of the Concession. The initial equity of just under £50M, plus an issue of warrants, was subscribed at that time by the founder shareholders. This was the so-called Equity 1 and effectively reimbursed expenditures to date.

6.   Equity 1 had been planned for earlier in the year but the timetable for Equity 1 and for Equity 2 slipped under the pressure of a mass of prerequisite negotiations. A decision was taken to defer Equity 2, a private placement of about £200M among French and British investment institutions, from the summer to the autumn of 1986. Behind this setback lay the difficulties in concluding a Construction Contract by the ten founder shareholder contractors to the satisfaction of both the banks and a core group of independent shareholders.

7.   In the meantime, work had been going on to draft the term sheet, or outline terms, for the proposed £5 billion credit facilities. By this time, there were forty banks in the potential underwriting syndicate. The negotiations were conducted by the five arranging banks. They were supported by a "technical" group of banks, who reviewed the Construction Contract, and by a "financial" group of banks, who monitored the development of the Term Sheet. In addition, the banks had consulting engineering firms as their technical advisers, as well as an external traffic and

revenue adviser to review the forecasts prepared by Eurotunnel's own independent traffic and revenue consultants.

8.   Agreement on the Term Sheet was reached early in September 1986 and it was on this basis that the forty banks reconfirmed their commitment in principle to underwrite the proposed credit facilities. Thus reconfirmation was essential for Equity 2, as were the Construction Contract and the heads of agreement that had been exchanged with British Rail and SNCF.

9.   Despite this underpinning, new investors at Equity 2 were being asked to make an investment with considerable risks. Apart from project risks themselves, the necessary legislation to ratify the Treaty and bring the Concession into force had not been approved by the French and UK legislatures. In the UK there was still considerable public and parliamentary opposition to the project. In addition, the Railway Usage Contract had still to be negotiated in detail and signed and the Credit Agreement also had to be negotiated, underwritten, syndicated and signed. Their subscription of Equity 2 would indeed be "venture capital".

10.   Equity 2 was only completed late in October 1986 with some difficulty in both the UK and France, some of the British problems well publicised at the time. In the end venture capital amounting.to just over £200 million was obtained, all of it likely to be spent or committed by the time the go-ahead would be available in mid-1987 - following completion of the legislation and of negotiations with the railways and banks!

11.   The project was able to progress to the next stage.

AN INDEPENDENT EUROTUNNEL

12.   Equity 2 also accomplished a vital step in the development of the project. The entry of the investment institutions was conditional on the promoters (contractors and banks) giving up control of Eurotunnel. From October 1986 an independent majority on the board could get on with building up a competent bi-national management. Eurotunnel acquired "a mind of its own".

THE STEPS TO EQUITY 3

13.   All the focus moved on to the steps necessary to achieve a successful public flotation. Nothing could be confirmed before the Treaty was ratified by the British and French parliaments and the Concession brought into force. The heads of agreement with British Rail and SNCF, Eurotunnel's two largest single customers, also had to be turned into a signed Railway Usage Contract and the Term Sheet approved by the banks had to be turned into a syndicated and signed Credit Agreement. The project itself had to be brought to a state of design, and readiness to

start main construction, that would allow a full public prospectus to be issued. The structure of the equity that would be offered to French and UK investors had to be agreed and, not least, the detailed equity marketing plan had to be developed and implemented. It was not contemplated that £750 million equivalent in equity could be raised except by public offering, requiring the issue of a full prospectus, necessitating a high level of information on a multitude of aspects of the project, all of which needed to be sufficiently certain or positive!

14. All this had to be done against a background in the UK of well orchestrated opposition from various groups to the project such as some ferry companies and Dover Harbour Board and, at least early in 1987, of a very doubtful and uncertain image in the British media. In addition, there was the near certainty of a UK General Election at some stage in 1987 - with the implications this could have on the passage of the Channel Tunnel Bill - and there was also the ever present possibility that the long-running bull market in equities would come to an end.

15. The programme was challenging. Following the (also well publicised) board changes in February 1987, it was agreed that the original plan to float Equity 3 in July 1987 should be delayed to the autumn despite the risk of the bull market fading. S.G. Warburg, and additional financial, legal and public relations support were added to Eurotunnel's team. This team, working with the existing UK merchant banks and stockbrokers and with Banque Indosuez and their advisers in France, spearheaded the Equity 3 campaign.

16. All legislative, contractual and project matters had to be co-ordinated with the requirements of Equity 3 and the prospectus on which we would offer equity to the public. The responsibilities on the Eurotunnel directors were extremely heavy and meeting the requirements of two stock exchange authorities to enable the simultaneous listing to take place in Paris and London was a formidable task. It is a great credit to all those involved that it was successfully achieved. A great deal of new ground was opened up in British and French securities practice.

17. A key feature in the process was some very effective research and analysis of the potential value of Eurotunnel units by Warburg Securities.

THE TREATY AND THE CONCESSION
18. While the passing of the necessary legislation in France by June 1987 was uneventful and unanimous, given the almost universal support for the project across the political spectrum, the situation in the UK was very different. There were strongly held differences of opinion on the project and on the procedures for approving it and for granting the

necessary powers to proceed. Despite this the views of those opposed to the project were thoroughly aired and changes were made. The timetable was tight and the General Election in June 1987 did delay passage of the Channel Tunnel Bill by almost two months before, under the hybrid bill procedure, it could be reintroduced in the new Parliament at the stage it had reached in the old Parliament. Royal assent, ratification of the Treaty and the coming into force of the Concession were achieved in July 1987.

THE RAILWAY USAGE CONTRACT

19. The Railway Usage Contract was signed on the day the Treaty was ratified and the Concession came into force on July 29. This followed tough and complex negotiations with British Rail and SNCF, whose own interests were not always in closest harmony. Eurotunnel agreed to assure the railways half the capacity of the tunnel in exchange for an increase in the level of revenues committed by the railways. All parties committed themselves to the start date.

THE CREDIT AGREEMENT

20. The Credit Agreement negotiations were also tough and complex. The Term Sheet had been 74 pages long. The final Credit Agreement was about 1400 pages long. The banks wanted a considerable degree of control over the project and close involvement in the project for themselves and their advisers. Eurotunnel needed freedom and flexibility to manage the project and the assurance that the money would be available when we wanted it. This was vital for the Equity 3 prospectus. The resulting agreement was an uneasy compromise, since it needed to balance the requirements of a public equity issue, which had rarely, if ever, been so closely linked to a project financing, with the need to syndicate the credit facilities internationally.

21. The Credit Agreement was underwritten in August 1987 by fifty banks, following the addition of ten major international banks to the underwriting group. This was followed by the syndication, itself a major marketing exercise for the arranging banks, with Eurotunnel participating in roadshows around the world in teams, usually led by the Co-Chairmen and including technical, financial and commercial presentations.

22. The target of syndicating 50% of the banks' original commitments was achieved with a margin to spare, providing a valuable vote of confidence by the banking community in the project and an essential prelude to Equity 3 - upon which the syndication was conditional. No equity, no bank loans.

23. The European Investment Bank played a valuable role in the syndication since it had earlier agreed to lend Eurotunnel up to £1 billion within the framework of the main £5 billion facility. Its independent and careful appraisal

of the project was recognised by other banks. The EIB facility was not additional finance, but was designed to enable Eurotunnel to borrow funds for longer maturities than were available from the banks; these loans would be secured by letters of credit from the banks until EIB were able to take project risk, expected to be shortly after operations commenced.

24. The Credit Agreement was signed simultaneously in London and Paris in November 1987 by 198 banks, once the issue of the preliminary prospectus was assured.

THE PROJECT STATUS AND MANAGEMENT

25. Also vital to the prospectus were the status of the project itself and the readiness to start main construction. The £200 million from Equity enabled work to proceed during 1987 on the preliminary tunnelling works on both sides of the Channel. On the French side, the access shaft at Sangatte had been largely completed and in the UK work had started to redevelop access via the tunnel abandoned in 1974. The first tunnel boring machines were being assembled.

26. Considerable progress was made in building an independent, credible and effective management team in Eurotunnel. A number of key managers, such as Pierre Durand-Rival, who played a vital role in the first stages of construction and is now Technical Adviser to the Co-Chairman, Alain Bertrand, our Managing Director - Operations and Safety, now working alongside Dr. Tony Ridley, Managing Director - Project, and Graham Corbett, now our Finance Director, were recruited from outside the ranks of the founder shareholders, augmenting a number of excellent people from the original sponsors who opted to stay with the project. An initial organisation structure was put in place. The creation of a new entity able to deal at arm's length with the founder shareholders in their capacity as contractors and banks was crucial to further progress.

TGV-NORD

27. Another very important contribution to the successful flotation of Eurotunnel was the commitment by the French government in October 1987 to build the TGV-Nord, the high speed rail line from Paris to the Belgian border, with a spur from Lille to the Channel Tunnel. This commitment enabled the railways to offer a journey time between Paris and London, city centre to city centre, of 3 hours from 1993. This would be competitive with the journey time by air and considerably increased the appeal of the project to the public and to investors.

THE EQUITY STRUCTURE AND MARKETING PLAN

28. The final ingredient for the Equity 3 flotation was the structure of the equity to be offered to investors and

the marketing plan. It was recognised that the issue had to be successful primarily in France and the UK. There were plans for 20% of the £750 million to be raised by way of an international placing linked to the main public issue of £300 million in each primary country. The issue would succeed or fail in Britain and France.

29. Eurotunnel's French advisers stressed the attraction to French investors of warrants and in the UK a strong case was made for the value of travel privileges or "perks"; there was a precedent with one of the cross-Channel ferry companies. Other proposals were reviewed, but these suggestions were accepted as providing an equity package with considerable marketing appeal and potential.

30. The marketing plan in the UK involved a number of elements from earlier "privatisation" issues, including television and other advertising, a share information office in London and a mass enquiry office in Bristol, three circulars prepared by Warburg Securities on traffic, construction and returns, meetings with institutions, press briefings and also briefings for broker's sales forces. In France, there was a television advertising campaign and brokers circulars. From mid-October to late November there were ceaseless national and international roadshows in support of the issues and placing. One already exhausted team travelled around the world in 125 hours, giving full presentations on the way in New York, Boston, Montreal, Toronto and Tokyo, and claimed that the flights had enabled them to get more sleep than they had had recently. Another team "did" nine Western European cities in six days after a weekend visit to the Middle East!

EQUITY 3
31. All of these elements - the legislation, the Treaty, the Concession, the Railway Usage Contract, the Credit Agreement, the EIB Credit Agreement, the readiness of the project, the management team, the TGV-Nord, the equity package and the marketing plan - were necessary to launch the priced prospectus on, to use the UK term, Impact Day - November 16, 1987.

32. One last event could have undone all that work - Black Monday, October 19. The long running bull market broke and the international stock markets collapsed. This led to the postponement or cancellation of almost all new equity issues and a virtual cessation of international equity commitments. In the UK, the Chancellor's decision not to postpone the BP issue did not help the market. Despite this, the decision was taken to proceed, at the price per unit already planned. Postponing the issue would have been fatal for the project and the investment risk was much longer-term than any current market situation. In spite of the atmosphere, the issue was

successfully underwritten. Eurotunnel was assured of the money needed to start main construction.

33. The rest of the story is well-known. Although a portion of the issue was left with underwriters in both countries, the issue attracted 300,000 investors. The initial price fell to about the level of the Equity 2 issue, but recovered to the Equity 3 level again in early March 1988, less than three months after dealings started. Any "stale" underwriters were either out or converted by then.

34. Seven months later, in October 1988, demand began to increase as the potential of the Channel Tunnel began to gain acceptance. In the late spring of 1989, the price of one unit and one warrant, floated 18 months earlier for £3.50, or 35 FF, exceeded £12 or 120 FF.

LESSONS

35. What lessons can be learned from Eurotunnel's experience? No two major projects are exactly alike and it could be argued that Eurotunnel is unique. A project first proposed almost two centuries ago and one that has aroused strong emotions in the UK in particular, it has had a number of false starts. This time it had effective political backing from President Mitterrand and Mrs Thatcher, determined promoters, courageous banks and founder institutions and single-minded management. Mrs Thatcher's determination that it would succeed undoubtedly helped ensure the passage of the necessary legislation in the UK, and the commitment to the TGV-Nord in France was crucial to the financing.

36. It is impossible to overestimate the importance of the decision to grant a long-term "Concession". This mercantile concept from Europe's Middle Ages, entitling the Concessionaire to develop and exploit a geographic and/or economic "territory" is fundamental. The Governments', the operator's and the investors' rights are spelled out and safeguarded.

37. The project does have characteristics of a general nature - it is large, it is costly and it has a long lead time. To help reduce the lead time, it was decided that a 7-year design and build contract should be used. This was intended to help the financing, but it is clear that such an approach also has its drawbacks in project finance. It leaves design and other uncertainties to be resolved during construction and makes it that much more difficult to estimate costs. One banker commented recently, that in his experience, the uncertainties were not usually resolved until half way through such contracts, a stage Eurotunnel reaches around the end of this year. We can testify to the difficulties that uncertain cost estimates have on a financing programme.

38. Linked to this point is the stage at which equity is raised and from whom. In the case of Eurotunnel the promoters took shares in Equity 1 to reimburse themselves for their major outlays to launch the project. It was only two months later that institutional investors were asked to subscribe Equity 2, at a stage when the risks were still very considerable. Although Equity 2 had the positive effect of putting the founder shareholders in a minority, the difficulties the issue encountered, with their adverse effect on the financial image of the project, might have been avoided if the promoters had financed part of the next stage as well. They were best placed to take those risks. It might have meant a slower start to construction and a longer contract period, but could have allowed design to proceed further. Fifteen of them could perhaps have gone to £100 million, not £50 million.

39. The financial structure adopted by Eurotunnel was of a straight-forward variety, essentially a simple combination of debt and equity, with the only flavour added by warrants and the introduction of a tier of quasi-equity such as a convertible and redeemable stock can permit a more discriminating allocation of risk and allow the "equity" to be reduced in size once the project is up and running. In cases where some public sector support is still required, this can be an effective way of providing a suitable instrument for public sector participation without affecting the private sector character of the project.

40. One of the key lessons highlighted by Eurotunnel's experience is the need to recognise the different roles of promoter and owner. The promoters of Eurotunnel had no desire to be involved in the future operation of the system; they were obviously interested in the contracting and banking opportunities the project presented. There is nothing wrong in that, except that contracts negotiated among a promoting group are not necessarily best suited to the requirements of the future owner, particularly an owner that needs to raise equity from the public. The creation of Eurotunnel as a separate entity, independent of both the contractors and banks, should have happened much earlier after award of the Concession. Eurotunnel was endowed at its eventual "birth" at Equity 2, nine months after the Concession with a Construction Contract and a Term Sheet for a Credit Agreement that it had not negotiated. Negotiating a workable Credit Agreement that supported the public equity issue was made that much more difficult. Eurotunnel is still living with the consequences of its delayed birth, and yet we shall always pay tribute to the vision and determination of the promoters.

41. This need for the early establishment of an independent board and management group in major stand-alone

projects, particularly where there is a future operating role that is of no interest to the promoters, is probably the single most important lesson from Eurotunnel's experience to date.

42.   It is now just over four years since the sponsors of Eurotunnel came together to seek the Concession. It is just under four years to opening. We are at the half-way point in the great European venture launched by President Mitterrand and Mrs Thatcher with their Invitation to Promoters.

Ce rapport résume l'évolution du programme de financement privé de la Société Eurotunnel, ainsi que les trois étapes qui ont permis la mise en place de fonds propres d'un montant de 10 milliards de Francs. Ce rapport se concentre plus spécialement sur les différentes étapes à franchir avant de pouvoir réaliser l'augmentation de Capital 3: l'adoption des projets de loi en France et au Royaume-Uni, la ratification du Traité et l'entrée en vigueur de la Concession, la conclusion de la Convention d'Utilisation avec les Chemins de Fer, la syndication et la signature de la Convention de Crédit d'un montant de 50 milliards de Francs, le développement du Projet, la constitution d'une équipe de direction indépendante à la tête d'Eurotunnel, la décision de la mise en service du TGV-Nord, la mise en place des éléments de l'Emission Publique et du programme de marketing associé. Ce rapport tire ensuite quelques leçons de l'expérience d'Eurotunnel pour d'autres projets.

4. Le système de transport. Point de vue de l'opérateur

A. H. BERTRAND, Directeur Général, Opérations et Sécurité, Eurotunnel

SYNOPSIS. En plus des trains "normaux" des compagnies ferroviaires nationales, le tunnel sous la Manche devra acceuillir des navettes, trains spéciaux de 750 m de long, chargés de véhicules routiers, qui parcourront deux galeries à 117 km à l'heure de moyenne: une véritable autoroute roulante. L'exploitation de ce double trafic et la sécurité exigeront la mise en oeuvre de techniques de pointe.

1.   Le système Eurotunnel, composé de ses trois tunnels et de ses gares terminales est conçu pour offrir deux prestations très différentes:
    -pour les usagers de la route, une véritable autoroute roulante grâce à l'exploitation régulière des trains-ferry circulant à intervalles très rapprochés;
    -le chaînon marquant entre les réseaux ferroviaires britanniques et continentaux, ce qui bouleversera l'équilibre existant entre les modes de transport transmanche.

2.   La ligne Calais-Folkestone se situera, dès sa mise en service, parmi les plus chargées du monde, avec 360 trains et 210 000 t en moyenne par jour. L'exploitation de cette ligne, qui superposera des trafics très divers, en circuit fermé aussi bien que de transit international, posera des problèmes d'exploitation originaux. Il sera fait appel à la technologie de pointe de l'industrie ferroviaire mondiale.

3.   Mais le fait que la ligne soit presque entièrement souterraine, sous-marine en vérité, induit d'autres originalités. C'est ainsi que les problèmes de sécurité, notamment à l'intérieur des trains-ferry, amènent à imaginer des solutions spécifiques faisant appel à des technologies quelquefois plus proches de l'industrie pétrolière que du chemin de fer.

4.   Ce lien fixe est aussi un point frontière, la première frontière terrestre entre la Grande-Bretagne et le continent. Là encore, des solution originales sont à trouver pour permettre à la fois des contrôles efficaces, et adaptés aux flux de trafic attendus.

5.   L'exploitation du système reposera sur une organisation humaine originale, binationale jusqu'au niveau des "cols bleus"; 2 200 personnes environ seront bientôt sélectionnées, recrutées et formées pour faire fonctionner le système Eurotunnel dès le 15 mars 1993.

LA CIRCULATION DES TRAINS

6.   L'exploitation de l'autoroute roulante reposera sur la circulation des trains-ferry se succédant en période de pointe toutes les 12 minutes.

7.   Chaque train, composé de 12 ou 24 wagons, est une véritable section de route articulée. Chaque wagon est une sorte de caisson de 25 m de long et de 4 m de large, permettant aux véhicules de s'embarquer à une extrémité du train, et de rouler à l'intérieur.

8.   Les véhicules accèderont à la zone d'embarquement par des routes et des ponts affectés à chaque type de trafic: voitures légères, voitures avec caravane, autocars, camions. Les navettes seront spécialisées par trafic: tourisme et fret. Un personnel bien formé guidera les conducteurs vers les zones de classement. Une fois le véhicule garé, les passagers resteront à bord. Ainsi sera-t-il possible de réaliser les opérations de chargement et de déchargement en quelques minutes à peine.

9.   Les trains navettes auront des dimensions inusitées, très supérieures au gabarit ferroviaire européen: 750 m de long, 4 m de large, 5 m de haut. Ces trains pèseront de 1 500 à 2 000 t selon leur chargement. Ils seront tirés par 2 ou 3 locomotives - elles de dimensions classiques - développant chacune 4 000 à 6 000 kW.

10.   Nous démarrerons l'exploitation avec 18 de ces trains - 9 pour le fret, 9 pour le tourisme. Mais il faudra en commander d'autres pour faire face à la croissance du trafic des années suivantes.

11.   Les trains parcourront la distance entre les deux terminaux en 33 minutes, à 117 km/h de moyenne avec des vitesses de pointe pouvant atteindre 160 km/h.

12.   Les terminaux ressembleront à des gares ferroviaires classiques, à ceci près que les quais seront réservés au trafic automobile. Il n'est pas prévu que des piétons ou des cyclistes s'embarquent directement dans les navettes.

13.   Compte tenu du caractère "fermé" de cette partie du système, il convient de réaliser une intégration fine des fonctions d'exploitation et de maintenance - du fait des gabarits très importants du matériel. Il est, par ailleurs, impossible d'assurer l'entretien des installations des

réseaux ferroviaires. C'est pourquoi les terminaux seront flanqués de zones de maintenance très bien équipées, pour répondre aux besoins sur toute la durée de vie du matériel roulant.

14. Les trains internationaux, TGV, trains de nuits, trains de marchandises rejoindront le réseau Eurotunnel par un ensemble de raccordements permettant soit l'accès en vitesse (cas des trains de voyageurs) soit l'accès après escale dans un faisceau de préparation et de contrôle (cas de trains de marchandises).

15. Ces trains trouveront un équipement fixe compatible avec leurs propres équipements embarqués. La caténaire (25 kV) sera du type classique mais son alimentation devra tenir compte de l'impossibilité de prévoir des sous-stations intermédiaires sous la mer. Pour fournir aux batteries très denses de train un courant de bonne qualité, il faudra probablement alimenter chaque demi-tunnel depuis l'extrémité à l'air libre, et en un point intermédiaire grâce à des câbles à basse impédance. Il reste à régler les problèmes délicats posés par des phénomènes de déséquilibre entre phases, spécialement critiques en Grande-Bretagne, du fait de la qualité des réseaux haute tension.

16. La signalisation permettra de réaliser un débit élevé. Elle sera du type Cab Signal. Il est, en effet, exclu de recourir à la signalisation lumineuse classique, qui, outre sa faible capacité, peut prêter à confusion dans les longs alignements que les conducteurs rencontreront en tunnel. Si le Cab Signal est un système bien connu, notamment à la SNCF, il faudra l'adapter pour répondre à des contraintes d'exploitation originales et sévères, et notamment la variété des types de train qui se succèderont. L'industrie ferroviaire, pour répondre à la demande des exploitants (de réseaux urbains ou interurbains) fait beaucoup de progrès en ce moment, et nous bénéficierons de cette avancée spectaculaire. Bien entendu, la signalisation intègrera un contrôle de vitesse très élaboré, outil essentiel des chemins de fer modernes.

17. Enfin, les télécommunications seront assurées entre les trains et les centres de contrôle, par l'intermédiaire d'antennes conçues sur le principe des câbles à fuite.

18. Le contrôle et la gestion de cette ligne seront assurés par un poste de commandement regroupant les fonctions de surveillance, de communications et de commande pour les mouvements ferroviaires, d'une part, mais aussi pour tous les équipements que l'on trouvera dans les tunnels. Des postes secondaires assureront le contrôle des opérations dans les zones des terminaux et de la maintenance.

19. Les contrôleurs en service seront aidés par des systèmes d'aide à la décision. Ils devront, par ailleurs, évidemment être bilingues.

20. Il faudra, en effet, gérer à la seconde près les arrivées et les passagers des trains, qui obéiront à des horaires précis.

21. Il est probable que British Rail, tant qu'ils ne disposeront pas d'infrastructures spécialisées pour le trafic transmanche, ne seront pas capable d'assurer une régularité sans faille. La tâche de nos opérateurs sera donc essentielle, pour permettre le passage dans les meilleurs conditions, tout en assurant le meilleur usage de la capacité du lien. Les règles du jeu entre Eurotunnel et les chemins de fer ont été mises au point sous forme d'une convention d'usage qui lie les trois exploitants pour la durée de la concession.

ASSURER LA SECURITE

22. Si la capacité offerte et la qualité de l'exploitation sont deux arguments de vente importants, la sécurité des biens et des personnes est, elle, fondamentale.

23. Les gouvernements ont d'ailleurs choisi le projet Eurotunnel, il y a deux ans, essentiellement parce qu'il était le seul à avoir la sécurité intrinsèque qu'offre la technologie ferroviaire.

24. Eurotunnel sera l'un des systèmes de transports les plus sûrs au monde. Trois facteurs principaux assurent ce haut niveau de sécurité. Tout d'abord, le tunnel lui-même est un ouvrage très complexe. Il est, en fait, composé de deux galeries principales assurant le passage des trains; chaque galerie est bordée d'un quai continu. Entre ces galeries principales, on trouvera une galerie de service, véritable route intérieure, reliée tous les 375 m par un rameau de communication aux tunnels ferroviaires. Cette galerie de service assure trois rôles essentiels: l'air frais aspiré de l'extérieur est distribué par le système de ventilation via cette galerie. La surpression en résultant donne à cette galerie le caractère d'espace refuge: en cas d'incendie grave dans les tunnels principaux, les passagers et le personnel seront évacués dans la galerie de service, où ils resteront à l'abri des émanations de gaz et de fumées, dans l'attente des secours. Enfin cette galerie sert de voie d'accès routier pour les véhicules de maintenance et de secours.

25. Ensuite, le système de transport repose avant tout sur la technologie ferroviaire. Eurotunnel adoptera des spécifications très sévères pour les composants de base de système - voie, matériel roulant, signalisation, télécommunications.

26.   Il convient de remarquer ici que les risques les plus fréquents sont éliminés par le fait même de la circulation des trains dans des tubes monodirectionnels. Les eaux de ruissellement seront canalisées vers des stations de pompage. Le réseau hydraulique est largement surdimensionné.

27.   Pour lutter contre le risque d'incendie se déclarant dans une navette transportant des véhicules avec leurs passagers, une attention toute particulière a été portée au matériel roulant.

28.   Chaque navette - train-ferry - sera conçue en réalité comme une suite de caissons roulants qui pourront être individuellement isolés, après évacuation des passagers par des portes coupe-feu. Leur structure résistant au feu autorisera le maintien de la marche du train vers l'air libre, où le sinistre sera traité sur une voie d'intervention. Des dispositifs de détection et d'extinction utilisant des technologies variées permettront aux membres de l'équipage d'un train d'agir de façon efficace dans tous les cas. Dans chaque navette, l'ensemble des contrôles d'environnement seront installés dans un poste central relié par radio aux centres de contrôle principaux d'Eurotunnel.

29.   Ce concept permettra ainsi de garantir un espace refuge à proximité immédiate du wagon sinistré, tout en permettant au convoi de s'évacuer par ses propres moyens vers l'air libre.

30.   Il convient de noter enfin que par sa présence permanente, le personnel d'Eurotunnel veillera au confort et à la sécurité des passagers; il s'assurera aussi que les règles élémentaires sont respectées: interdiction de fumer, de mettre en marche le moteur avant arrêt de train, etc. D'une façon générale, le personnel d'Eurotunnel sera formé de façon très approfondie, notamment vis-à-vis de la sécurité. L'ingénieur sécurité pour la phase d'exploitation est déjà en place, ainsi que les responsables du service ferroviaire et des terminaux. Tous ont une très grande expérience, acquise sur le terrain, qu'ils mettent aujourd'hui au service de l'ingénierie d'Eurotunnel.

31.   Des exercices seront organisés à intervalles réguliers, avec les services locaux de secours et de lutte contre l'incendie, pour vérifier l'efficacité des formations, et tester les procédures mises en oeuvre. Il est important de noter que les contacts sont déjà établis avec les brigades de pompiers de Kent et de Pas-de-Calais, et c'est ensemble que nous mettons au point les données techniques des ouvrages ou de matériel ferroviaire, garantissant ainsi les meilleures conditions de sécurité pour nos passagers.

LE PREMIER POSTE FRONTALIER TERRESTRE DE LA GRANDE-BRETAGNE

32. Il peut sembler paradoxal d'aborder le sujet des contrôles de frontière alors qu'Eurotunnel doit être mis en service après la date prévue pour la mise en place du marché unique européen; pourtant les incertitudes qui planent sur cette échéance et sur ses conséquences obligent Eurotunnel à envisager un système souple et évolutif.

33. Pour la première fois de son histoire, si on excepte la frontière avec la République d'Irlande qui reste un cas très particulier, la Grande-Bretage va avoir, marché unique ou pas, une véritable frontière terrestre avec le continent. Les conséquences en seront d'autant plus importantes que le scénario, classique ailleurs mais quasiment révolutionnaire en l'occurrence, des contrôles binationaux juxtaposés avec sortie libre après le tunnel a été retenu.

34. Concrètement cela entraîne que des douaniers et policiers français et britanniques travailleront côte à côte aussi bien en France qu'en Angleterre. Il faut donc s'attendre à des changements profonds d'habitude. Eurotunnel a d'ailleurs des contacts très nombreux avec les autorités administratives françaises et britanniques.

35. Dans ce contexte Eurotunnel étudie une organisation de contrôle qui permette sans difficulté une circulation aussi fluide que possible pour les véhicules légers aussi bien que poids lourds. Des installations adaptées permettent de traiter efficacement les cas nécessitant un contrôle approfondi.

CREER LA PREMIERE VERITABLE COMPAGNIE BINATIONALE DE TRANSPORT

36. Pour conduire les trains, guider et accompagner les voyageurs, percevoir les billets, contrôler les accès, gérer les circulations en liaison avec les chemins de fer et des centres d'information routiers, entretenir les installations et le matériel roulant, Eurotunnel devra recruter environ 2 200 personnes, qui vivront l'expérience originale d'appartenir à la première société réellement binationale.

37. L'intégration ira dans certains cas presque dans les détails: par exemple, pour les équipages à bord des navettes. Dans d'autres cas, des équipes nationales travailleront en alternance, ou en appui, par exemple dans l'entretien des installations fixes.

38. Les prochains mois verront le démarrage du processus qui permettra de définir la stratégie sociale et l'organisation générale du secteur exploitation.

39. Il faudra ensuite affiner les descriptions de poste, en s'attachant à créer une structure efficace donnant aux hommes une responsabilité réelle de tâches enrichies. C'est

la seule voie pour garantir un service de grande qualité à
notre clientèle. Nous commencerons la sélection et le
recrutement des premières équipes suffisamment tôt pour
qu'elles participent aux essais de réception et aux mises au
point du matériel. Ces noyaux de base assureront aussi
l'armature sur laquelle Eurotunnel s'appuiera pour mettre en
place les équipes nécessaires le 15 juin 1993.

SUMMARY

1. The Eurotunnel system will link the British and Continental road and rail systems, with a novel rolling road shuttle service for road vehicles and a more conventional rail link for passenger and freight trains.

2. The combination of these two systems and the frequency of the service will raise novel operational problems and require the most advanced technology of the world's railway industry to be brought into play.

3. There are also some novel safety problems to be devised for a system which is almost entirely underground.

4. As if this were not enough, the system also forms a border crossing and here again novel solutions are being developed.

5. The Shuttle trains are generally of two types, Tourist Shuttles or Lorry Shuttles; for safety reasons lorries are not carried on the same trains as passenger carrying vehicles.

6. The Tourist Shuttles will consist of two rakes each of fourteen wagons; a loading/unloading wagon at each end and twelve carrier wagons. OnE rake will comprise double deck carrier wagons for cars, carrying four or five cars on each deck; the second rake of the same composition but the wagons are single deck to carry vehicles over 1.85m in height, such as coaches, minibuses.

7. The freight trains are also composed of two rakes of fourteen wagons, but they will be single deck and above 20m long, compared with the 25m of the Tourist Shuttles wagons. All wagons are 4m wide and 5m high.

8. The trains will be about 750m long and be hauled by two locomotives developing between 4 and 6 MW. We shall start operations with at least 9 Tourist Shuttles and 9 Lorry Shuttles.

9. The trains will take 33 minutes to travel from terminal to terminal and be capable of reaching a maximum speed of 160 kph.

10. The terminals will resemble conventional railway stations except there will be vehicles instead of pedestrians getting on and off the trains.

11. Once inside the trains, drivers of tourist vehicles will remain in or alongside them for the journey.

12. The power supply for the tunnel trains is duplicated, being capable of supply from sub stations in either Britain or France, though normally each half tunnel will be supplied from one side.

13. The signalling system will be of the cab signal type and capable of being developed to handle a high flow rate of mixed traffic.

14. There will be telecommunication systems between the trains and the control centres, one of which will be the Operational Control combining all control functions for the rail traffic and the equipment in the tunnels. Secondary control centres will be responsible for operations in the terminal and maintenance zones.

15. Safety has been one of the highest priorities in the development of the system - by the time the Operating Certificate is granted by the Special Safety Authority advising the Intergovernmental Commission it will be a very safe system of public transport.

16. This high level of safety is due to three main factors: the system is based on well tried rail technology with single track rail tunnels; the presence of specially trained attendants on all shuttles; the tunnel arrangement with continuous walkways in the running tunnels and connecting passages at regular intervals linked to a continuous service tunnel.

17. The frontier control arrangements will be based on the free exit principle, with any formalities which Britain and France require being carried out at the departure terminal. These novel arrangements which are still being developed reflect the construction of Britain's first land frontier with the Continent.

18. Considerable human interest must however come from the recruitment of staff to the first genuinely binational transport company and then integration into a single working organisation.

5. Expected traffic demands

A. DICK, Chairman, Alastair Dick & Associates, planning adviser to Eurotunnel; A. BLANQUIER, Directeur Général, SETEC Economie, and D. DANFORTH, Principal Associate, Wilbur Smith Associates

SYNOPSIS

The expected traffic flows for the transport services that will use the Tunnel have been determined by independent consultants. The procedure they have adopted is to establish the likely share of the expected total cross-Channel traffic that will be captured by the Tunnel services. In addition an allowance is made for traffic that will be created both by the fixed link and the relaxation of frontier controls, etc. as Europe moves towards a single market. The paper describes the various procedures used and assumptions adopted.

INTRODUCTION

1. **Background.** Traffic estimates for the likely usage of the Tunnel's services are primarily required to support the financing plan for the Project. A secondary requirement was to produce an indicative level of demand to develop the transportation systems. The traffic estimates and their associated revenues were produced by independent consultants commissioned by Eurotunnel and acceptable to the syndicate of lending banks. The consultants who were appointed in May 1985 were SETEC Economie of France, in association with Wilbur Smith Associates of USA and Alastair Dick & Associates of the UK. This work has continued with SETEC Economie and Wilbur Smith Associates when Alastair Dick became Planning Adviser to Eurotunnel in February 1988.

2. The Consultants have produced a series of forecasts:

October 1985 - used for the submission by the Promoters.

April 1986   - which contained the same traffic estimates as in the submission document, used for raising equity from institutional investors and as indicator of likely demands within the Construction Contract.

February 1987- used as a basis for syndicating the Bank Loan and the public equity subscription.

June 1988   - first of the annual updating estimates as required by the syndicate of lending banks.

3. The traffic estimates produced by the Consultants should be considered as being indicative of the possible traffic volumes. The actual volumes of traffic using the transport services will depend upon the marketing policies of Eurotunnel and the companies promoting the services using the Tunnel, e.g. British Rail/SNCF, the coach operators, road freight companies, etc. Also the traffic forecasts, produced so far, are those capable of being justified to the banks and their consultants as being probable users of the system. The created traffic included in the current forecasts has been estimated on conservative grounds to allow for the uncertainties attached to this type of traffic as far as bankers are concerned.

4. **Assumptions**. The following assumptions were adopted by the Consultants for the development of their forecasts:

(a) current international agreements and conventions regarding trade and passenger movements between the United Kingdom and the Continent of Europe would remain in effect during the forecasting period. (This assumption has been progressively developed to include the moves towards a Single Internal Market proposed for 31 December 1992);

(b) no national or international emergencies will arise which will abnormally affect cross-Channel traffic;

(c) no additional fixed link across the Channel will become operational during the Concession period. (It is recognised that the Concession requires Eurotunnel to bring forward a proposal and allows for others to promote schemes after 2020);

(d) ferry companies will continue to operate cross-Channel services both before and after the commencement of Tunnel services;

(e) the Tunnel services and facilities will be adequately marketed, signposted and maintained to encourage usage;

(f) sufficient capacity will exist to cope satisfactorily with the forecast level of traffic;

(g) the operational characteristics of the Shuttle services will be compatible with the current information provided by Eurotunnel;

(h) average tariffs for Eurotunnel Shuttle services will be the same as those charged for ferry services on the Dover-Calais route in the future;

(i) the pricing policy of the Railways will be similar to the present, although in the most recent forecast some supplements have been assumed on particular services.

5. **Methodology**. The estimates of the traffic volumes likely to use the Tunnel and hence the likely revenues are based upon the following steps:

1 **What is the existing cross-Channel traffic market?**

2 How big will this market be in the future?

3 What will be the market share gained by the various
   services using the Tunnel?

4 How much additional traffic will be "created" by the
   Tunnel?

5 How much revenue will Eurotunnel receive?

An additional step is required to convert these estimates of
likely passenger and freight volumes into vehicles and trains
for a specific time period, for example per day or per hour.

6. Computer-based models were developed to implement this
methodology. Separate **global demand models**, each incorporating
many sub-models, were established for passengers and freight,
based upon the analysis of approximately twenty years of
cross-Channel traffic volumes and macro-economic and social
parameters. **Mode of travel/route allocation models** were also
established using data on passenger and freight travel patterns
that had been especially collected. The **European road, rail
and air transport networks** were also modelled in terms of
travel time, cost and frequency. These networks are compatible
with the travel demand matrices produced by the demand models.
The passenger matrices utilise a 71 zoning system while the
freight matrices consist of 6 zones in Great Britain and 24
zones for continental Europe. In addition, a model of the
cross-Channel ferry operations has been developed so that the
relationships between operating costs, traffic occupancy and
tariffs could be employed.

7. The approach adopted for each of these steps is described
in the following sections.

CROSS-CHANNEL TRAFFIC MARKET

8. **Definition:** The potential market for the Tunnel traffic
services is defined in terms of:

(a) **catchment area:** passenger and freight movements between
    Great Britain and the appropriate Continental countries;
    and
(b) **traffic categories:** traffic has been categorised into
    different types because of different growth rates,
    criteria for choice of mode of travel, routes, etc.

9. The catchment area of the potential market is as follows:

(a) for **passengers**
    (i) **without vehicles:** distinction is made for travel
         between Great Britain and
         - the **"eight principal countries"**, France, Belgium/
         Luxembourg, Netherlands, West Germany, Italy,

Switzerland, Austria, Spain;
- the **"four other European countries"**, Portugal, Greece, Yugoslavia, Denmark;
(ii) travelling by **car or coach** between Great Britain and the above 12 countries plus the rest of Europe (excluding Ireland).

(b) for **freight**
movements of commodities, other than bulk fuel, between Great Britain and the eight principal countries plus other movements that use the Channel or South North Sea ferry services (essentially this is British traffic to other Continental countries plus transiting traffic to/from Ireland).

10. Passenger trips are categorised according to:

(a) **country of residence:** United Kingdom, each of the eight principal countries, four other European countries, rest of the World;
(b) **reason for travel:** business or leisure which is further sub-divided into independent travel and inclusive tours;
(c) **mode of travel:**
 (i) by air,
 (ii) by sea without a vehicle, which are subdivided into "normal" sea foot (sometimes described as "classical" passengers) and excursionists (i.e. day-trippers),
 (iii) occupants of vehicles, subdivided into car and coach,
 (iv) by through rail (this mode of travel will only be available when the Tunnel is operational).

11. Freight traffic is categorised according to:

(a) **commodity:** 14 principal groups with 39 sub-divisions;
(b) **direction of trade:** import to, or export from the UK;
(c) **type of transport:**
 (i) unitised which is subdivided into freight carried in trucks (i.e. RoRo), containers, rail wagons;
 (ii) non-unitised which is subdivided into bulk (excluding bulk fuel) and new vehicles presently transported by ferry.

12. Size. An extensive data base of travel information has been compiled from:

(a) the International Passenger Survey (IPS) which has been continuously conducted by the UK Government since 1962 although useful time series data commenced in 1964;
(b) the UK Customs & Excise data on import and exports;
(c) statistics produced by the British Civil Aviation Authority;

(d) statistics produced by the Organisation for Economic Cooperation and Development (OECD);

(e) ferry and port statistics;

(f) supplementary surveys specially commissioned by Eurotunnel.

13. None of the regular surveys (i.e. a-d in the previous paragraph) record the total cross-Channel passenger volumes. The surveys are not for transport data but for other purposes. IPS, for example, is primarily for establishing the balance of payments in the tourism sector. It records country of residence of visitors rather than from where they travelled to the UK. This means that cross-Channel travel by, for example, residents of non-European countries have to be derived from other sources. The number of trips made by UK residents to Western Europe (excluding Eire) and Continental residents to the UK is given in Table 1.

Table 1 - Cross-Channel Passenger trips (excluding lorry drivers) (thousand per year)

Year	UK residents			Continental residents			Total
	Business	Leisure	Total	Business	Leisure	Total	
1965	964	7394	8358	730	2206	2936	11294
1970	1704	10084	11788	1342	4550	5892	17680
1975	2310	15562	17872	1986	7938	9924	27796
1980	3522	22154	25676	2878	10720	13598	39274
1985	4362	30280	34642	3316	10096	13412	48054
1986	4410	36152	40562	3666	10554	14220	54782
1987	4848	39026	43874	3966	11898	15864	59738

Source: IPS

14. Table 2 summarises the mode of transport used for freight movements between the UK and the eight principal countries.

Table 2: UK Trade Volume (Non-Fuel) through GB Seaports with the 8 Principal Countries (million nett tonnes)

Year	Unitised						Bulk	Total
	RoRo*			Rail Wagon	Container	Total		
	Acc.	Unacc.	Total					
1975	*	*	6.5	**	4.5	11.0	19.6	30.6
1980	*	*	11.2	0.7	3.2	15.2	20.2	35.4
1985	8.8	8.2	17.0	0.9	4.8	22.7	25.9	48.6
1986	9.5	8.7	18.2	0.8	5.1	24.1	29.4	53.4
1987	10.2	9.9	20.2	0.8	5.6	26.6	28.4	55.0

Source: Customs & Excise

Notes: Totals may not agree due to rounding.
*   Accompanied and unaccompanied RoRo were only separately
        identified from 1982.
**   Rail Wagon traffic was included within Bulk until 1977.

15. Global Demand Models. The passenger global demand model
consists of two families of sub-models together with a
procedure to instil non-modelled elements, e.g. trips made by
"rest of the World" residents. The **passenger trip generation
sub-models** estimate cross-Channel trips made by UK and
residents, for leisure and business purposes. The **passenger
type sub-models** of the "eight principal countries" then
allocate these trips to the various modes of transport and the
other sub-categories such as excursionists. These are;

    (a) inclusive tour sub-model,
    (b) coach passenger sub-model,
    (c) car passenger sub-model,
    (d) excursionist sub-model.

The residual trips, i.e. after allocation to Coach, Car,
Excursionist are thus "foot" passengers travelling by Air or
Sea (but excluding excursionists). The IPS time series data
was used in various forms of multiple regression techniques to
develop, systematically, the best relationships. The principal
explanatory variables were found to be GDP, Consumer
Expenditure per head, Purchasing Parity Index and Car
Ownership. The initial relationships have been further refined
as additional data has been added to the original database.

16. The freight global demand model allows estimates to be
made of the effect on foreign trade of economic growth of each
country and such as changes in customs duties. The model
consists of two sequential steps; firstly a forecast of **UK
foreign trade** in value and in tonnage, by 14 commodity groups,
then a sub-division of these groups into **unitised traffic** and
further disaggregation into 39 sub-groups of commodities.
There are three sub-models within the UK foreign trade
procedures;

    (a) **value sub-model**, which estimates foreign trade, in value
        terms, solely on the basis of the economic growth of the
        countries concerned;
    (b) **purchasing power (price effect) sub-model**, which
        incorporates the effects of exchange rates and inflation;
    (c) **value/weight sub-model**, which converts freight in value
        terms to weight terms.

The 14 commodity groups produced by the foreign trade model are
not refined enough to be used for estimating the tonnages that
could be transported by unitised transport. Therefore, the
main commodity groups were subdivided and the propensity for
being unitised determined. This relationship has been further

refined as data has become available to support a time
dependent relation which reflects the increasing tendency
toward unitised modes of transport.

17. These models have been used to estimate the global demand
for cross-Channel traffic for 1993, the expected opening year
for the Tunnel, 2003 and 2013.

18. <u>Global Demand Forecasts</u>. The estimates of the future
total cross-Channel traffic demands are derived by adopting
values for the explanatory variables for the global demand
models. The most important of these variables is the UK GDP
and its closely associated consumer expenditure per head. The
assumed values for the explanatory variables have varied
slightly since 1985, as an example Table 3 shows the values
that have been used for UK GDP growth rates.

Table 3: Assumption for UK GDP Growth Rate
(% per annum, constant prices)

Period	Year of Forecast		
	1985/6	1987	1988
1983 - 1993	2.15	-	-
1985 - 1993	-	2.15	-
1986 - 1993	-	-	2.50
1993 - 2003	2.75	2.15	2.15
2003 - 2013	-	2.00	2.00

19. The actual and forecast global cross-Channel traffic
demands are shown in Table 4.

20. The principal reasons for the increased traffic forecasts
are the updating of the model foundations by incorporating the
latest data and the higher GDP growth rates that have been
assumed for the latest forecasts. In this context it is of
interest to note that the total 1988 cross-Channel traffic
volumes were;

- passengers 58.4 million (7.4% higher than 1986),
- unitised freight 35.9 million gross Tonnes (24.6% higher
  than 1986) and,
- bulk freight 33.9 million gross Tonnes (7.4% less than
  1986).

That is the actual unitised freight flows for 1988 have already
exceeded the initial forecasts for 1993 volumes.

Table 4: Actual and Forecast cross-Channel Traffic within the Catchment Area

	Base Year: Actual			Forecast								
				1993			2003			2013		
	1983	1985	1986									
Year of Forecast	1985/6	1987	1988	1985/6	1987	1988	1985/6	1987	1988	1985/6	1987	1988
Passenger (millions)												
Foot (incl. excursionists)	33.2	35.6	39.6	49.3	46.4	56.2	75.0	62.6	80.9	–	80.3	109.2
Car	6.7	7.3	8.7	9.5	9.6	10.6	11.9	12.6	13.7	–	14.7	16.4
Coach	6.1	5.2	6.1	8.4	8.3	10.1	11.7	12.9	15.1	–	16.9	20.9
Total	46.0	48.1	54.4	67.2	64.3	76.9	98.6	88.1	109.7	–	111.9	146.5
Freight (mgT)												
Unitised	23.4	27.0	28.8	32.1	39.8	42.6	47.9	59.7	65.8	–	85.4	97.1
Bulk	30.0	33.4	36.6	41.8	44.6	48.6	64.5	62.4	69.4	–	84.4	94.5
Total	53.4	60.4	65.4	73.9	84.4	91.2	112.4	122.1	135.2	–	169.8	191.6

mgT = million gross Tonnes

TRANSPORT NETWORKS - CHOICE OF MODE OF TRANSPORT

21. The transport networks used for traffic movements between Great Britain and the Continent are;

(a) Road used by car, non-scheduled coaches and trucks.
(b) Coach used by scheduled coach services.
(c) Rail used by passengers and freight.
(d) Air used by passengers.

In addition there is a sub-network incorporating the cross-Channel multi-purpose and container ferry services. This sub-network also incorporates the Eurotunnel Shuttle services. These networks enable the compilation of the matrix of zone to zone journey times and costs. The journey times being derived from the estimated running speeds for the different types of highways (with allowances for stops for refreshments, frontiers, etc.) or from timetables for scheduled air, rail and coaches services. The travel costs for the base year networks are based upon perceived costs for car travel and actual costs or fare tariffs for the other modes of travel. The networks for future years incorporated committed improvements e.g. new motorways, railways, etc and assumed future travel costs.

22. Future travel costs and fares. Future costs for journeys, and to a major extent future fares and tariffs, between Great Britain and the Continent depend upon real increases in unit costs and productivity improvements. Significant factors in transport costs are payroll and fuel costs. These indicate a probable real increase in transport costs and thus fares for various modes of travel. In addition to these general trends in transport costs, there will be competitive responses to the new transport services using the Tunnel; in particular, by the ferries and the airlines. During the initial forecasts a fairly simplistic view was taken of these competitive responses in ferry tariffs and airline fares. However, since that time;

(a) a comprehensive study has been made of airline pricing incorporating consideration of moves towards deregulation and abolition of duty free sales, etc.

(b) a computer based model of ferry operations has been constructed.

The conclusion of these two exercises, together with the general review of transport costs, was that ferry travel and air travel between London and the near Continental cities will be cheaper in real terms after the commencement of the Tunnel transport services. This has been accounted for in the forecasts.

23. Future transport services. The most significant new transport service will be the through passenger trains between

61

London and Paris and London and Brussels. There has been a variation in the assumed journey times as the details of these services have been developed.

Table 5: Assumed Journey Times

Service	Year of Forecast			
	Conventional Train*	High Speed Train		
	1985/6	1985/6	1987	1988
London-Paris	4hr 30mins	3hr 15mins	3hr 00min	3hr 00min
London-Brussels	4hr 15mins	2hr 47mins	2hr 30mins	2hr 40mins

* In 1985/6 the TGV Nord line had not yet been approved.

24. The other new transport service is, of course, the Eurotunnel Shuttles which transport motorway traffic between the M20 at Folkestone and the A26 near Calais. The assumed characteristics of the Shuttles are summarised below.

Table 6: Motorway to Motorway Journey Times using the Shuttles
(Units: minutes)

Year of Forecast	Tourist Shuttle			Freight Shuttle		
	1985/6	1987	1988	1985/6	1987	1988
Transit through Tunnel	35	33	33	35	33	33
Toll payment, queueing, official checks	9*	15	9+	)	15	9+
Loading/unloading and waiting times	25**	23	22	) 72)	43	32
Total	69	71	64	107	91	74

```
*    Excludes payment time
**   Includes payment time
+    Assumes limited checks compatible with "Europe without
     frontiers".
```

25. Mode/Route Choice Model. The choice of mode of travel and then route depends upon the perception of the traveller of several factors including:

(a) origin to destination cost of a journey including the cross-Channel element;
(b) the overall journey time;
(c) frequency of service;
(d) perceptions of comfort, reliability and other qualitative factors.

The fundamental assumption underlying the calculation of market shares for the different modes of transport and route is that travellers will endeavour to minimise a combined function of journey times, cost and frequency. The relative importance attached to each of these factors has been ascertained by examining the existing pattern of cross-Channel movements. In particular, the perceived value of time savings by travellers of the various nationalities was established, separately for business and leisure trips, to ensure that the modelled estimates for the different modes and routes closely matched the observed pattern. The procedure also ensured that with the introduction of more attractive services (in terms of time and cost savings), there would be the creation of additional trips ("generalised price induced trips").

FORECASTS OF TUNNEL TRAFFIC
26. The estimates of Tunnel traffic flows consists of two elements;

(a) the market share of the global cross-Channel traffic flows derived by using the mode/route choice model and the details of the future networks. The procedure, as stated above, also calculates the generalised price induced trips created by the improved services and the expected price cuts in the ferry and airline fares.

(b) traffic volumes that are not taken into account in the modelling procedure. These models are based upon past trends in cross-Channel traffic and current perceived values of the existing services. They cannot, therefore, fully reflect the effects of changes in perceptions, behaviour, industrial structure and land-use patterns likely to occur when the Tunnel opens and as Europe progresses towards a single market.

27. The estimates for created traffic have been developed and included in the forecasts when sufficient evidence has been estalished. So far the full potential for created traffic has not been included in the forecasts, because, as stated in paragraph 3 above, the prime purpose of the forecasts is to establish "bankable" forecasts of likely revenues.

28. The traffic forecasts for the Tunnel services are given in Table 7. The 1993 traffic forecasts are stated as if the Tunnel were operating for the full year.

Table 7: Tunnel Traffic

(Units: million passengers/million gross tonnes)

Category	Forecast Year	1993 Diverted	1993 Created Price-induced	1993 Created Other	1993 Total	2003 Diverted	2003 Created Price-induced	2003 Created Other	2003 Total	2013 Diverted	2013 Created Price-induced	2013 Created Other	2013 Total
Shuttle:													
by car	1985/6	6.0	0.3	-	6.3	7.0	0.3	-	7.3	-	-	-	-
	1987	6.0	0.3	-	6.3	7.5	0.4	0.4	8.3	8.3	0.4	0.5	9.2
	1988	7.5	0.3	-	7.8	9.2	0.4	0.5	10.1	11.2	0.5	0.7	12.4
by coach	1985/6	6.4	1.1	-	7.5	7.8	1.1	-	8.9	-	-	-	-
	1987	6.4	0.5	-	6.9	8.6	0.7	0.5	9.8	9.9	0.8	0.6	11.3
	1988	7.0	0.5	-	7.5	9.8	0.8	0.8	11.4	12.8	1.1	1.1	15.0
freight	1985/6	6.0	-	-	6.0	7.5	-	-	7.5	-	-	-	-
	1987	7.5	-	-	7.5	10.3	-	0.2	10.5	13.0	-	0.2	13.2
	1988	8.1	-	-	8.1	11.6	-	0.6	12.2	15.4	-	0.8	16.2
Through trains													
Passenger	1985/6	14.5	1.4	-	15.9	18.8	2.0	-	20.8	-	-	-	-
	1987	14.5	2.0	-	16.5	17.9	2.5	1.0	21.4	21.6	3.1	1.4	26.1
	1988	14.0	1.4	-	15.4	17.1	1.6	1.1	19.8	19.4	1.7	1.3	22.4
freight	1985/6	7.2	-	-	7.2	11.4	-	-	11.4	-	-	-	-
	1987	7.3	-	-	7.3	10.3	-	0.3	10.6	14.2	-	0.4	14.6
	1988	7.4	-	-	7.4	10.9	-	0.5	11.4	15.6	-	0.8	16.4

29. In order that a general estimate could be established to assist in specification of the shuttle transportation system the 1985/6 traffic forecasts were converted in vehicles per hour. This procedure incorporates two steps:

(a) conversion of annual passenger and freight flows into vehicular volumes based upon average occupancy/load per vehicle. These were; 2.55 passengers per car, 40 passengers per coach, 12.4 gross tonnes per truck.

(b) conversion of annual flows to hourly flows. This was based upon the existing pattern of cross-Channel traffic through Dover and Calais. The actual pattern of hourly flows throughout the year when the Tunnel is operating will be different. However, until the commercial policy for the Shuttle services is determined it is premature to derive the likely pattern.

The hourly flows incorporated within the Construction Contract, which were based upon the 1985/6 forecasts are given in Table 8. These are given for the thirtieth busiest hour and are for a single direction.

Table 8:   30th highest hourly flows

Category	At Opening		2 0 0 3	
	Tourist Peak	Freight Peak	Tourist Peak	Freight Peak

Cars
    less than 1.85m high
    1.85m high and over
	570
	100
		185
		30
			665
			120
				220
				40

Coaches	65	20	74	25
Trucks	40	100	55	130

TUNNEL SOUS LA MANCHE - Prévisions de trafic

RESUME

1. CONTEXTE DES ETUDES

Les prévisions de trafic concernant le tunnel sous la Manche
ont eu pour premier objectif de contribuer à l'élaboration des
plans de financement du projet. Un second objectif était de
fournir les éléments d'information concernant le niveau de la
demande de trafic à satisfaire par le système de transport
développé par Eurotunnel.

Les estimations de trafic et de revenus ont été élaborées par
des consultants indépendants nommés par Eurotunnel et agréés
par le Syndicat des banques chargé de réunir les prêts.

Les consultants qui furent appointés en mai 1985 étaient
SETEC-Economie pour la France, en association avec Wilbur Smith
Associates des Etats-Unis et Alastair Dick & Associates pour le
Royaume-Uni.

Les études ont continué avec SETEC-Economie et Wilbur Smith
Associates lorsqu'Alastair Dick devint Conseiller d'Eurotunnel
en février 1988.

2. ETUDES REALISEES

Plusieurs études de trafic et de revenus ont été réalisées
depuis 1985. Les principales étapes des études ont été :
octobre 1985 pour la soumission aux deux Gouvernements par les
promoteurs du projet, avril 1986 pour la seconde émission de
capital auprès des investisseurs institutionnels, février 1987
pour la syndication du prêt bancaire et l'émission d'actions
auprès du public et enfin juin 1988 pour la première mise à
jour des prévisions de trafic à la demande d'Eurotunnel et du
Syndicat bancaire.

3. CONTEXTE DES PREVISIONS DE TRAFIC

Chacune des prévisions de trafic décrites précédemment a été
établie dans un certain contexte économique concernant la
croissance au Royaume-Uni et sur le Continent ainsi que les
temps, les fréquences et les prix de transport caractérisant
les services offerts par Eurotunnel et par les modes
concurrents.

Une liste détaillée des hypothèses dans lesquelles les
prévisions de trafic ont été régulièrement fournie dans les
divers rapports rédigés par les consultants.

4. DESCRIPTION DES ETAPES DE LA PREVISION

L'estimation des volumes de trafic susceptibles d'utiliser le
tunnel et la prévision des revenus correspondant à ces trafics
ont été fondées sur une procédure séquentielle qui est la
suivante.

(i)   Quel est le trafic existant actuellement à travers la Manche ?

(ii) Quelle sera la croissance de ce marché dans le futur ?

(iii) Quelles seront les parts de ce marché qu'Eurotunnel pourra obtenir pour chacun des services qu'il offrira ?

(iv) Quelle sera l'importance du trafic additionel qui sera "créé" par le tunnel ?

(v)   Quels seront les revenus perçus par Eurotunnel pour l'ensemble des services qu'il offrira ?

5. METHODOLOGIE

Pour traiter les cinq points précédents, un ensemble de modèles de trafic a été développé. Parmi les principaux d'entre eux, on peut mentionner :

(i)   les modèles de demande globale pour les passagers d'une part et le fret d'autre part, ces modèles se décomposant en de nombreux sous-modèles correspondant à chacun des segments du marché ;

(ii) les modèles d'affectation du trafic par route et par itinéraire entre les services offerts par Eurotunnel et les modes concurrents ;

(iii) les modèles de concurrence intermodale permettant de réaffecter les trafics d'Eurotunnel par mode de transport (concurrence entre le mode voiture particulière, le transport en autocar et le transport ferroviaire) ;

(iv) le modèle particulier consacré à l'exploitation des ferries sur les lignes Trans-Manche qui permet d'éclairer les relations existant entre les trafics, les fréquences offertes, les coûts d'exploitation et les tarifs des traversées maritimes.

L'ensemble des modèles décrit précédemment permet de déterminer à chacun des horizons étudiés les trafics captés par Eurotunnel dans chacun des segments du marché Trans-Manche et les revenus correspondant à ces trafics. De nombreux tests de sensibilité des résultats aux hypothèses concernant la croissance économique ou l'offre de transport, ont été réalisés à l'aide de ces modèles.

6. Engineering management and transportation system design

B. D. BROWN, BSc, FIMechE, FIMarE, FINucE, Engineering Director, Transmanche

SYNOPSIS. The cross Channel fixed link Project represents a milestone for engineering in many respects. Firstly the sheer size and complexity of a project where the design, construction and commissioning of the multi billion pound undertaking must be completed within 7 years. An entirely new, integrated road/rail transportation system has to be created, that functions to strict operational and safety criteria and that accommodates trains from the national railway networks. Finally, the project truly is bi-national, and brings French and British Engineering expertise together to meet the challenge of the cross Channel fixed link Project in terms of both organisation and methodology.

INTRODUCTION

1. Overall Project Concept   The basic concept of the cross Channel fixed link Project is a simple one. It is to transport passengers, their vehicles and goods by rail, under the Channel by means of a fixed link. It is therefore, a transportation system.

2. In order to fulfil this function, the system requires major infrastructure works and these together form the cross Channel fixed link project. It is shown in schematic form in Fig. 1.

3. Transportation System Concept   The concept and subsequent engineering of the project have been determined by the needs of the transportation system. The key network groupings of the system are shown in Fig. 2 together with the resulting major elements of the fixed link. The engineering organisation for the project evolved around these major elements.

4. The engineering of the Tunnels and Terminals are dealt with in detail in later papers. The present paper deals more specifically with the engineering of the transportation system and overall engineering management philosophy.

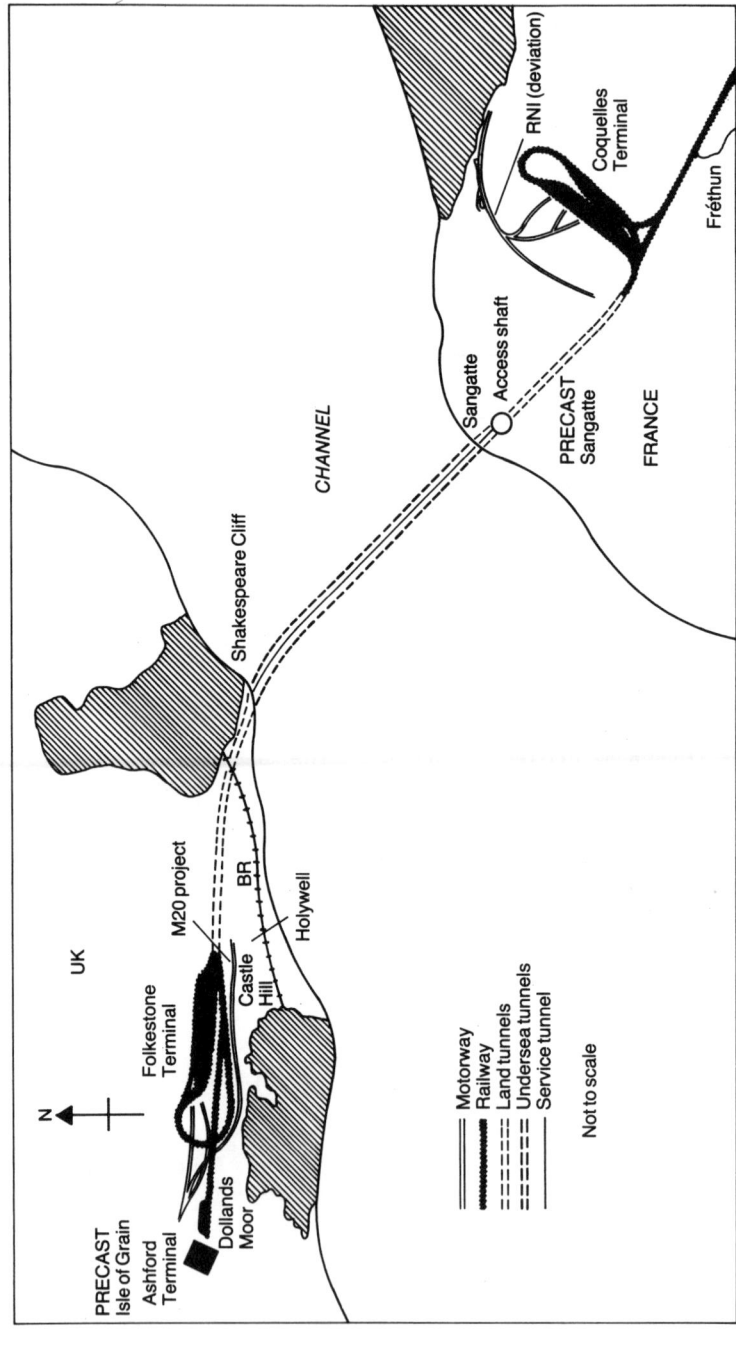

Fig. 1. Transportation System Configuration

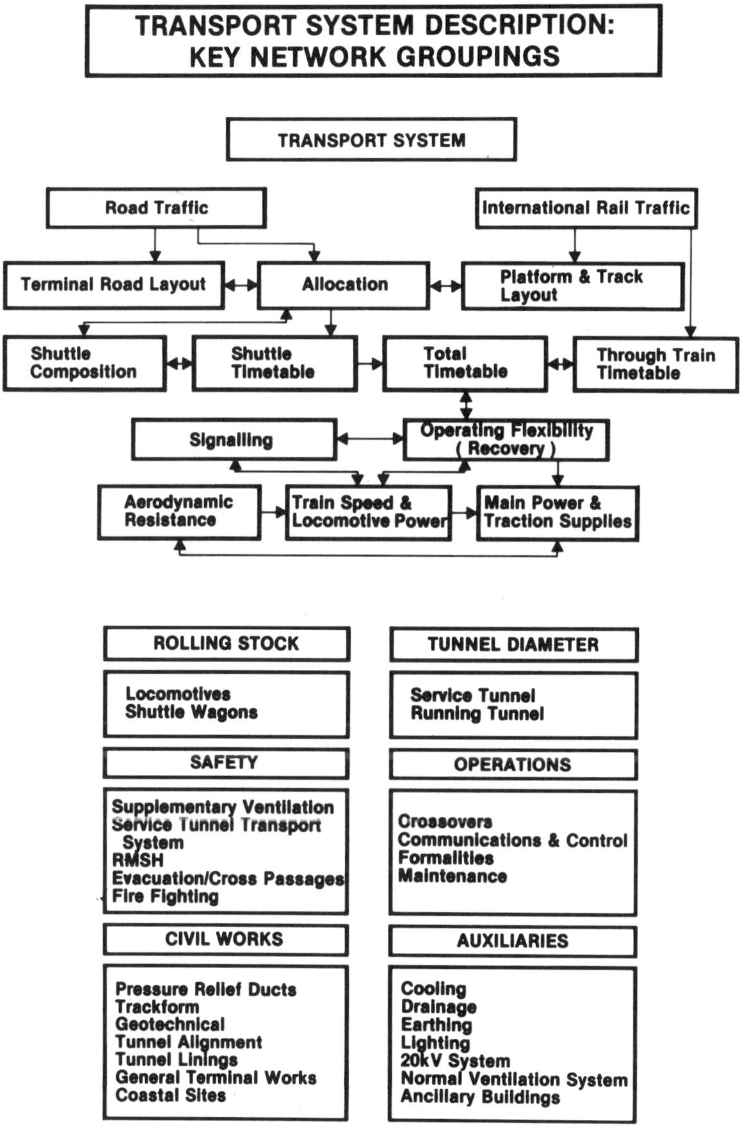

TRANSPORT SYSTEM DESCRIPTION: KEY NETWORK GROUPINGS

TRANSPORT SYSTEM

Road Traffic

International Rail Traffic

Terminal Road Layout

Allocation

Platform & Track Layout

Shuttle Composition

Shuttle Timetable

Total Timetable

Through Train Timetable

Signalling

Operating Flexibility (Recovery)

Aerodynamic Resistance

Train Speed & Locomotive Power

Main Power & Traction Supplies

ROLLING STOCK

Locomotives
Shuttle Wagons

TUNNEL DIAMETER

Service Tunnel
Running Tunnel

SAFETY

Supplementary Ventilation
Service Tunnel Transport
 System
RMSH
Evacuation/Cross Passages
Fire Fighting

OPERATIONS

Crossovers
Communications & Control
Formalities
Maintenance

CIVIL WORKS

Pressure Relief Ducts
Trackform
Geotechnical
Tunnel Alignment
Tunnel Linings
General Terminal Works
Coastal Sites

AUXILIARIES

Cooling
Drainage
Earthing
Lighting
20kV System
Normal Ventilation System
Ancillary Buildings

Fig. 2. Transportation System
Key Network Groupings

ENGINEERING MANAGEMENT OF THE PROJECT
Technical Basis of the Project

5. Although the cross Channel fixed link has a much longer history, the present project was first defined by a Proposal to the Governments of France and UK in October 1985, which included a schedule of descriptive drawings and outline specifications.

6. The drawings and specifications were subsequently included in the Construction Contract finalised between the newly formed Concessionaire - Eurotunnel (ET) and the contractor - Transmanche Link (TML) Joint Venture comprising 10 of the leading UK and French construction companies.

7. The technical basis of the project is the performance of the transportation system and is defined under each of the following five categories:

(a) Throughput of the system
(b) Performance of the shuttle trains
(c) Environment
(d) Safety and evacuation
(e) Scenarios of operational procedures.

The criteria defining this performance are included in the specifications. The drawings and the material quantities derived from them were intended as a guide to possible solutions for the project.

8. The drawings and specifications together with the Proposal to Governments were defined as the preliminary outline design of the project. This preliminary outline design represented a concept generally acceptable to the 'Employer' (ET) and summarised the employer's performance requirement for the fixed link transportation system. It was representative of a viable cross Channel fixed link but in need of reappraisal and design development. All key areas of the project design were to be included in this reappraisal.

Engineering Policy Objectives

9. To take into account the complexity and originality of the Project The channel tunnel project is one of unique nature and complexity.
(a) it is a major civil and building engineering infrastructure linked closely to geotechnical conditions and construction methods.
(b) it is also a complex transportation system
- road/rail
- shuttle/national rail networks
- bi-national regulations.

10. To take into account the unusually fast track nature of the Project The project schedule is some 12 months short from being able to fully optimise all solutions and to carry out in succession the various phases of design.For example,

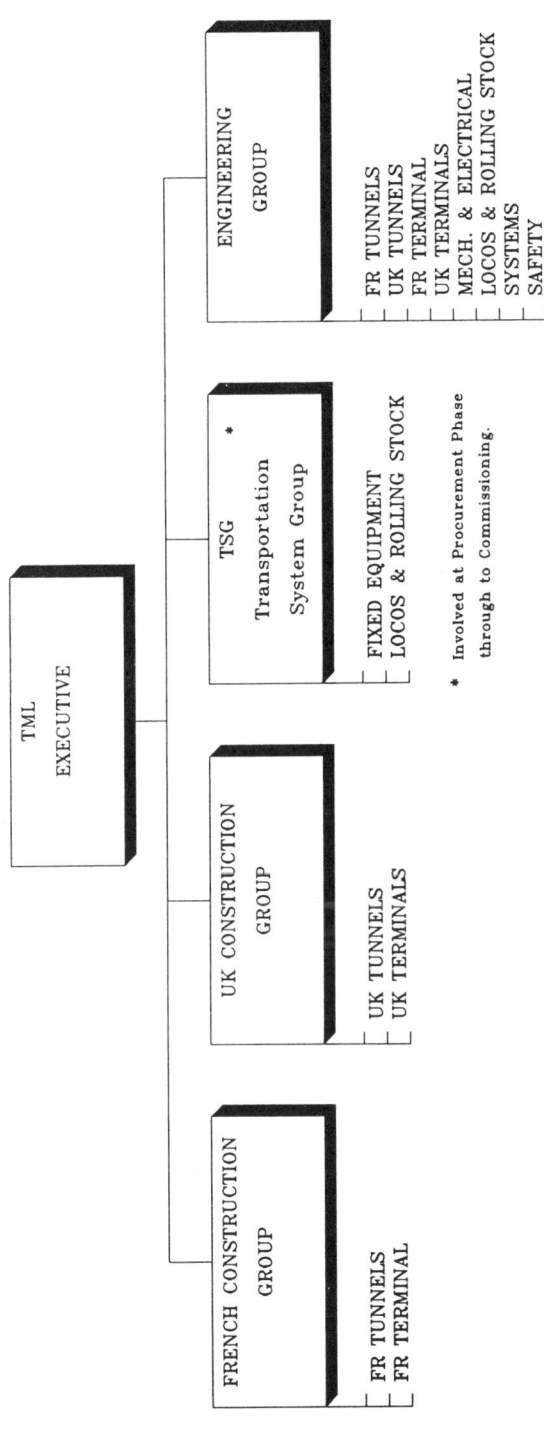

Fig. 3. TML Operational Groups

the functional studies are carried out at the same time as the civil structures design. The management of the engineering has also to take account of the design and construct nature of the project in order to maintain the overall objective to remain on schedule and within budget.

Autonomous Engineering Group

11. The engineering policy objectives led directly to the decision to create an 'autonomous' engineering group within TML. This group has responsibility for total project design and is one of the four major operational groups within TML Refer to Fig. 3. The 'autonomy' of the group was made necessary principally due to the overall design development to be carried out from the start of the project. If a fully developed transportation system had been defined from day one, then separate engineering design teams may have been a viable option. These could have been integrated into the other TML operational groups with a lightweight interfacing co-ordination organisation dealing with cross discipline issues.

Structure of the Engineering Group

12. The engineering group is a joint French - UK organisation currently based at the TML project liaison office at Sutton, Surrey. All engineering departments are located together to ensure total integrated management and a maximum common approach to project design.

13. The engineering group is managed by a single engineering director to successfully combine the multi-disciplinary nature of the project. He is responsible for all project design and is a member of TML executive committee.

14. The engineering group has been structured around three principal activities.

(a) Locomotives, rolling stock and operations.
(b) Electro-mechanical fixed equipment.
(c) Civil and building works.

15. Engineering departments have been formed to manage the design of these activities and are shown in Fig. 4. In essence, they also reflect the general organisation breakdown structure of TML thus ensuring that the engineering managers responsible for the departments, have direct opposite numbers in the other TML operational groups.

16. The next level of organisation within a department is an engineering section. This generally corresponds to an elemental or primary sub-system breakdown of the transportation system. The engineering department sections are shown in Fig. 4. A section manager is responsible to the engineering manager at this level, for the project design.

Design Management Principles

17. The central theme of engineering management has been to operate a centralised TML organisation to effectively

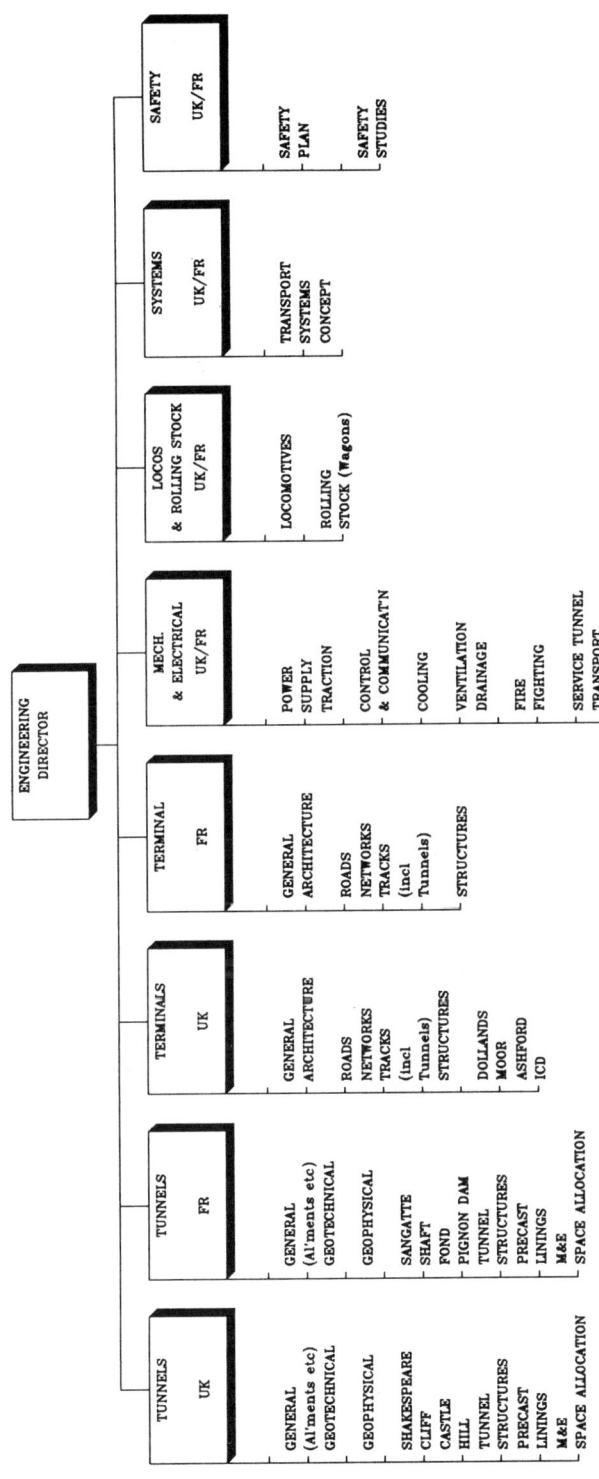

Fig. 4. TML Engineering Group

TML DEPARTMENT	Principal External Design Organisation
FR Tunnels	BETU (Bureau d'Etude Tunnels) Spie Batignolles, SGE - overall design IUR - Alignment BRGM, Geocean - Geotechnical/Geophysical
UK Tunnels	Mott Macdonald (MM) - overall design MM/Wimpey - Geotechnical/Geophysical ILF - NATM Taylor Woodrow Balfour Beatty Bush & Rennie - Structural Risk Analysis Hydraulics Research-Marine/Wave Studies
FR Terminals	BETER (Bureau d'Etude Terminal) Dumez, Bouygues, SAE - overall design Aeroports de Paris-General Architecture Simulog - Traffic simulation model CSTB Nantes - Wind Studies
UK Terminals	MM+BDP (Building Design Partnership) - overall design BMT - Wind Studies Terrasearch - Land Site Investigation
Mechanical & Electrical	MM+ EPDC (Engineering Power and Development Consultants) + Trouvin+ SGTE (Societé Générale de Techniques et d'Etudes) overall design Birmingham University - Supply Harmonics UMIST - Corrosion St Cyr Labs France - Aerodynamics model tests
Locomotives & Rolling Stock	Kennedy Henderson Ltd (KHL)/SGTE Alsthom Atlantique, GEC, BREL (ABG) Bertin-Noise NEU-Ventilation
Systems	Sofrerail, Sofretu, Transmark, KHL
Safety	SGTE/KHL CERCHAR - Fire Tests

Fig. 5.   Principal External Design Organisations

manage and co-ordinate all aspects of design, together with clear, controlled, delegated design authority to external design organisations.

18. <u>External Design Organisations</u> The volume and diversity of the development and design workload and the range of specialities necessary to develop the engineering solutions required, has meant that with very few exceptions, all work is carried out by external design organisations. (Refer to Fig. 5).

19. In the UK this has been done by several major independent design consultants and also TML member company design offices, and in France by utilising TML member company design office capacity in specially created design teams. In all cases these external design organisations report directly to the TML engineering department concerned.

20. <u>Respective Roles of TML and External design Organisations</u> Whilst the elaboration of the design is the principal task of the design office, the TML engineering group is responsible for the following main functions:

(a) overall responsibility for project design
(b) control and co-ordination of design offices
(c) interface with ET
(d) liaison with third parties (InterGovernmental Commission (IGC), MDO (maitre d'oeuvre), utility authorities, local and county authorities, national rail networks, etc,..), via ET.
(e) liaison with other TML groups
(f) acceptance on behalf of TML, of the project design.

21. The design organisations although chosen principally for their proven design track records, were also those known to be able to integrate the entrepreneurial role of TML for cost and programme management. The following considerations have therefore ranked alongside those of technical design quality, when developing the design of the project.

(a) working to an agreed programme and within an agreed budget
(b) continually seeking optimum design solutions for optimum project cost.

Particular Organisations in Engineering

22. <u>Infrastructure design Co-ordination</u> The role for co-ordination of infrastructure design has been conferred to the civil-building engineering departments. This entails organising all co-ordination and interface data and producing and distributing the subsequent co-ordination drawings. The area most affected by this role is the civil-building/mechanical-electrical interfaces. This has a special significance in the tunnels where space requirements

for mechanical and electrical equipment are the most sensitive. It is intended that once the mechanical and electrical detail design is underway, the M&E engineering department will assume this role.

23. Safety and System Design Two Engineering Departments have been established for engineering safety and systems.

24. The safety engineering department deals with all safety matters in design and liaises closely for all Avant Project submissions (document submissions for Approval under the Concession Agreement) made to the IGC.

25. The (transport) system engineering department was formed to reinforce the management of interfaces across all engineering departments and has specific responsibility for:

(a) developing and maintaining a coherent concept of the Transportation System.
(b) reviewing, checking and managing key performance criteria.
(c) design checks on selected documents for compliance with the above.
(d) Preparation of commissioning tests, to confirm system performance.

26. Technical Committees In addition to the normal day to day technical liaison carried out by each engineering department, a series of technical committees have been organised to accelerate the process of integration of engineering design, with regard to the other TML groups and ET acceptance. In particular the technical committees for tunnel construction and M&E fixed equipment have been invaluable in bringing together the various work disciplines within the engineering design teams. This has combined expertise and experience in the search for optimum solutions to technical issues related to design and construction methods.

27. Liaison with ET Although TML are responsible for the design of the project, ET comment on the TML designs and solutions proposed. An independant MdO also carries out audits and gives technical advice to ET under the terms of the Concession Agreement. There is therefore a constant process of collaboration in determining optimum solutions and criteria resulting from the requirements of ET as ultimate operator of the transportation system.

THE ENGINEERING DESIGN PROCESS
Summary of the Design Process

28. It has been seen that the preliminary outline design forming the technical basis of the concession for the cross Channel fixed link is a transportation system performance requirement at opening and for the year 2003. All design development must respect this performance requirement and the design process is structured to ensure this, as shown in Fig. 6.

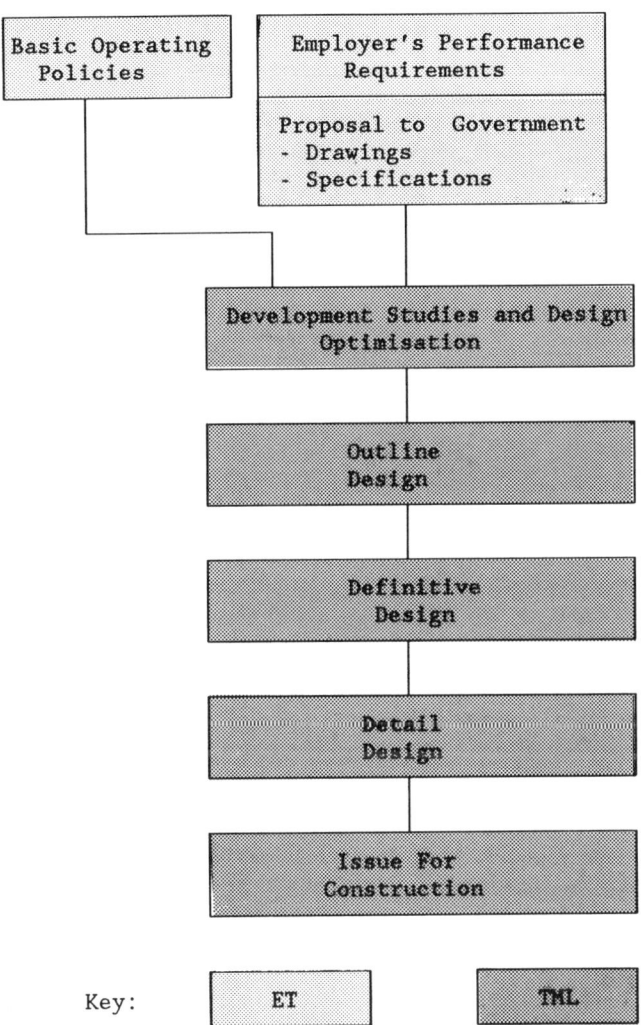

Fig. 6   Summary of Design Process

Phases of Design
29. There are four main phases of design
(a) Development Studies - Etudes de Development
(b) Outline Design   - Avant Project Sommaire (APS)
(c) Definitive Design - Avant Project Detaillé (APD)
(d) Detail Design   - Plans d'Exécution d'Ouvrages
                        (PEO)

30. **Development Studies** The reappraisal of the preliminary outline design takes place in the development study phase. This ensures that a balanced solution is reached in terms of passenger safety and comfort etc; transit time Terminal to Terminal etc; capital and operating costs.
31. In addition to the performance requirements of the preliminary outline design, certain basic transportation system operating policies were necessary to commence the design process. These included volume and traffic pattern together with operating, maintenance and safety conditions.
32. The development studies in reassessing the preliminary outline design also served to define the design criteria and also to optimise the project within the framework of the performance requirement, such that the design is
(a) cost effective
(b) contains no unnecessary features that would be unattractive to ET or the user
(c) acceptable to the IGC.

33. **Outline Design Phase - (APS)** The outline design phase is intended to build upon the design criteria and optimisation given by the development study phase in order to amplify the preliminary outline design. It also serves as the basis for submissions to the IGC for approval in accordance with the terms of the Concession Agreement.
34. **Definitive Design - (APD)** The definitive design phase is intended to act as a check on the development of the outline design phase including any modifications made at the comment stage of this design phase. Where design and construct packages are awarded, the definitive design documentation is used as the basis for the technical content of the tender documentation.
35. **Detail Design Phase - (PEO)** The detail design phase results in the production of detailed documentation (drawings, specifications, criteria, etc,..) necessary for the construction/fabrication of the finished works. In all it is estimated that over 100,000 design documents will have been produced by the end of the design phase. A specially developed technical documentation register available at any authorised PC or computer terminal, allows instant access to all information concerning design documents, whether this be

in the design offices, at the joint liaison office or at the various job sites. This ensures that the latest up to date design document listing is automatically available to all concerned.

Engineering Group Scope in the Design, Construct Commission Process

36. Although the engineering group retains responsibility for all project design, some engineering design may be carried out under the auspices of other TML operational groups. Fig. 7 summarises the management of the overall design process across the main project disciplines

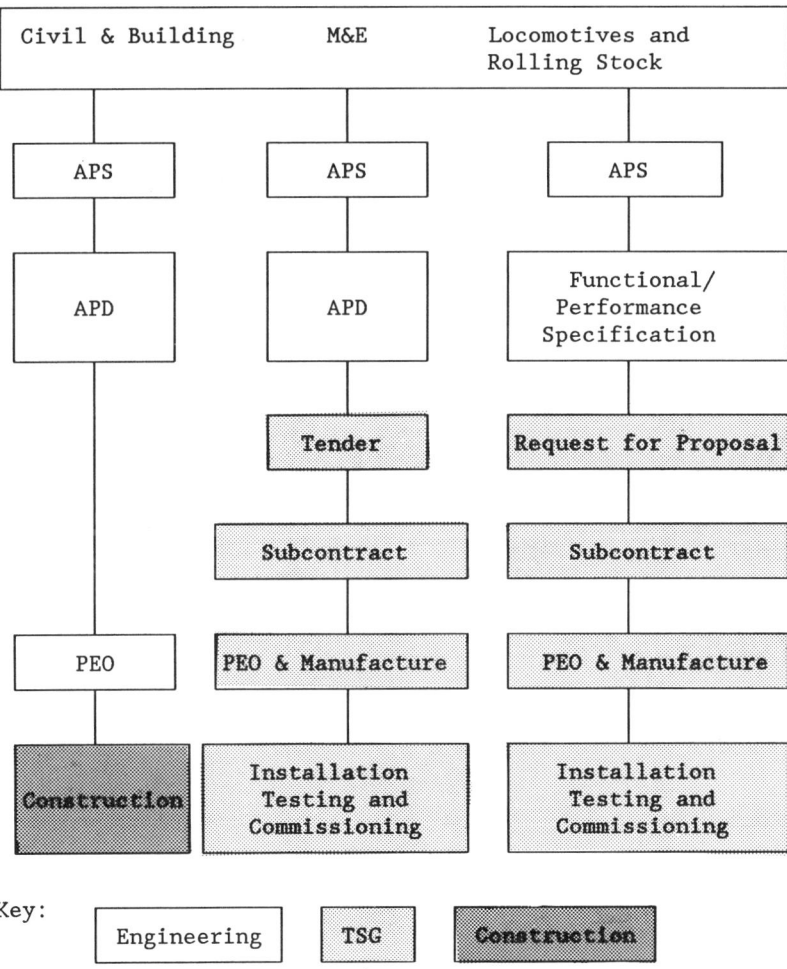

Fig. 7.   TML Engineering Group -
          Overall Design Process

TRANSPORTATION SYSTEM DESIGN

37. The transportation system consists of a tunnel rail link between two terminals one at Coquelles in France and the other at Folkestone in the UK. The users transit with their vehicles in special trains (shuttles), that move round a basic figure of eight rail configuration. For users, both terminals operate on the free exit principle. Tolls, customs and other formalities and allocation/directing of road vehicles is carried out prior to loading. An 'on demand' level of service is being designed for, avoiding the necessity for pre-booking.

38. Figure 8 lists the forecast number of different types of road vehicles to be designed for at the 30th peak hour at opening and at 2003. Under normal operating conditions platform to platform transit times of between 33 and 35 minutes are achieved . With platform dwell times of 20-22 minutes for loading, unloading and purging etc, a shuttle completes a figure of eight round trip in under two hours.

Vehicle	Opening		2003	
	At Tourist Peak	At Freight Peak	At Tourist Peak	At Freight Peak
Cars less than 1.85m high	570	185	665	220
Cars 1.85m high and over	100	30	120	40
Coaches	65	20	74	25
Lorries	40	100	55	130

Fig. 8.   30th Peak Hour Traffic Levels (One Direction)

Locomotives and Rolling Stock

39. The studies of railway operation lead to 2 distinct types of shuttles that are up to 770m long and can weigh as much as 2370 tonnes.

(a) Tourist Vehicle Carriers
(b) Heavy Goods Vehicle Carriers (HGV)

40. Single Deck Carriers The tourist shuttle comprises 24 carrier wagons and two sets of loading/unloading wagons. Twelve of the wagons are termed 'single deck' and are for transporting coaches, light vans or cars with trailers (Refer to Fig 9).

41. Double Deck Carriers The remaining twelve carrier wagons are termed 'double deck' and are for transporting cars. (Refer to Fig 10). The loading/unloading wagon for the double deck has an internal loading ramp thus avoiding the necessity for separate loading facilities at the terminal sites. (Refer to Fig 11). On average each full shuttle can carry 135 vehicles.

42. Single Deck Carriers HGV The HGV shuttle consists of 24 single deck carrier wagons and two sets of loading/unloading wagons. One HGV up to 44 tonnes may be carried per carrier wagon ie. 24 total per shuttle train.

43. Locomotives Two electric locomotives will be provided per shuttle train, each rated at 5.6 MW and powered from an overhead 25kV catenary. The two locomotives are located at the front and rear to ensure the capability of shuttle movement out of the tunnel in either direction, for a range of incident scenarios that have been extensively evaluated, to ensure a very low probability of system risk/failure. Locomotives will be equipped with an asynchronous drive system with gate turn off (GTO) convertors. All systems will be designed to maximise their 'tunnel proof' characteristics. This is basically necessary due to the effects of a 'cold' locomotive entering a 'hot' and humid tunnel with inevitable potential for condensation on the electrical equipment. With the levels of adhesion forecast for tunnel conditions, 12 powered axles per shuttle are required for both normal and abnormal (restarting) operating conditions.

44. Shuttle Wagons The single and double deck wagons are designed to create and maintain acceptable environmental transit conditions for passengers and goods. Principal amongst these are:

(a) temperature
(b) air quality
(c) ride comfort
(d) noise level
(e) air pressure transients.

45. Prototyping The nature of the design development of the rolling stock detailed in Fig. 7 has meant that certain tests, trials and prototyping work became the most cost effective way of determining optimum solutions and validating the various design and performance criteria adopted.

(a) Full Scale loading and unloading trials of all types of vehicles and passenger evacuation has taken place.

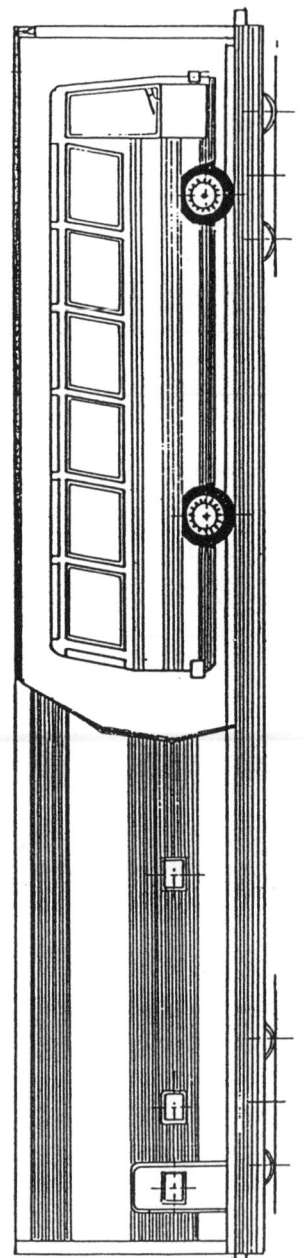

Fig. 9. Single Deck Wagon

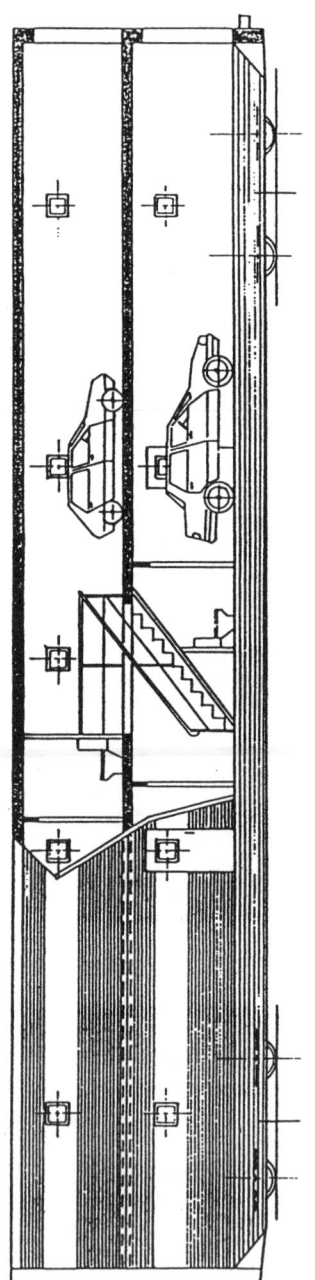

Fig. 10. Double Deck Wagon

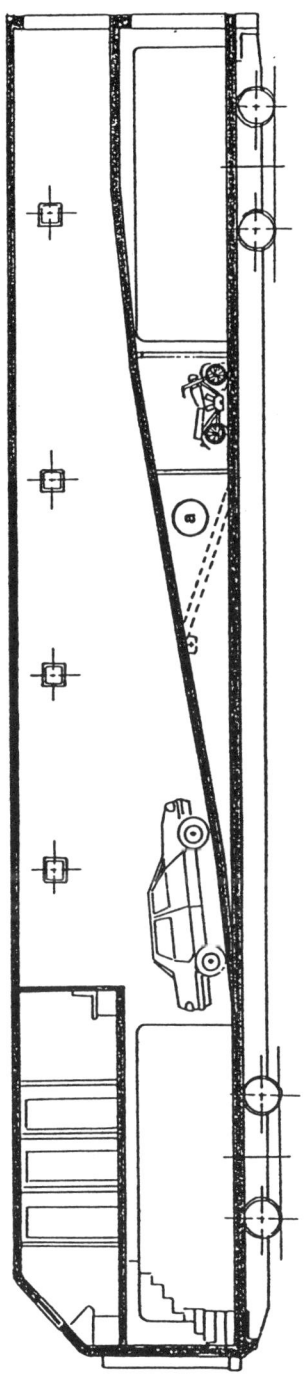

Fig. 11. Double Deck Loading Wagon

This has validated many aspects of personnel movement, vehicle movement and manoeuvrability.

(b) Full scale fire tests to check fire detection and suppression.

(c) Prototypes of the more unique features of the rolling stock have been made and tested including shuttle wagon gangway end connections, passenger vehicle side loading doors, HLV telescopic loader hood and wagon bogies.

Mechanical and Electrical Fixed Equipment

46. The overall development and design of the M&E fixed equipment has been managed by identifying the main systems. These may be divided broadly into three groups as shown in Fig. 12:

(a) Systems requiring 'in depth' technical evaluation before design commenced.

(b) Systems requiring optimisation before design commenced.

(c) Systems requiring refinement.

```
┌─────────────────────────────────────────────────────┐
│ SYSTEMS REQUIRING IN DEPTH EVALUATION                │
├─────────────────────────────────────────────────────┤
│   Tunnel Aerodynamics                                │
│   Tunnel Cooling                                     │
│   Tunnel Ventilation                                 │
│   Traction (Tunnels/Terminals)                       │
│   Signalling (Rail Network)                          │
│   Control Centres                                    │
│   Data Transmission                                  │
└─────────────────────────────────────────────────────┘

┌─────────────────────────────────────────────────────┐
│ SYSTEMS REQUIRING OPTIMISATION                       │
├─────────────────────────────────────────────────────┤
│   Fire Fighting                                      │
│   Drainage                                           │
│   Tunnel LV/Lighting                                 │
│   Edf/SEEB Main Power Supply Connection              │
│   Telecommunications                                 │
│   Tolls and Road Signalling                          │
└─────────────────────────────────────────────────────┘

┌─────────────────────────────────────────────────────┐
│ SYSTEMS TO BE REFINED                                │
├─────────────────────────────────────────────────────┤
│   Tunnel Earthing                                    │
│   Maintenance Facilities                             │
│   Fire Detection                                     │
│   Access Control                                     │
│   Terminals Power Supply                             │
│   Main Substations                                   │
└─────────────────────────────────────────────────────┘
```

Note that there will be a progression from type a) through b) to type c) as design work progresses.

Fig. 12   Principal M&E Systems

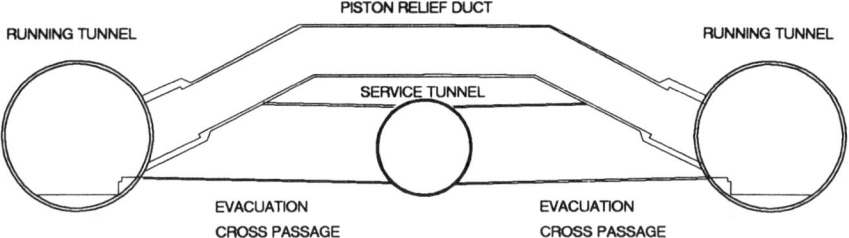

Fig. 13.   Typical Tunnel Cross Section

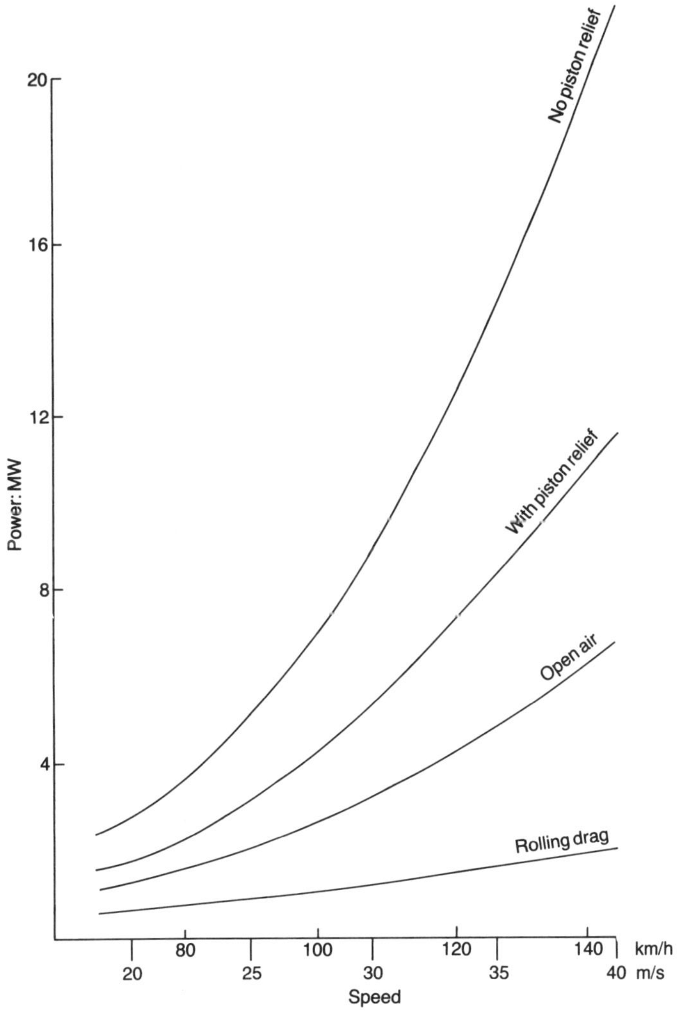

Fig. 14.   Effect of Piston Relief
Ducts on Power Demand

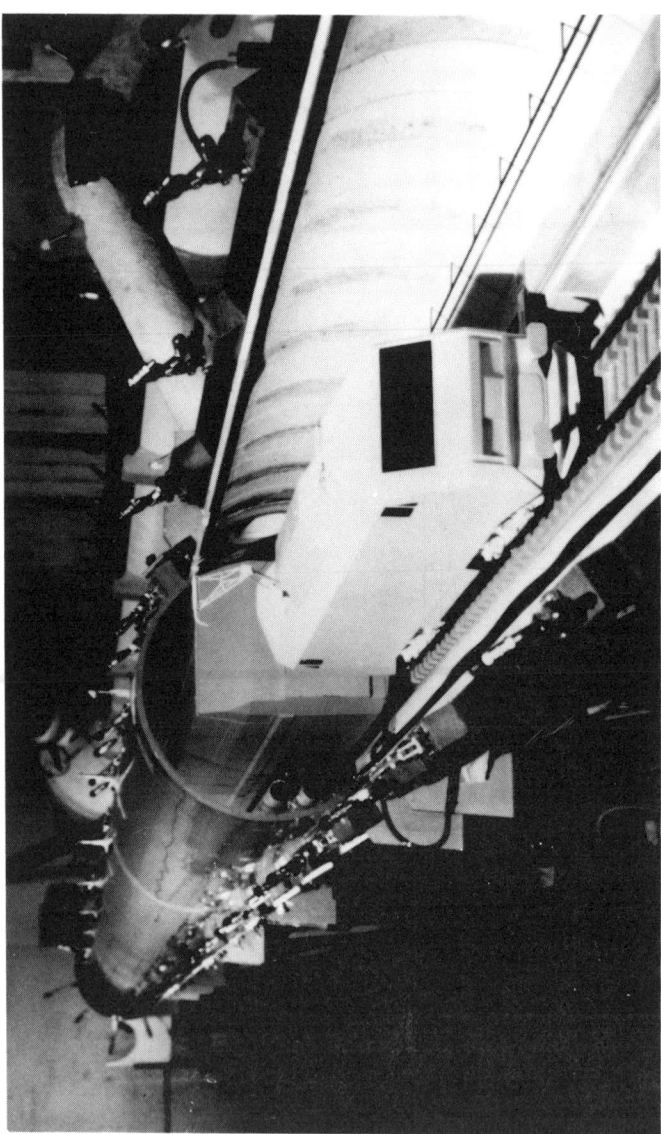

Fig. 15. Aerodynamic Test Model

47.<u>Aerodynamics</u> One of the key technical elements in the successful evaluation of the transportation system has been tunnel aerodynamics. With a direct impact on shuttle speed and locomotive power, main power and traction supplies, plus an effect onto signalling, operating flexibility, and the transportation system timetable.

48. <u>Cause and Effect of Aerodynamics</u> The shuttles generate a relatively high aerodynamic resistance in the tunnel due to their high 'blockage ratio' - the ratio of shuttle cross-sectional area to the empty tunnel cross-sectional area. The effect of the high resistance, would cause if not alleviated, unacceptable electrical power demands and low operating speeds.

49. <u>Piston Relief Ducts</u> The major effects of aerodynamic drag are significantly reduced by providing piston relief ducts between running tunnels. (Refer to Fig. 13). The ducts approximately 2m in diameter and 250m pitching pass from one running tunnel to another, over the central service tunnel. (Refer to Fig 14). They function by connecting and thereby reducing the high and low pressure regions at the front and rear of a train by means of air flow to and from the other running tunnel. However, complex pressure patterns are established as a by-product of train movements in both tunnels and these required detailed investigation.

50. <u>Aerodynamic Testing and Modelling</u> In order to comprehensively study and evaluate the aerodynamic effects of shuttle movements an exhaustive schedule of tests, mathematical and physical modelling took place including:

(a) Aerodynamic forecasting by validated computer model using the Method of Characteristics to give
    - aerodynamic resistance values
    - pressure regimes
    - air flow velocities

(b) Scale model tests of the tunnel piston relief ducts and shuttles have been made at the St Cyr laboratories in France (Refer to Fig 15). These measured the lateral force effects of piston relief duct discharge into the tunnels and its effect on the shuttles.

(c) Assessment of the acceptability of small, frequently repeated pressure pulses to the travelling public, made on volunteers in a special pressure chamber at BR Research Centre in Derby.

    (d) Tests in the Simplon tunnel in Switzerland where measurements gave the full scale results of piston relief duct effects on rolling stock.

    (e) Detailed mathematical analysis using a computational fluid dynamics computer code - to simulate certain dynamic effects.

51. Cooling The operation of high speed shuttles/trains throughout the tunnel will generate heat that will be removed by means of a chilled water bare pipe cooling system. With heat exchange at the surface of the epoxy coated steel pipe, the cooling water will circulate to heat discharge installations at Shakespeare Cliff and Sangatte before re-circulation into the tunnel.

52. Ventilation A normal ventilation system operating by the service tunnel will maintain tunnel air quality to prescribed levels. Operating at approximately 144 m^3/s it will provide an equivalent 26 m^3 per person per hour for the passengers. Supplementary ventilation of the running tunnels will control air movements during incidents (eg. to maintain the airflow over trains stopped in tunnels).

53. Traction The catenary supplying the electric locomotives at 25kV in each half tunnel is single end fed. It is capable of expansion to a more powerful 3 section catenary as rail traffic throughput grows, thus giving increased flexibility for operations.

54. Signalling The characteristics of the transportation system, provides a major technical challenge for the design of the signalling system

    (a) In tunnel maximum shuttle speed of 160 kph with average speed of 120 kph.
    (b) Through train maximum speed 200 kph. (projected).
    (c) 'Scissor' tunnel crossovers with 60 kph train speed.
    (d) 6 trains/shuttles per hour per direction during single line maintenance working between crossovers.
    (e) Mixture of shuttles and through trains together at close headways.
    (f) Long shuttles moving at close headways within relatively confined terminal track layouts.

The signalling system with automatic train protection (ATP) will provide for 20 paths (shuttle equivalent) at opening expandable to 24 paths.

55. Control Centres and Data Transmission The main system control centre will be located at the UK terminal with standby facilities at the French terminal for degraded mode operations. High use will be made of fibre optic technology (principally for avoiding interference problems)

to relay many thousands of remote control points information back to the central control centre.

56. <u>Drainage</u> The tunnel alignment between UK and France is largely fixed by geological considerations. It keeps within the railway operation gradients of not greater than 1 in 90 and to maintain minimum drainage, gradients of not less than 1 in 1000. Pumping stations located below the tunnel at low points take gravity drained seepage water and discharge it through pipes run in the Service Tunnel to treatment plants at the coastal sites.

57. <u>Trackform</u> Whereas the track in the terminals is traditional ballasted type, in the tunnels a support giving a overall dynamic resilience equivalent to ballast, is to be provided. The evaluation of trackform at the development phase led to consideration of almost 20 different configurations to be integrated into the running tunnel invert. The final trackform selection will be made on the basis of compliance with a comprehensive functional specification.

CONCLUSION

58. The overall engineering and co-ordination for the cross Channel fixed link to achieve such a unique transportation system, has been a challenge. Particularly to accomodate intensive road traffic into the two major UK and French terminals complexes with the minimum of transit time delays, then onto shuttles forming part of a high capacity, high speed, mixed traffic rail network in a tunnel configuration.

59. Design work to APD and PEO level is well in hand to achieve completion for all major areas by the end of 1989. The design is on target to meet the construction and procurement/installation phases of the project.

60. For fixed equipment and rolling stock, the emphasis on engineering activities will shift to the monitoring of designs produced by sub-contractors as well as interface management across all engineering disciplines.

61. Finally TML engineering group would like to express its appreciation to all organisations who participated with them during the engineering phase, without that assistance and cooperation it would have been difficult to achieve the overall project objectives.

RESUME

Concept du Projet Le concept de base du liaison fixe à travers la Manche est simple : il est de transporter les passagers et leurs véhicules par voie ferrée sous la Manche en utilisant un lien fixe ferroviaire. C'est donc un système de transport. L'ingénierie du projet s'est organisée autour des groupements de réseaux clés du système de transport.

Base technique du Projet La base technique du projet est représentée par les plans et les spécifications inclus dans la Proposition faite aux Gouvernements en 1985 sous le nom d'avant-projet sommaire préliminaire. Les spécifications comprennent les critères de performance auxquels le système de transport doit respondre. Ceci représentait une solution viable de liaison fixe à travers la Manche mais nécessitant une réévaluation.

Objectifs de la Politique d'Ingénierie Prendre d'abord en compte la complexité et l'originalité du projet qui représente un grand ouvrage de génie civil étroitement lié aux conditions géotechniques et aux méthodes de construction et qui est aussi un système de transport très complexe. Considérer ensuite la compression inhabituelle des délais. Il manque 12 mois pour optimiser le projet et franchir par étapes les phases d'études. Les études fonctionnelles sont menées de façon simultanée aux études de génie civil. Les méthodes adoptées sont principalement de structurer l'ingénierie autour de trois natures d'activités (matériel roulant et opérations, electromécanique, génie civil et bâtiment), et de mener avec les mêmes centres de décision et les mêmes bureaux d'études toutes les phases d'études.

Groupe Ingénierie Autonome Le groupe ingénierie est une organisation conjointe française/anglaise avec un seul directeur d'ingénierie capable d'intégrer l'aspect pluri-disciplinaire du projet.

Phases d'Etudes Depuis l'avant-projet sommaire préliminaire les études ont suivi quatre phases bien définies :

a) Les études de développement (optimiser la conception et fixer les critères d'études).

b) L'avant-projet sommaire pour confirmer les études de développement et représenter la base des soumissions de l'avant-projet à la Commission Intergouvernementale.

c) L'avant-projet détaillé pour compléter les choix et options et, en ce qui concerne les ensembles conception et construction, pour servir de base technique aux appels d'offres.

d) Les études d'exécution.

<u>Conception du Système de Transport</u> La conception du système de transport a commencé avec un nombre défini de véhicules routiers de types variés à transporter entre les deux terminaux. Les études de l'exploitation du système ferroviaire ont abouti à deux types distincts de navettes :

    a)   navette de tourisme (simple pont et double pont)
    b)   navette poids lourds (simple pont).

<u>Locomotives</u> Chaque navette comportera deux locomotives électriques d'une puissance unitaire de 5,6 MW alimentées par une caténaire de 25 kV.

<u>Prototypes</u> De nombreux prototypes ont été construits pour permettre de déterminer les solutions optimales et valider les différents critères de conception et de performance au moindre coût pour les équipements suivants : interconnections, barrières coupe-feu, bogies, porte de chargement, système de détection incendie etc,...

<u>Equipement Fixe M & E</u> L'ensemble du développement et de la conception des equipements fixes M & E a été géré en identifiant les systèmes principaux. Les systèmes nécessitant une évaluation technique approfondie avant le commencement des études sont l'aérodynamique, le système de refroidissement, la traction, la signalisation, les centres de contrôle, etc... En particulier les études d'aérodynamique ont utilisé les maquettes et les essais grandeur nature pour valider et confirmer les résultats d'étude. L'utilisation de rameaux de pistonnement a beaucoup réduit la résistance aérodynamique et ainsi la demande en alimentation des locomotives.

7. Tunnel design

G. S. CRIGHTON, BSc, FICE, FIHT, FASCE, FIEAust, Engineering
Manager—UK Tunnels, and L. LEBLOND, Ingénieur Civil de l'Ecole
Nationale des Ponts et Chaussées, Engineering Manager—French Tunnels

SYNOPSIS. This paper summarises the general aspects of the
design of some of the major civil elements of the tunnel
works. A brief description of the general geology, important
to any tunnelling project, is also given. The space
requirements provided by the tunnel civil design are dictated
by the rolling stock, mechanical, electrical and signalling
requirements and details of these designs are contained in
another paper.

INTRODUCTION
1. A design, construct, supply and equip contract was let
to TML in May 1986. The project programme dictated that
detailed engineering design had to virtually proceed in
parallel with construction. The design and construct type of
contract allowed close cooperation between the TML Engineering
and Construction Departments and helps to ensure that the
construction target dates are met.
2. Design work for the tunnel started in the Summer of 1986
and at the same time field work for additional geophysical and
geotechnical information, along the line of the tunnel was
started. The development of the precast concrete tunnel
lining mix designs and subcontracts for lining joint tests
were also initiated towards the end of 1986. Field survey
work to establish the main Channel Grid was also started at
this time.

GENERAL ARRANGEMENT
3. The geographical location of the main elements of the
tunnel works is shown in Figure 1. An overview of the tunnel
civil engineering works is given below.
4. Length of the underground works from the Beussingue
Portal to the Castle Hill Portal is 50.5 km made up as
follows:-

Underland French	3.3	km
Underland UK	9.3	km
Undersea	37.9	km

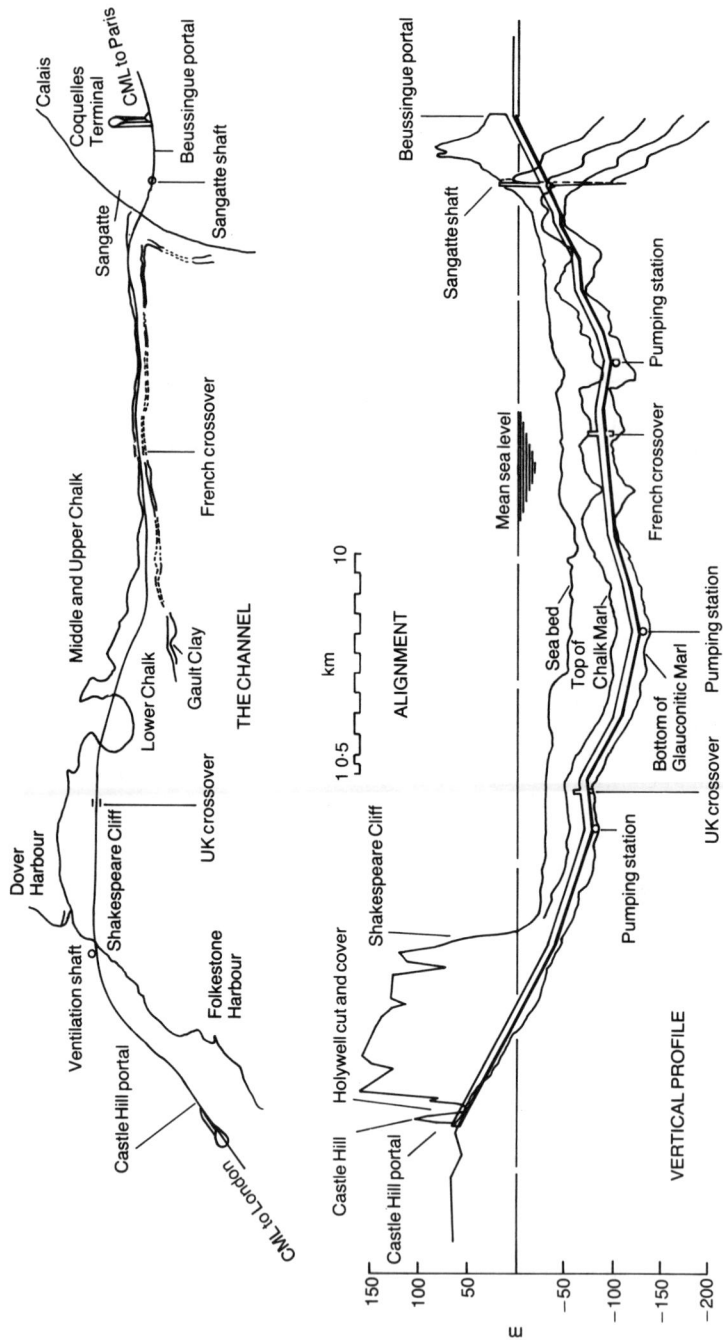

Fig. 1. Main elements of tunnel works - location

5. <u>Standard Tunnel Cross Section</u> consists of two runnings tunnels, 7.6 m internal diameter, linked to a central service tunnel of 4.8 m internal diameter.

6. <u>Linings</u> mostly consist of high strength precast concrete. The main features of these linings are described in detail later in the paper.

7. <u>Evacuation Cross Passages</u> of 3.3 m internal diameter, every 375 m, allow for passenger evacuation from one running tunnel to the service tunnel or the other running tunnel; they also house some fixed equipment.

8. <u>Electrical Equipment Rooms</u> which house the fixed equipment for control and communications and power supply, are 3.3 m internal diameter and generally adjacent to the evacuation cross passages.

9. <u>Electrical Substations</u> Six stations are located near specific installations (Shakespeare Cliff, Sangatte Shaft, Beussingue Portal and the three underground pump stations) and fifteen spread over the rest of the alignment. These are 4.8 m internal diameter, lined either with cast iron or concrete.

10. <u>Signalling Rooms</u> of 3.3 m internal diameter, are located at approximately 3 km centres and lined in cast iron.

11. <u>Neutral Section Supply Rooms</u> Three located within larger structures (Sugarloaf Hill, Undersea Crossovers) and three sited in 4.8 m internal diameter cross adits.

12. <u>Piston Relief Ducts</u> of 2 m internal diameter, at 250 m centres lined in cast iron.

13. <u>Pumping Stations</u> Three at the low points on the alignment and one at Shakespeare Cliff and one at Sangatte.

14. <u>Crossovers</u> Two situated at the third points of the tunnel. Details of the Crossovers are given later in this paper.

15. <u>Beussingue Portal</u> is the tunnel entrance/exit on the French side. The two railway lines emerge at the same level, with the service road emerging above and to the north of the railway lines. The portal structure also contains some technical equipment rooms.

16. <u>Sangatte Shaft</u> Details given later in the paper.

17. <u>Fond Pignon Dam</u> Details given later in the paper.

18. <u>Shakespeare Sea Wall</u> Details given later in the paper.

19. <u>Shakespeare Cliff Underground Works</u> Details given later in the paper.

20. <u>Holywell Cut and Cover</u> section consists of reinforced concrete box structures, 512 m long, built within an open cut.

21. <u>Castle Hill</u> section consists of three parallel tunnels, 493 m long, built using NATM primary lining and an in situ secondary concrete lining. These tunnels emerge in a 117 m long reinforced concrete box structure at the portal.

22. <u>Castle Hill Portal</u> is the tunnel entrance/exit on the British side. As with the Beussingue Portal, the two railway lines emerge at the same level with the service tunnel emerging above and to the south of the railway lines.

ALIGNMENT AND GEOLOGY
General

23. Establishment of the tunnel alignment was a critical stage in the design process. The basic geometric criteria applicable to a high speed railway was determined and other criteria applying to the minimisation of power consumption during operation of the railway, and the establishment of optimum drainage and pumping regimes were also established.

24. As well as complying with these criteria, the key issue in establishing and optimising the alignment was to locate the tunnels as far as possible within the most favourable tunnelling material, the Chalk Marl, and to avoid known fracture and fault zones or other adverse natural ground features which would reduced tunnelling progress.

25. The Chalk Marl is generally an impermeable, weak, homogeneous rock, which allows excavation to be carried out at speed. Nevertheless, fractured zones with high water inflows are experienced in the Chalk Marl. It is overlain by Grey Chalk, which in general is harder and more fractured material, giving higher overall water inflows, and requires special measures to allow tunnelling to be carried out at an acceptable rate of progress. The Chalk Marl is underlain by Gault Clay which, while essentially impermeable, could give rise to swelling action and significant time-dependent deformation on excavation, and is less favourable than the Chalk Marl for tunnelling.

26. Near the French Coast, the Gault Clay stratum is relatively thin and the proximity of the permeable Lower Greensand underlying the Gault Clay is an important consideration in determining the alignment.

27. It has been essential in determining the most suitable alignment to assess the accuracy of the geotechnical predictions of strata and feature locations. A detailed reliability analysis of the component elements of the geophysical survey work, including bathymetry, positioning accuracy, signal timing and velocity data, data accuracy weighting, and the quality of processing and data combination procedures, was carried out to attempt to maintain the tunnels within the Chalk Marl stratum.

28. The UK seaward tunnel drives start at the location of the previously constructed section of service tunnel at Shakespeare Cliff and leaves the Kent Coast at a depth of 45 m approximately below seabed, following a slight eastward dip of the Chalk Marl stratum. The tunnels reach their lower point 115 m below sea level, approximately 13 km offshore, at a point where the rock cover above the tunnels is as its greatest, approximately 75 m. Over this section the geological conditions are sensibly uniform, the strata are gently warped and the tunnel passes obliquely through the central axis of two minor folds. There is evidence of minor faulting associated with these folds which may result in the ground being fractured to a greater degree along their crests,

particularly in the more brittle chalks overlying the Chalk Marl; these are, however, well above the tunnel horizon.

29. From the low point in the UK section, the tunnels continue in the Chalk Marl and rise gently towards the mid-Channel position. Over this central section, the tunnels run parallel with, but at least a half kilometre away from, the Fosse Dangeard, a major buried valley. The extent of this buried valley was confirmed by recent surveys which do not provide any evidence of instability or discontinuity which might be associated with it at the location of the alignment.

30. The French seaward tunnel drives leave the coast at a depth of 35 m below seabed and remain within the Chalk Marl stratum wherever possible. However, geological folding, relatively high strata dip and minor flexures in the French section required careful alignment of the tunnel to maintain an optimum position.

31. The generally poorer nature of the ground over the inshore French section has been confirmed and the method of construction adopted to deal with these conditions embodies tunnelling machines which can be operated in permeable ground as fully closed pressure balance machines.

32. Several geological faults and one substantial fault were identified within the French section. The latter crosses the tunnel alignment obliquely and the fault is subject to deep penetrative weathering within the Grey Chalk, although current evidence indicates ·that this does not affect the Chalk Marl and is not expected to extend down to tunnel level.

Geology

33. Geotechnical investigations have been carried out on the Channel since 1875. In fact, over 100 boreholes have been carried out since that time.

34. Recent geotechnical and geophysical surveys were carried out in 1986 and again in 1988. The main objective of the 1986 survey was to improve, update and supplement the existing geological and geotechnical information across the Channel, particularly in areas where significant changes were made to the alignment from the previous schemes.

35. As a result of the 1986 survey it was recognised that site specific information was needed at particular structures associated with the tunnels. Hence a further survey was carried out in 1988 at the location of the crossovers and the Fosse Dangeard.

36. The stratigraphical sequences on either side of the Channel, together with typical values of thickness for the various formations are given in Figure 2 and the geological plan and section is shown on Figure 3.

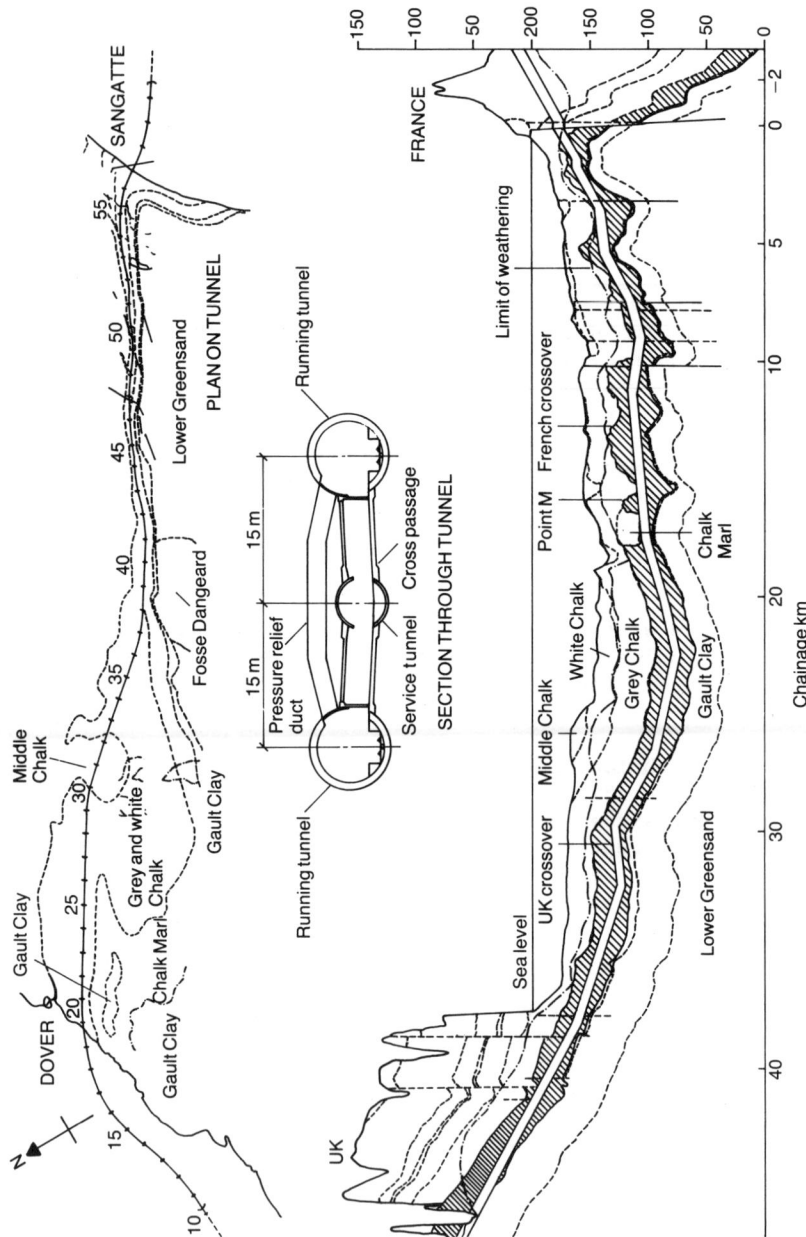

Fig. 2. Geological plan and cross-section

Formation	Typical Thickness (m) England	France	Brief Description
Middle Chalk	80 - 90	60 - 70	White marly chalk, Melbourn Rock (11m to 15m thick) and Plenus Marls (2m) at base of formation.
Lower Chalk	75	70	**Top:** Homogeneous yellow white chalk. **Middle:** Grey chalk-grey marly chalk. **Lower** : Chalk Marl-grey chalky marl. **Base** : Glauconite Marl -marly sands (2m to 3m)
Gault	45	15	Grey-blue calcareous clay/mudstone.
Lower Greensand	50	15	Alternations of weakly cemented sand and clays.

Figure 3. Simplified Geological Succession.

Geological and Geotechnical Criteria

37. The undersea section of the tunnels alignment generally complies with the following criteria, relative to the following geological horizons:-

(a) Distance to bottom of Glauconitic Marl - generally 5 m (minimum 2 m)
(b) Distance to top of Chalk Marl - generally 5 m (minimum 3 m)
(c) Cover above tunnel - 19 m minimum

The underland section of the tunnels are constrained in alignment by factors other than geology to such an extent that the scope of adjustment in respect of geology is extremely limited.

Main Alignment Operational Criteria

38. Track for Running Tunnels (Standard Section)

(a) Radii: Minimum horizontal 4200m
              Minimum vertical   15000m
(b) Gradients: Maximum   1.1%
               Minimum   0.18%
(c) Cant: Maximum   50mm

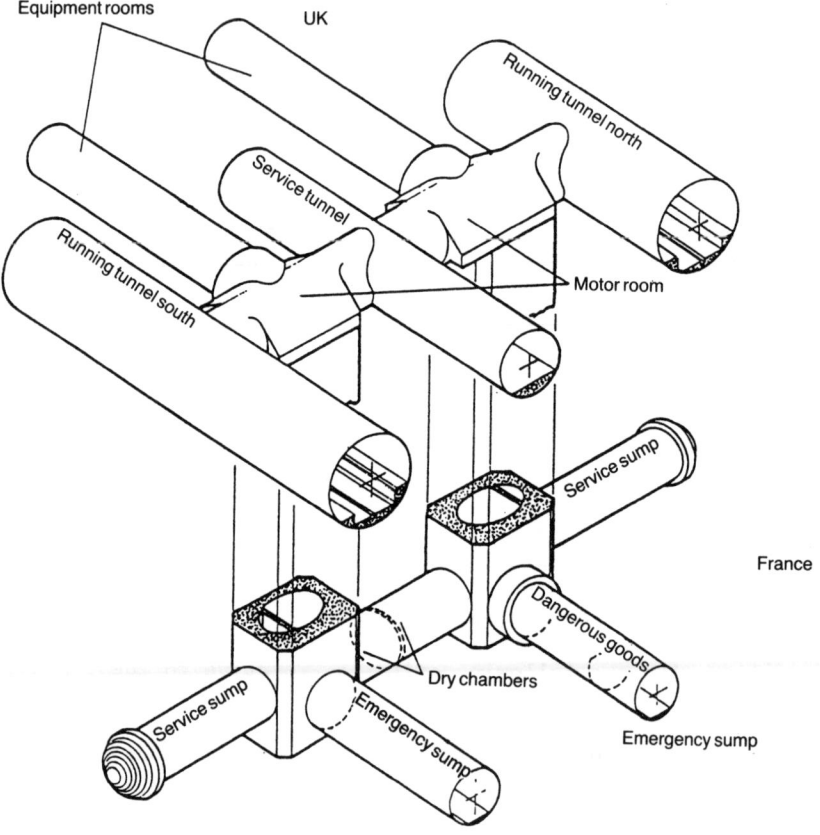

Fig. 4. French undersea pumping station and sump - isometric

39. Track for Service Tunnel

(a) Radii: Minimum horizontal 1000m
                Minimum vertical   1000m
(b) Gradients: Maximum   3.5%
                Minimum   0.18%

40. Crossover

(a) Spacing and Location: The spacing and location of the crossovers are based on operational, geological and construction criteria.
    Optimum operation spacing: 16.8 km
    Maximum operation spacing: 18 km
(b) Junction
    Speed   60 kph
    Minimum horizontal radii 475m
    Minimum vertical radii   15000m
(c) Gradient: Maximum   0.9%

TYPICAL UNDERGROUND STRUCTURES
Pumping Stations and Sumps
41. There are five main pumping stations and sump structures in the tunnel works. Three occur undersea at the tunnel low points and one at Shakespeare Cliff and another at Sangatte Shaft. The stations beneath the coastlines intercept water from the underland tunnels and prevent it from entering the undersea sections.
42. The principle functional requirements of these structures are:-

(a) collect and pump out seepage water
(b) contain and eventually dispose of liquids that may enter the tunnel system from leakages of the water service system and service pipe failures and waste water due to fire fighting
(c) impound and dispose of any accidental discharge of chemical liquids, which may enter the system.

43. An isometric of the French Undersea Station is shown in Figure 4. The UK Undersea Stations are similar.
44. Liquids are channelled from the main tunnels to sumps at the lower level. For normal operational flows, liquids are dealt with at the service sumps. Emergency sumps are provided to store water in the event of a partial or total loss of pumping capacity.
45. The entire station is symmetrical about the Service Tunnel structurally, and this allows a duplication between the two sides operationally. This is an important safety and maintenance feature.

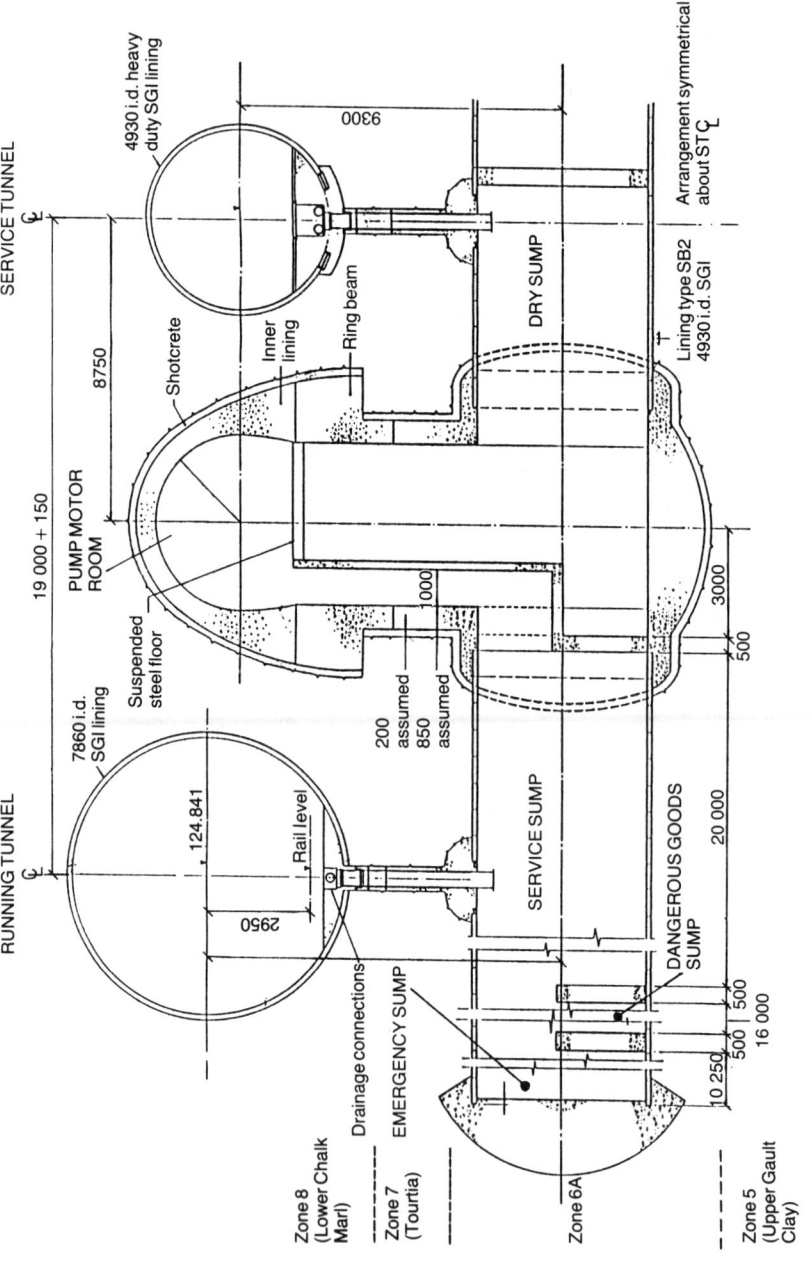

Fig. 5. UK undersea pumping station and sump – section

46. A further subdivision is made at the lower level into wet and dry chambers. The pumps are located in the dry side and driven by vertical spindles from motors at the high level.

47. Motor rooms, mechanical equipment and electrical substations are all at the high level in chambers parallel to and symmetrical about the Service Tunnel.

48. Figure 5 shows a section through a UK Undersea Pumping Station. The standard 30m spacing between Running Tunnels is increased to 38 m to accommodate the Pumping Station works between the Service and Running Tunnels.

49. The Pump Motor Rooms at the high level and the shafts are formed by NATM or classical excavation method. At the base of the shafts, eyes are formed and segmental sump adits driven perpendicular to and below the main drives.

Undersea Crossovers

50. General details of the UK Crossover are shown on figures 6, 7 and 8.

51. Crossing facilities are required between the two Running Tunnels to facilitate maintenance to sections of the tunnel. The UK Undersea Crossover is to be constructed at a position approximately 7km from the Shakespeare Cliff in an area of minimum faulting. It also satisfies the operational criteria for position and tunnel alignment.

52. The construction programme dictates that the crossover should ideally be constructed prior to the arrival of the Running Tunnel TBM's at the crossover location. A multi-disciplined development study in early 1987 concluded that this would be best achieved by the use of a single crossover structure incorporating a scissor track crossing. The track geometry for a 60 kph crossing speed determined the length of this single cavern structure which is now proposed with internal dimensions of 156m long and 18.1m wide.

53. The two Running Tunnels are deflected from a standard 30m spacing to within 10.5m centres at the cavern ends. The Service Tunnel is diverted beneath the Running Tunnels, away to the northern side of the crossover location at an offset distance of approximately 40m.

54. After the passage of the Service Tunnel TBM support train, the ST is enlarged and sloping cross adits are constructed to facilitate access to the crossover location. Design and construction of these adits and the cavern are based on the principles of NATM.

55. After the two Running Tunnel TBM's have passed, the secondary lining will be installed and internal structures completed. The construction adits are used in the permanent state to accommodate electrical rooms. Passenger evacuation crosspassages link the cavern to the Service Tunnel.

56. Numerous schemes were studied during the development of the design and the results of the definitive stage were subject of an independent review.

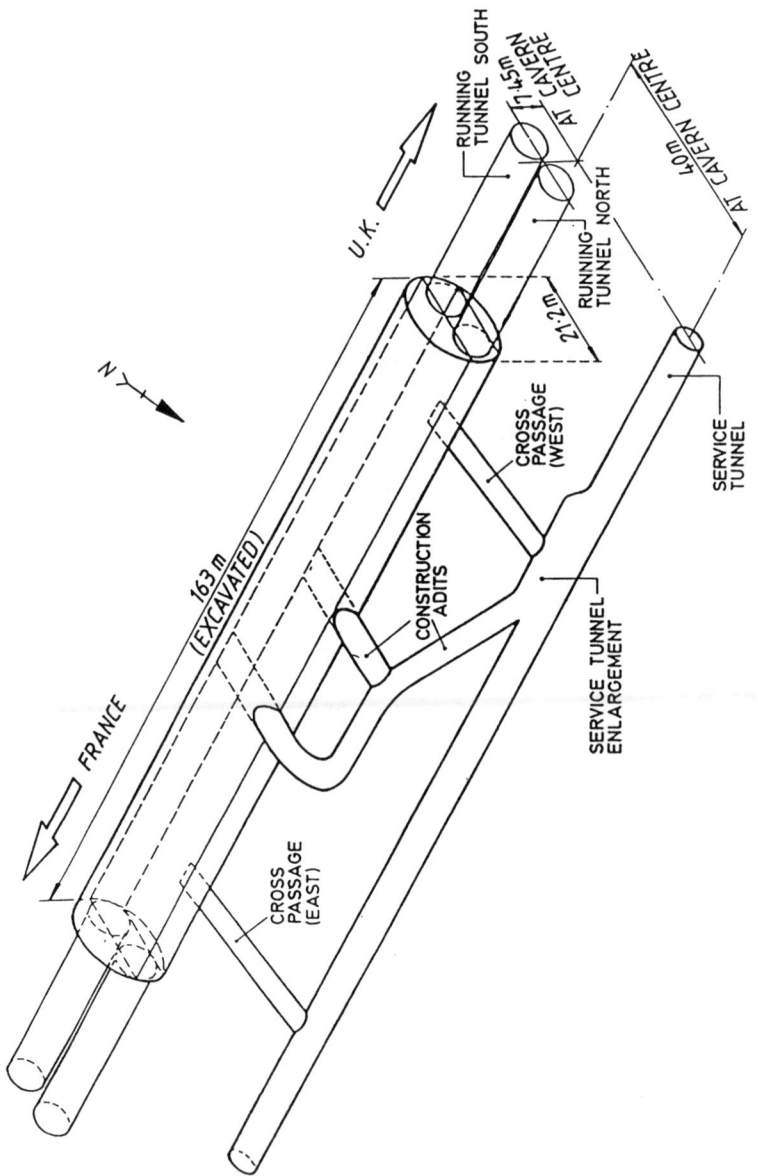

Fig. 6. Undersea crossover – isometric

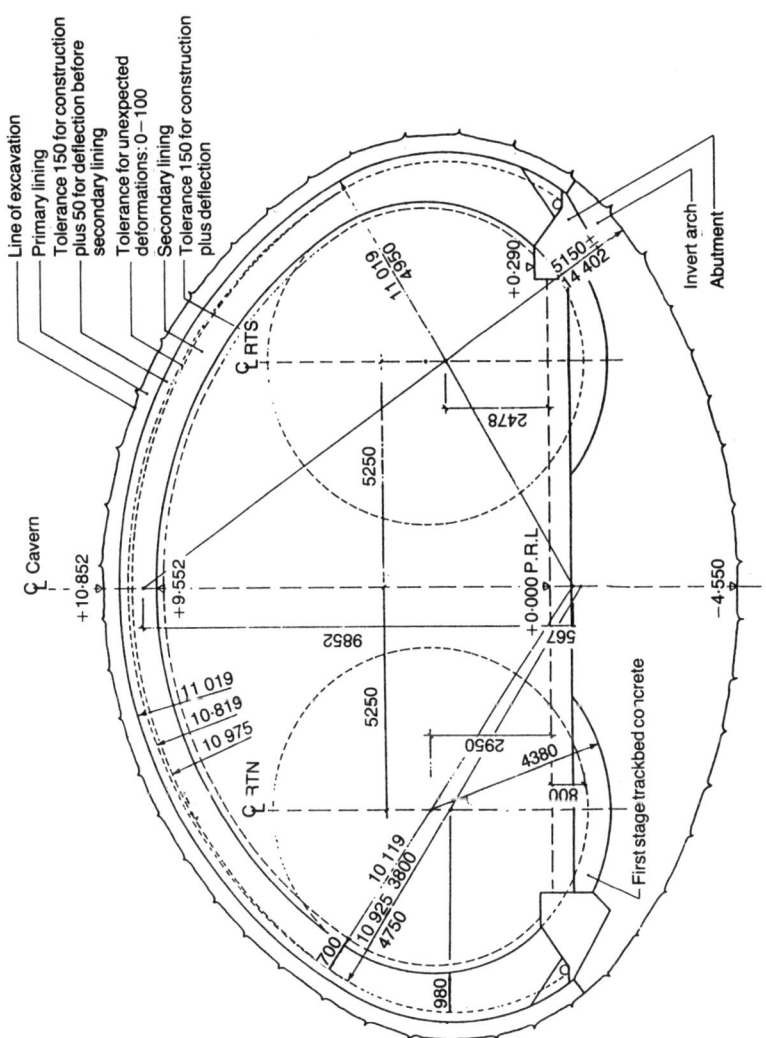

Line of excavation
Primary lining
Tolerance 150 for construction plus 50 for deflection before secondary lining
Tolerance for unexpected deformations: 0 – 100
Secondary lining
Tolerance 150 for construction plus deflection

Ȼ Cavern

Ȼ RTS

Ȼ RTN

+10·852

+9·552

11 019
10·819
10 975

11 019
4 950

+0·290

5150 ±
14 402

Invert arch
Abutment

2478

5250

5250

+0·000 P.R.L.

−4·550

9852

567

2950

4380

800

First stage trackbed concrete

10 119
3600

10·925
4750

700

980

Fig. 7. UK tunnel crossover cross-section – space proofing

107

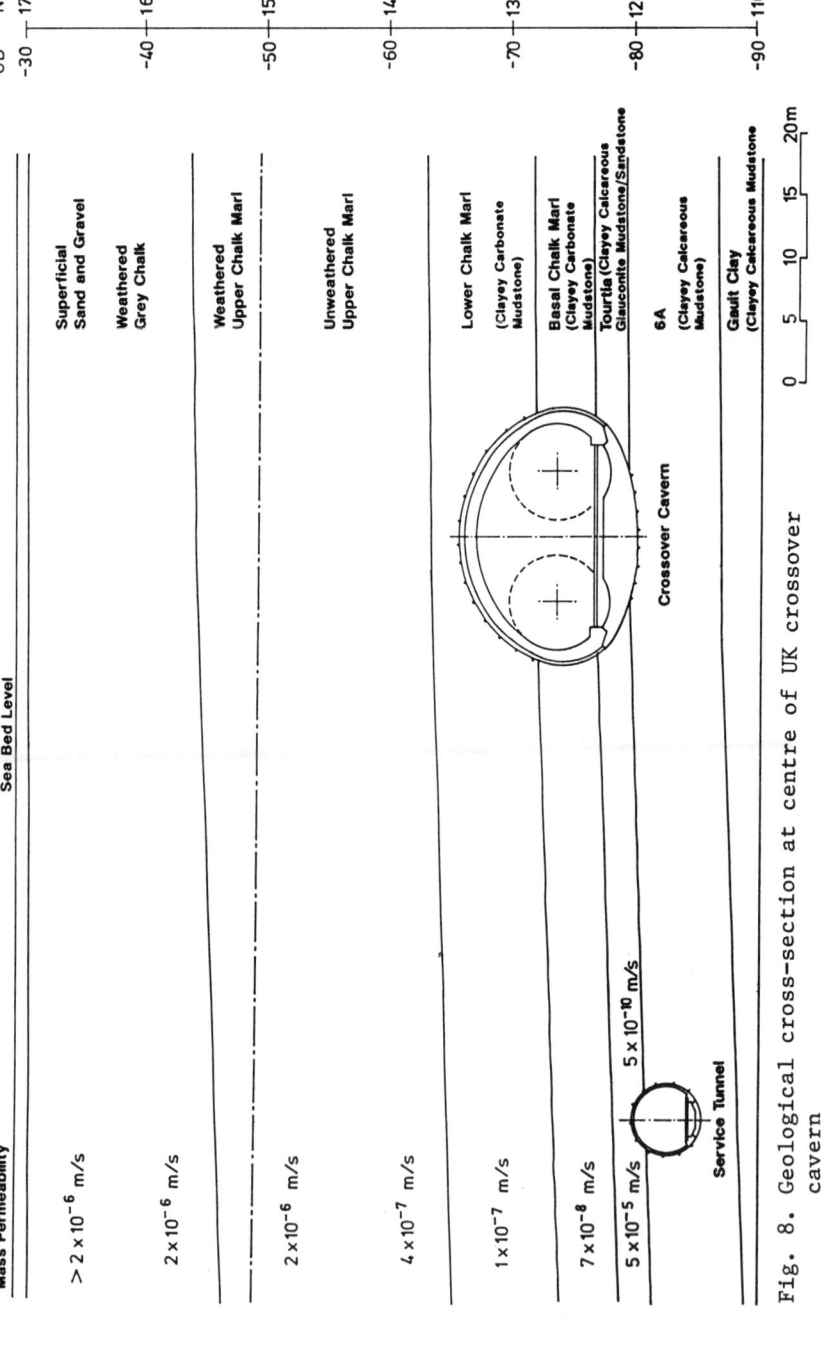

Fig. 8. Geological cross-section at centre of UK crossover cavern

57. A probabilistic risk assessment of the excavation and primary support stages of the proposed detail design scheme has been produced.

58. The design development of the crossover is illustrated graphically in Figure 9. This design development process is typical of the process undertaken for all design work on the Channel Tunnel Project.

59. At the date of writing this paper, the French Crossover design has not been finalised, as the construction date for this crossover is later than the UK one. The overall length of the French Crossover is similar to that of the UK Crossover. The construction method will take into account the programme constraints resulting from the location of the French Crossover near the meeting point of the tunnels, thus the construction method is critical. In particular, the construction method will allow the crossover to be built either from the Running Tunnels or from the Service Tunnel. The inside diameter is about 20 m. The construction method will be based on the use of small diameter tunnels adjacent to each other, filled with concrete.

Cross Adits

60. Cross Adits between the Service and Running Tunnels have two principal uses:-

(a) as Passenger Evacuation Passages and
(b) as Equipment Rooms.

61. Figures 10 and 11 show standard 3.3 ID arrangements for evacuation passages, for UK and French tunnels respectively. Spheroidal Graphite Iron linings are used. Openings from the main tunnels are formed by special hybrid concrete/cast iron sets and junctions formed with insitu concrete.

62. Equipment Rooms are of either 3.3 or 4.8m diameter depending upon the space provision required. The French arrangement is constructed after ground treatment from the main tunnels. Temporary support is placed as necessary and permanent lining formed with insitu concrete. The UK arrangement is similar to the smaller evacuation passage with SGI linings, reinforced concrete junctions and special lining sets in the main tunnels to form the openings.

Piston Relief Ducts

63. These ducts reduce aerodynamic drag on the trains, and hence traction power requirements, by allowing relief of the 'piston effect' generated by the moving train within the tunnel.

64. Figures 10 and 11 show the UK and French arrangements respectively. The configurations have been determined by consideration of both aerodynamic and construction requirements. Smooth bore SGI linings have been used in both cases. The high air velocities envisaged within the ducts would spread any water entering them as a mist through

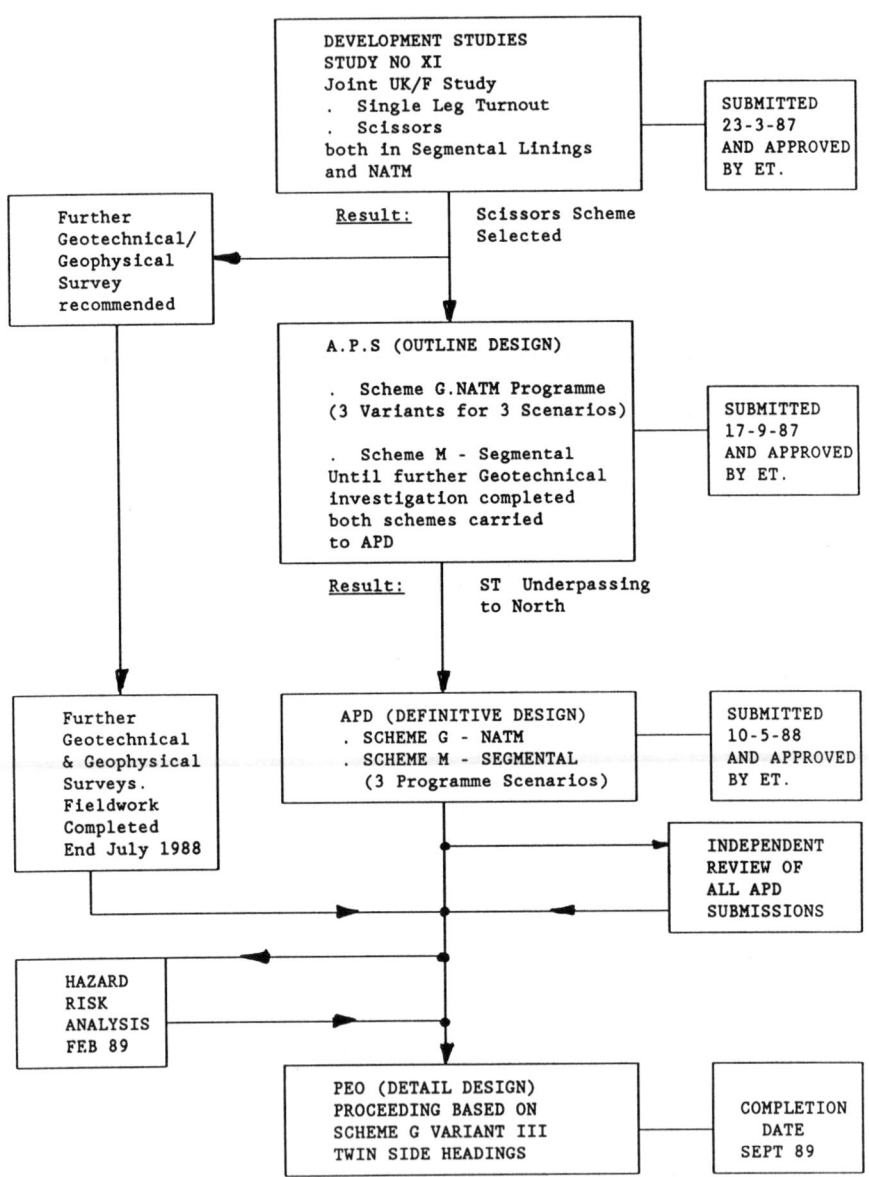

Fig. 9. UK crossover design development

UK ARRANGEMENT

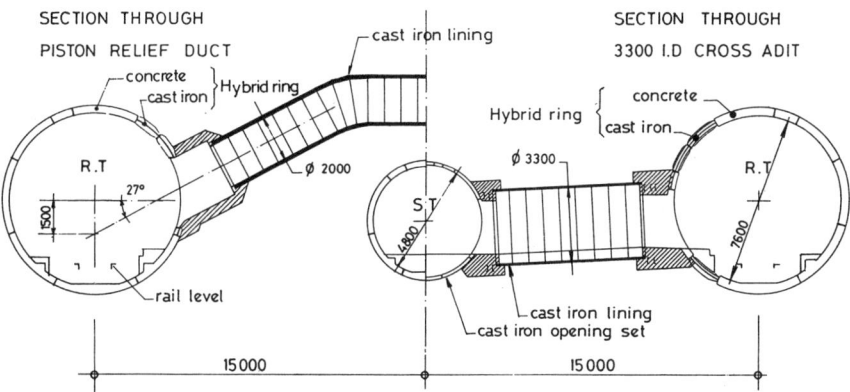

Fig. 10. Evacuation passages – UK arrangement

FRENCH ARRANGEMENT

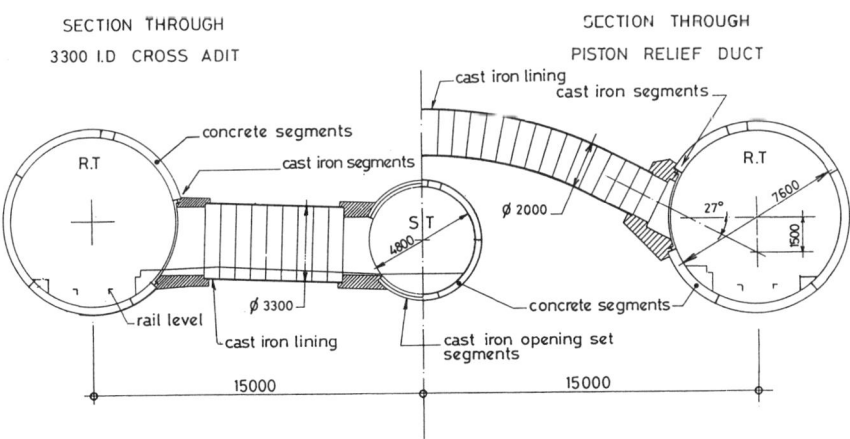

Fig. 11. Evacuation passages – French arrangement

the tunnels. Provisions have been made therefore to prevent water reaching the inside face of the linings.

SEGMENTAL TUNNEL LININGS
General
65. Segmental tunnel linings in precast reinforced concrete or cast iron are used for most of the tunnel length in normal sections and in sections where junctions occur with the cross works (cross passage, pressure relief ducts).

66. Differing geological conditions on the British and French sides of the Channel, leading to different construction methods, have meant that bolted watertight concrete segments are being used on the French side whilst an expanded articulated lining is suitable for the UK conditions.

67. Features of tunnel construction common to all of the principal drives include:

   (a) high-speed excavation and construction of circular tunnels
   (b) one-pass lining installation
   (c) probing ahead with ground treatment as necessary
   (d) ground support and protection close to the shield for safety.

68. All linings are grouted to ensure:

   (a) uniformity of contact between ground and lining
   (b) limitation of ground movements and fissures
   (c) an alkaline environment to promote lining durability
   (d) limitation of water inflow.

Design Criteria
69. Codes and Standards. There are no French, British or International codes relating explicitly to the design of tunnels. Structural design criteria have been developed specifically for the project, and as far as possible made common for the linings designed in both countries. The criteria are based upon the following main design Codes:

   (a) Concrete: BS 8110: 1985 and BAEL (83)
   (b) Steelwork: BS 5950: 1985 and CPC fasicule 61 titre 5
   (c) Spheroidal Graphite Cast Iron: As no suitable design codes are available for the use of cast iron, specific design criteria have been developed for its use.

70. Loads and Load Factors.

   Permanent Gravity Loads:
      Load factor 1.35 Ultimate Limit State
      Self weight of tunnel structures
      Ground pressures

Water pressures
Loading superimposed upon the ground
Loading from internal equipment generally.
Variable Gravity and Acceleration Loads:
Load factor generally 1.5 Ultimate Limit State
Construction traffic loads
Service loads
Braking, nosing and centrifugal forces
Handling loads upon preformed components
Tunnel Boring Machine forces during construction.
Imposed Deformations:
Temperature variation
Shrinkage and creep of concrete
Rotation and misplacement at radial joints in the
linings.
Accidental Loads:
Load factor 1.1 Ultimate Limit State
Fire
Earthquake (checked at Serviceability Limit State)

71. Materials. Partial material safety factors are
generally similar to those in the relevant national codes.
Spheroidal graphite cast iron is grade 500/7 or 600/3 to BS
2789 1985 or the equivalent in NF A32-201. Design rules for
cast iron have been derived from various recommendations,
and checked by a series of large-scale tests upon segments.
72. Design Methods. The estimation of the ground and water
loads acting upon the linings is made by analysis of the
interaction between ground and lining. Values for the
geotechnical parameters which have the largest effect upon
the loads, Young's Modulus and "creep" of the ground, are
based upon interpretation of the results of geotechnical and
geophysical surveys. The outcome of in-situ measurements made
in ground and linings in the 1974-5 scheme was also taken
fully into account.
73. Durability. Specific design criteria relating to the
achievement of long durability were evolved. The durability
of cast iron tunnel linings is provided for by a small
additional thickness to sections, and the application of a
barrier coating.
74. Very extensive studies have been made upon the subject
of durability of the reinforced precast concrete tunnel lining
segments. Dense, high strength concrete mixes have been
developed whose coefficients of permeability and diffusivity
are as low as can be found amongst standard production
concretes. (See Figure 12).
75. Performance in the event of a fire is adequate to permit
the safe evacuation of personnel over a 90 minute period.
76. In order to achieve maximum rate of progress at an
economic cost, the main type of tunnel lining chosen for the
TBM drives to build the Service and Running tunnels is
comprised of precast concrete segmental rings.

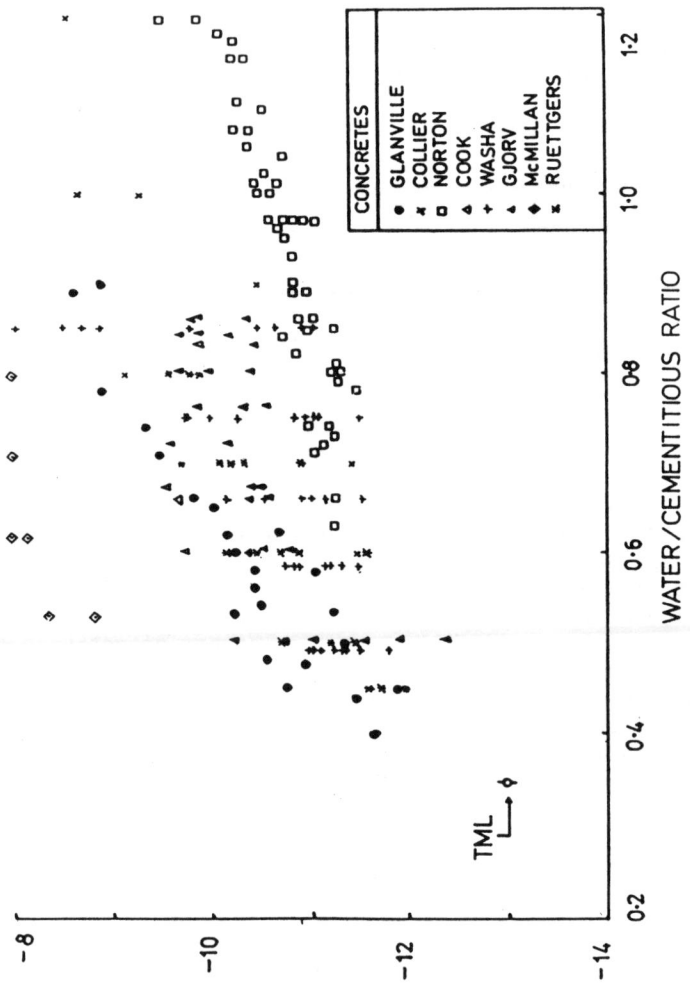

Fig. 12. Precast concrete linings. Test results and published data on water permeability coefficients for concrete

UK LININGS
Precast Concrete

77. The TBM on the UK side is an open faced machine (which can be closed in emergencies if necessary) consisting of two main assemblies, the cutting head and trailing 'gripper' section. The two sections have a telescoping action with each other.

78. Standard precast concrete lining is placed in an area immediately behind the 'gripper' section and expanded to bear against the newly excavated unlined chalk. The remaining small voids between the lining and excavated tunnel are grouted to provide uniform contact.

79. Where cast iron linings are required they are built either behind the gripper section or alternatively if required for reasons of safety can be built within the tailskin of the TBM. These linings are designed to withstand full thrust loads from the TBM.

80. The number of segments in a ring is determined by comparing design and construction requirements for efficiency. Design is efficient when bending moments in a ring are minimised; this is achieved by introducing rotation joints as often as possible. Construction is efficient if the number of precast units is minimised, if the specific tolerances are not too tight, and if the erector equipment can handle segments without difficulty.

81. Six segments plus a key were chosen for the Service Tunnel drives. Eight segments were appropriate in the larger Running Tunnels.

82. An articulated lining ring expanded against the excavated ground annulus can be erected faster than any other type of lining as it does not require the insertion and tightening of bolts. Such a lining cannot easily be made watertight during erection, as it must be erected behind the TBM shield, and not within it. It also requires that the excavated annulus behind the shield is sufficiently circular and of predictable circumferential length to permit closure of the ring using practical methods.

83. The ground conditions of the UK side of the Channel allow the use of an expanded ring over most of the length of the drives. A 'wedge key' was the preferred means of ring closure.

84. The thicknesses of the linings necessary to carry the potential loadings imposed by the surrounding ground and water in conjunction with the preferred segmentation etc, call for steel reinforcement in the segments, for two reasons:-

(a) safety during handling and erection
(b) the large stress concentrations which would occur at radial joints

85. Lining thicknesses were minimised, and resulted in dimensions ranging from 270mm to 540mm. Curved radial joint contact surfaces were needed to accommodate the anticipated rotations. The design of the radial joints is based upon results from large and small scale laboratory tests.

86. Practical design of tunnel construction equipment as well as considerations of durability of concrete tunnel linings undersea required concrete of a very high quality. A grade P60 to BSS 8110 and 5239 was specified. This is being achieved with a substantial margin.

87. Control of water inflows is firstly by grouting an annular space between the ground and the lining, formed by pads 20mm thick cast onto the extrados of the lining, and secondly by the provision of a channel at the intrados. Where water enters a joint, an insert is placed in the groove to direct the water to the drainage system.

88. The tunnels linings are shown in figures 13 and 14.

FRENCH LININGS
Precast Concrete

89. The geological conditions expected and experienced on the French side of the tunnel dictated the following general concepts:-

(a) Use of tunnel boring machines (TBMs) capable of working either in open or closed mode to control water inflow and withstand full hydrostatic pressure if necessary.
(b) Possibility in closed mode of introducing pressure between the excavation face and the seal.
(c) Segments placed directly from inside the pressure seal, the thrust forces of the TBM jacks being taken up by the lining already installed.
(d) Use of compressible sealing gaskets between segments and between rings.
(e) Provision of articulated radial joints between segments for improved centring of the compressive efforts.

90. Bolted, watertight precast concrete segments were designed to comply with the choice of the construction equipment and to limit the water inflow during construction and operation.

91. The reinforced concrete precast segments are of uniform thickness for the whole length of the French part of the tunnel; ie 320mm thick for the Service Tunnel segments, 400mm thick for the Running Tunnel segments.

92. Segments are manufactured in the dedicated precast plant in Sangatte. They are cast horizontally. After thermal treatment, neoprene gasket are fitted and the segments then automatically directed to the storage area.

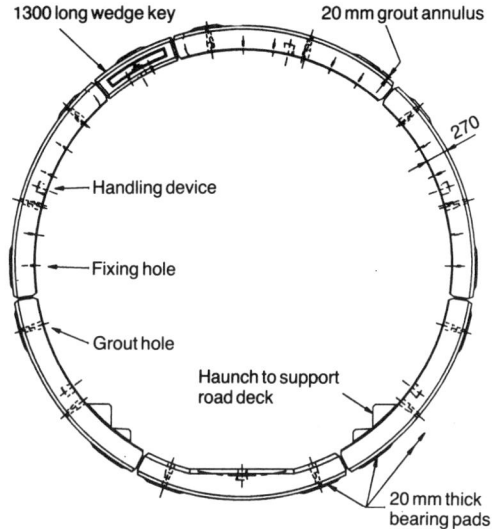

Fig. 13. UK side 4.8.m i.d. service tunnel lining, ring length 1.5 m

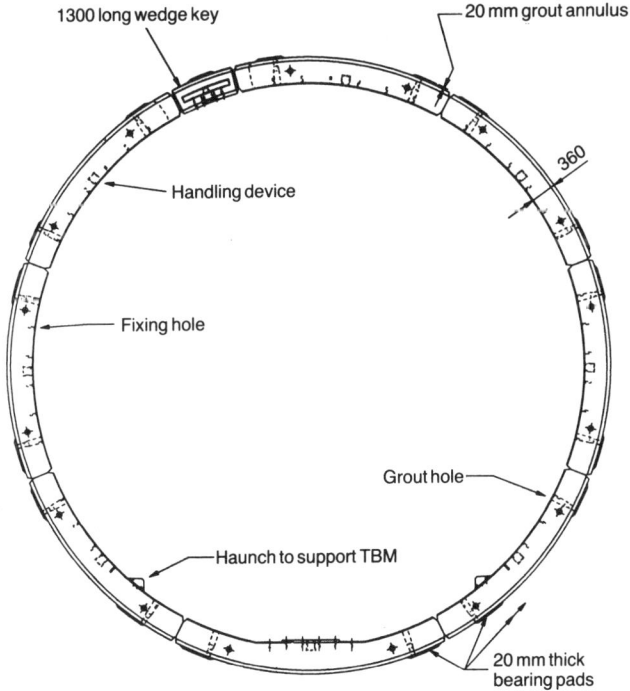

Fig. 14. UK side 7.6 m i.d. running tunnel lining, ring length 1.5 m

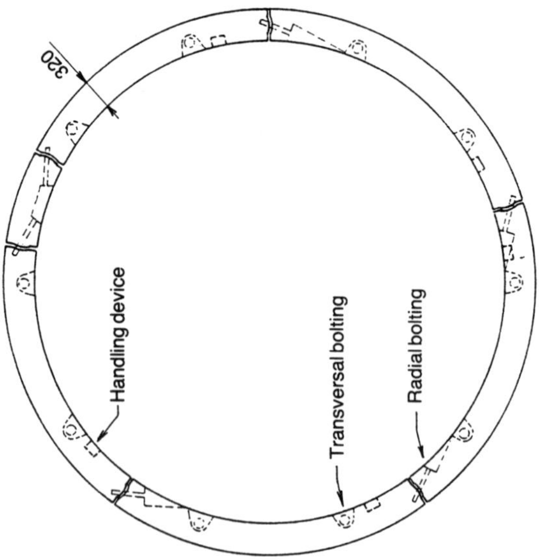

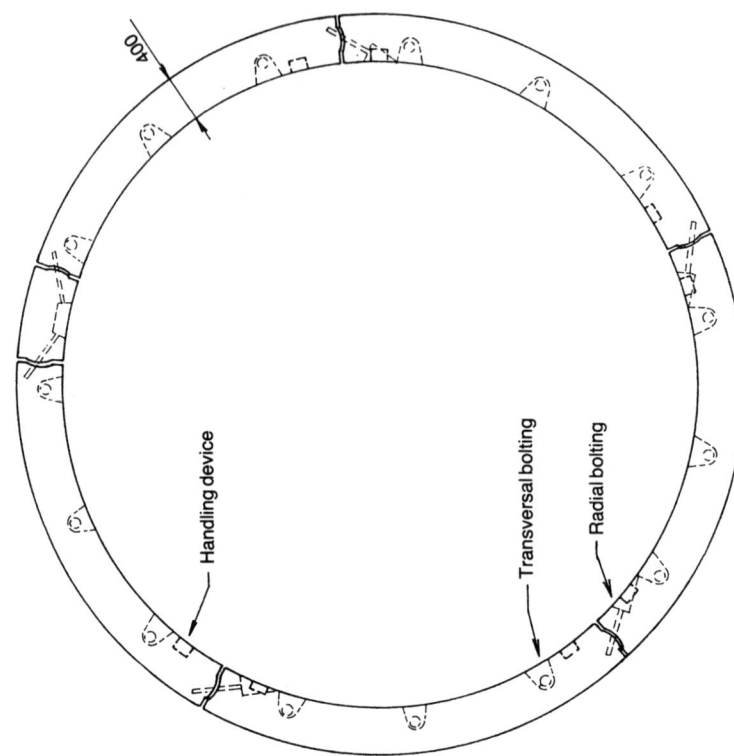

Handling device

Transversal bolting

Radial bolting

400

Fig. 16. French side 7.6 m i.d. running tunnel lining, ring length 1.60 m

Handling device

Transversal bolting

Radial bolting

320

Fig. 15. French side 4.8 m i.d. service tunnel lining, ring length 1.4 m

93. For transportation to TBM location, the segments are placed on special supports grouped in sixes to constitute a complete ring.

94. The segments are erected and bolted one by one inside the TBM shield. Depending on the working mode chosen for the TBM (open or closed), the TBM can bore (or not) during erection of segments.

95. The strength of segments was tested in CEBTP laboratory and has proven to be in accordance with design calculations.

96. The layout of typical segments and ring is shown on figure 15 and 16.

CAST IRON LININGS

97. Cast iron linings are 2 to 5 times more expensive than their precast equivalents. Cast iron segmental linings are employed in three types of locations:-

(a) at openings between tunnels, primarily in the Service and Running Tunnels at intersections with crosspassages and other tunnels;

(b) in the Service and Running Tunnels where ground conditions are too poor to erect the expanded precast concrete lining safety or sufficiently quickly, or at special chambers, such as parts of the Crossover and the Pumping Station complexes.

(c) in hand-excavated tunnels, such as cross-passages and piston relief ducts.

98. The intersection of two tunnels will impose additional loads upon the linings adjacent to their junction. It is not economic, in the UK drive conditions, to call for increased excavation diameter or lining thickness everywhere to cope with localised extra loads, and cast iron linings can be made shallower than concrete.

99. On the UK side, where ground conditions are poor, linings may need to be made watertight. The lengths of tunnel where this is the case are expected to be only a small proportion of the total drive on the UK side.

100. The savings in construction costs in hand-excavated tunnels if cast iron linings are employed come from the reduced volume of excavation as linings are thinner, and the ease of erection of segments which weigh only a quarter of their segmental concrete equivalents.

101. As far as possible linings at openings are designed to:-

(a) obviate the need to dismantle the previously erected linings

(b) provide linings at openings which can be erected in the same way as the standard precast concrete segments

(c) Hybrid cast iron opening sets were designed for the cross passage and piston relief duct openings in the UK

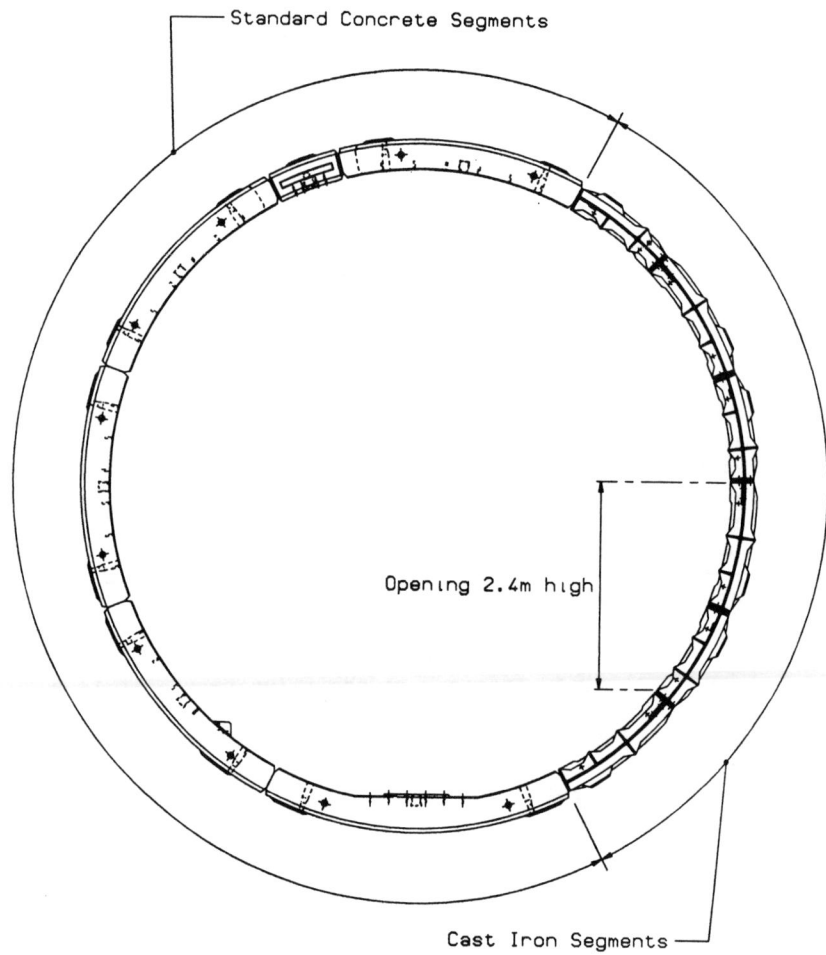

Fig. 17. Section through hybrid opening

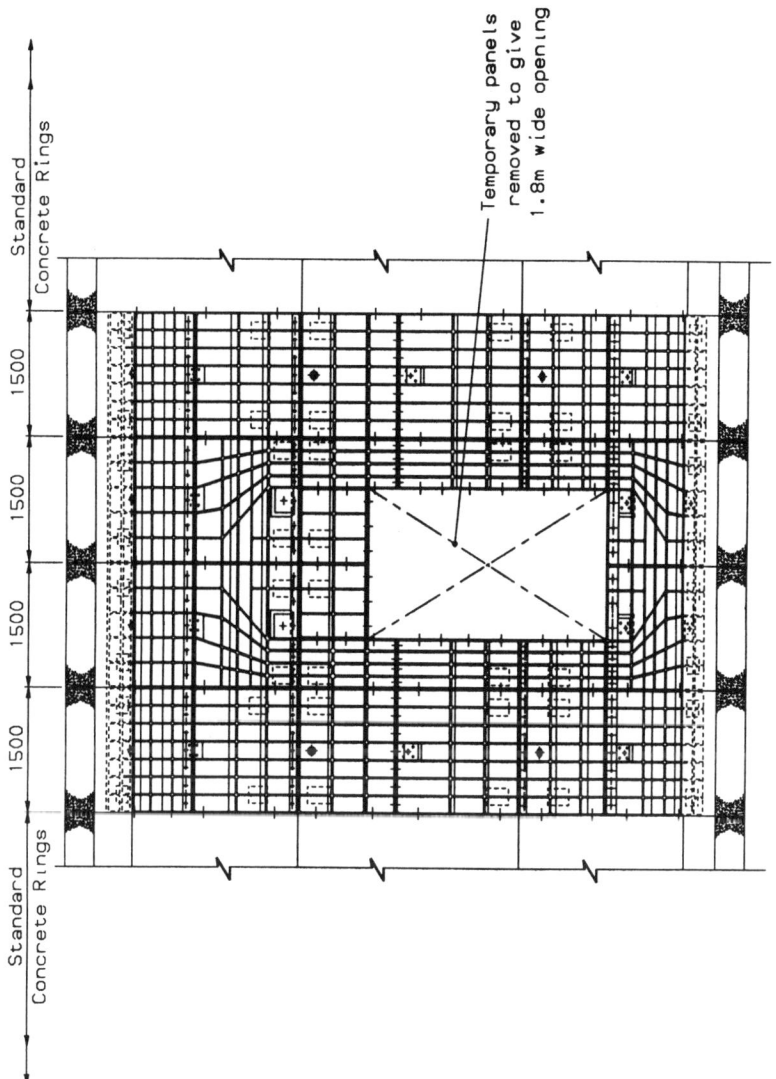

Temporary panels removed to give 1.8m wide opening

Standard Concrete Rings

1500 1500 1500 1500

Standard Concrete Rings

Fig. 18. Elevation of hybrid opening

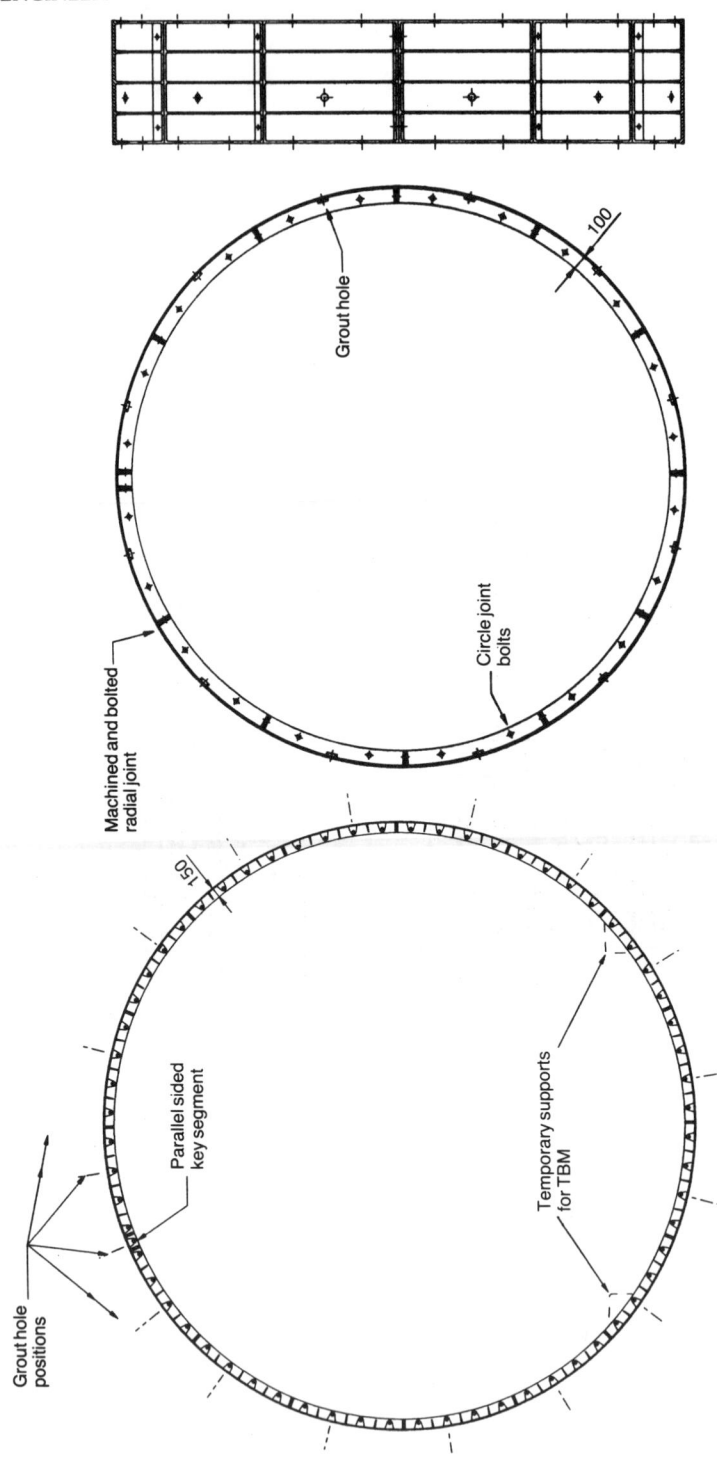

Fig. 19. Running tunnel – bolted cast iron lining, ring length 1.5 m

Fig. 20. Crosspassage lining, 3.3 m i.d., ring length 0.75 m

running tunnels. They were designed so that they could be built by the TBM in conjuction with cheaper standard precast concrete linings. See Figures 17 and 18 for details.
(d) Standard precast concrete segments are placed by TBM's at openings in the French Running Tunnels. These segments are then cut and cast iron elements placed to form the final opening.

102. These objectives call for expanded, articulated linings in the TBM drives (UK side) and bolted linings in hand-excavated chambers. A typical TBM lining is shown in Figure 19. Figure 20 shows a typical hand-excavated cast iron for the cross passage. Piston relief ducts are provided with a smooth internal profile for aerodynamic efficiency (See Figure 21 for details).
103. Spheroidal Graphite Cast Iron, grade 600.3 was chosen for these linings, as it possesses substantial ductility and high strength. Since a large proportion of the cost of cast iron linings is related to the weight of metal required, the high strength of this grade is beneficial, this benefit is enhanced because slender sections attract lower bending moments.

SHAKESPEARE WORKS
Adits and Marshalling Area
104. The UK tunnel works main construction site is at Shakespeare Cliff. The access adit A1 and the short stretch of Service Tunnel constructed in 1974 has been adopted for incorporation in the works. All the other underground structures at Shakespeare have been constructed under this contract.
105. The Shakespeare tunnels serve the following purposes:
(a) chambers for assembly and launch of six TBM's;
(b) supply and resourcing of all labour, plant and materials for the UK bored tunnels
(c) conveying, bunkering and removal of tunnel spoil
(d) supply and control of all temporary construction services (power, ventilation, water, drainage, locomotives, communications)
(e) permanent tunnel ventilation systems
(f) supply of permanent tunnel cooling system
(g) main permanent power distribution
(h) permanent tunnel drainage
(i) 500m long sections of the service and running tunnels (See Figures 22 and 23 for details).

106. The programme for the Channel Tunnel project demanded the fastest possible construction techniques to ensure that the TBM drives commenced on time. This programme require-ment led to the adoption of the New Austrian Tunnelling Method (NATM), with the secondary lining construction

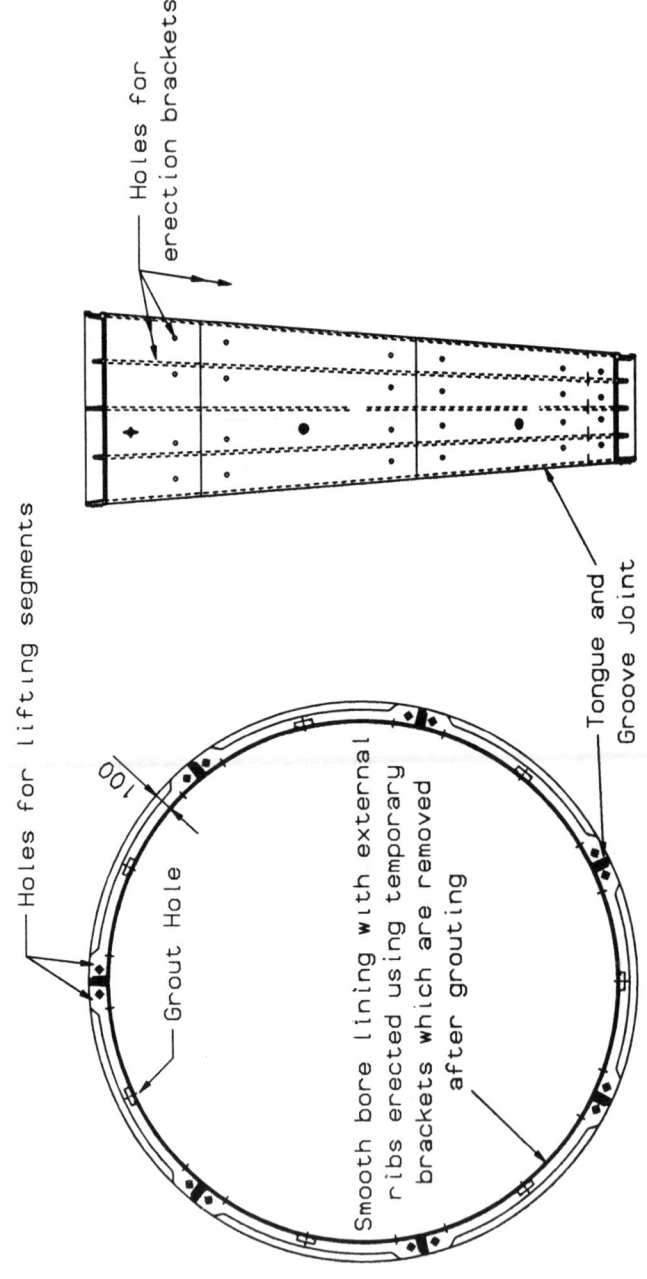

Fig. 21. Pressure relief duct, lining 2 m i.d., ring length
0.75 m

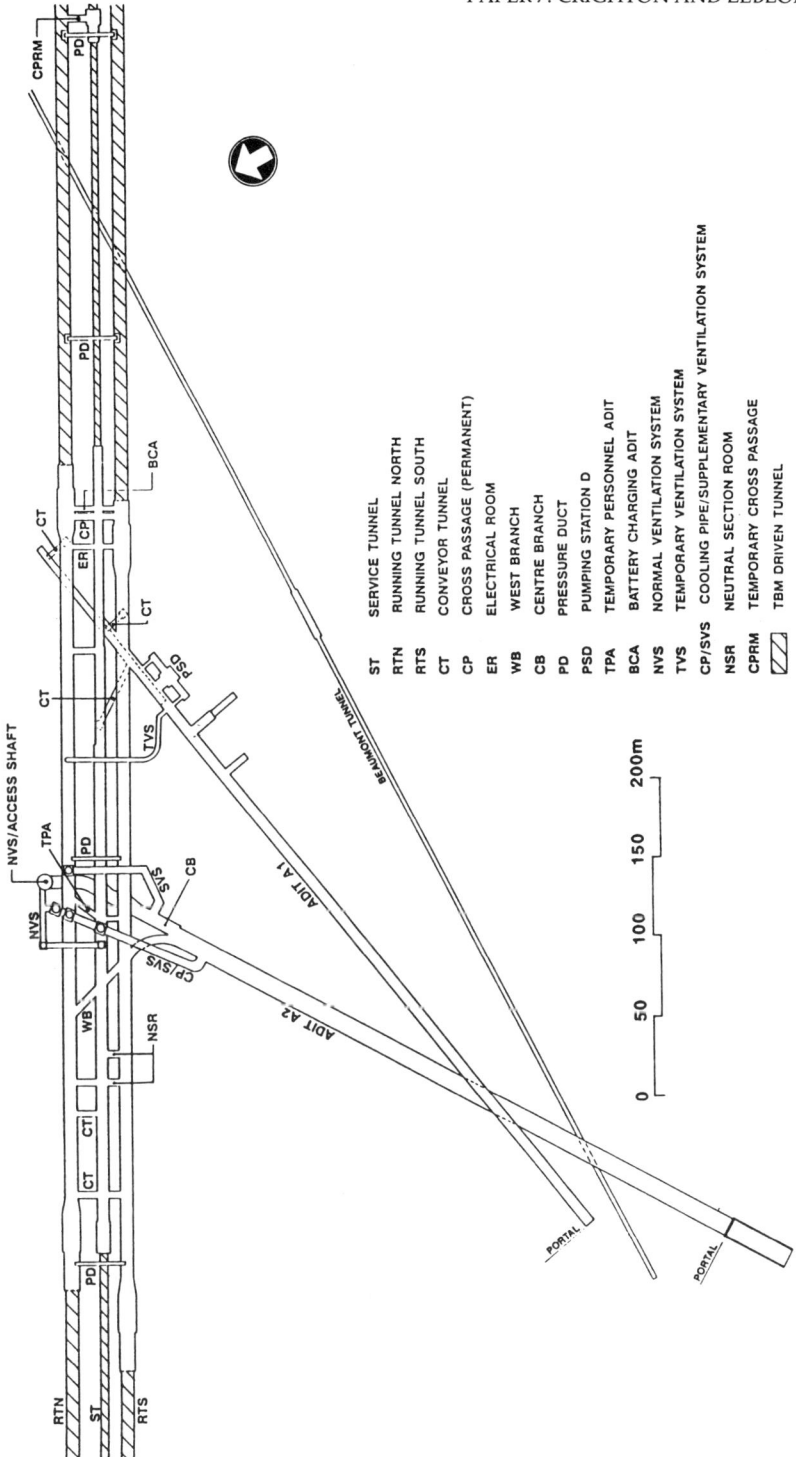

ST   SERVICE TUNNEL
RTN   RUNNING TUNNEL NORTH
RTS   RUNNING TUNNEL SOUTH
CT   CONVEYOR TUNNEL
CP   CROSS PASSAGE (PERMANENT)
ER   ELECTRICAL ROOM
WB   WEST BRANCH
CB   CENTRE BRANCH
PD   PRESSURE DUCT
PSD   PUMPING STATION D
TPA   TEMPORARY PERSONNEL ADIT
BCA   BATTERY CHARGING ADIT
NVS   NORMAL VENTILATION SYSTEM
TVS   TEMPORARY VENTILATION SYSTEM
CP/SVS   COOLING PIPE/SUPPLEMENTARY VENTILATION SYSTEM
NSR   NEUTRAL SECTION ROOM
CPRM   TEMPORARY CROSS PASSAGE
▨   TBM DRIVEN TUNNEL

Fig. 22. Shakespeare underground development – layout

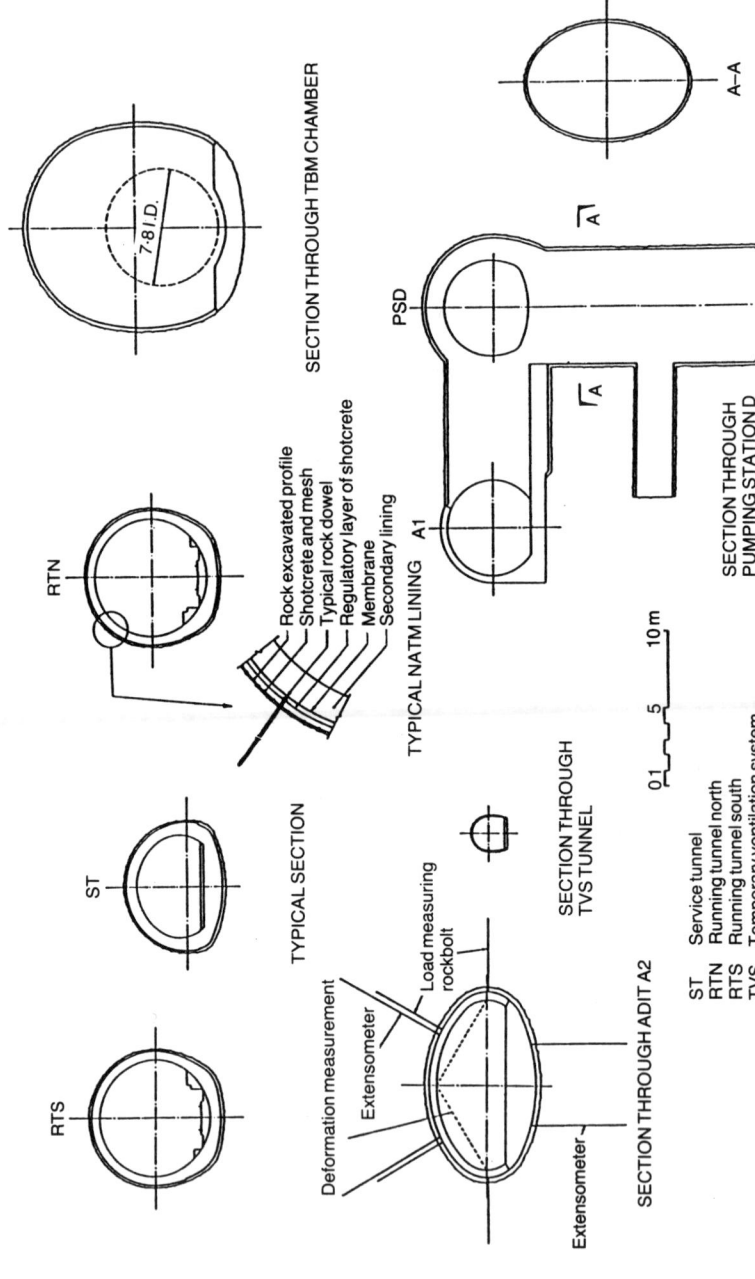

Fig. 23. Shakespeare underground development – section and details

delayed until completion of the TBM drives and removal of associated temporary construction plant.

107. NATM is an observational tunnelling method which uses the inherent strength of the ground to support the tunnel until casting of an in situ secondary lining. The favourable ground behaviour is mobilised by sealing with reinforced shotcrete the excavated tunnel walls behind the advancing face to prevent weathering. The ground is improved by placing of rock dowels drilled through this primary lining. The secondary lining is designed taking into account the in situ measurements of ground and primary lining deformation, rock dowel load and shotcrete stress.

108. The specific principles of NATM are as follows:

(a) Support of the excavation is provided by a compound structure consisting of the rock and various support elements such as shotcrete, anchors, steel lattice girders etc. The primary support is relatively flexible thereby allowing controlled deformations.

(b) Loosening of the original rock structure should be avoided to maintain the load capacity of the rock mass.

(c) The timing of installation of the primary support must be carefully controlled; the rheological characteristics of the ground, the advance rate of excavation, the stand-up time of the opening and the timing for completion of the invert arch must be taken into account.

(d) Monitoring and visual observation of conditions are an essential part of the NATM.

(e) The design of the final lining takes the time-dependent characteristics of the rock into account.

109. The Shakespeare tunnels lie predominantly in the Chalk Marl found to be favourable for NATM, and is generally dry. The geotechnical parameters adopted before construction were found to be conservative, following observation of the constructed primary linings. This has led to more economic and flexible designs for other NATM tunnels at Shakespeare. Indeed the success of the NATM at Shakespeare was a major factor in its adoption for the Castle Hill tunnels and, more significantly, the UK undersea crossover.

110. A particular feature of NATM is that the design, construction method, construction sequence and construction plant are closely interdependent. Successful implementation depends on continuous integration of these separate elements. Engineering personnel monitor the construction work continuously to confirm the design assumptions and provide a rapid on site response to varying construction demands.

111. The speed and facility of NATM is particularly suited to excavations of complex geometry. At Shakespeare,

multiple junctions, small and large tunnels, an elliptical shaft and various curved tunnel sections have been excavated and lined. To cater for complex secondary lining profiles in some of these areas, consideration has been given to the use of permanent shotcrete instead of conventional in situ concrete. Suitable specifications and materials for this are being developed at this stage.

112. The total length of the NATM tunnels and shafts at Shakespeare is about 2500m. Work commenced in September 1987 and has met the programmed starts for TBM launch. Primary lining construction is complete; secondary linings will be placed in 1991-1992.

Marine Works

113. At Shakespeare Cliff a maximum of 3.75 Mm^3 of spoil from the UK tunnels is being deposited in lagoons created by construction of temporary crosswalls from the shore to a new 1.5km long frontal sea wall, ensuring no loss of spoil to the sea. The reclaimed areas are used for supply and resourcing of the UK tunnel construction works.

114. The design chosen is advantageous in terms of simplicity, method, programme and cost. It consists of a mass concrete in situ wall placed in three lifts within 10m long Larssen 6 sheet-piled cells. The temporary crosswalls are similar but infilled with gravel.

115. An extensive set of studies was undertaken, examining wave, tidal, hydraulic and current conditions, beach processes, ground and spoil characteristics and environmental effects. The site is an area of high environmental, nature conservation and scientific interest.

116. The permanent sea wall is designed as a short-term breakwater and a long-term spoil-retaining structure. Overturning and sliding resistance are provided by the weight of concrete and the sheet-piled diaphragms. The wall is founded on sound Lower Chalk of high bearing capacity. The mass concrete is a grade 35 pfa/opc mix, having favourable heat evolution and strength-gain characteristics, giving the durability necessary for the exposure.

117. The seaward spoil slope is protected from erosion with concrete paving and energy-dissipating upstand units. Filtered drainage is provided for wave overtopping, rainwater and groundwater, and scour protection is placed at the toe of the seawall.

118. At the eastern end of the reclamation a new armoured slope prevents increased erosion of Shakespeare Cliff by wave reflection from the reclamation.

119. Construction of the marine works started in September 1987, is now well advanced and is on programme for completion in 1992.

120. Details are shown on Figure 24.

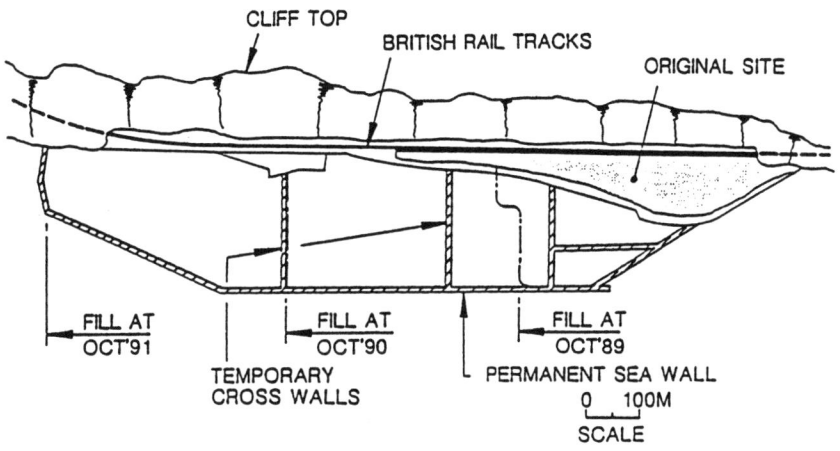

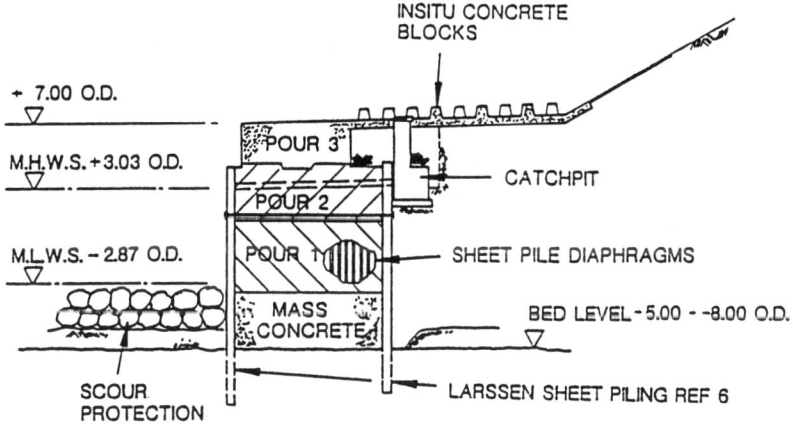

Fig. 24. Shakespeare marine works – plan and section

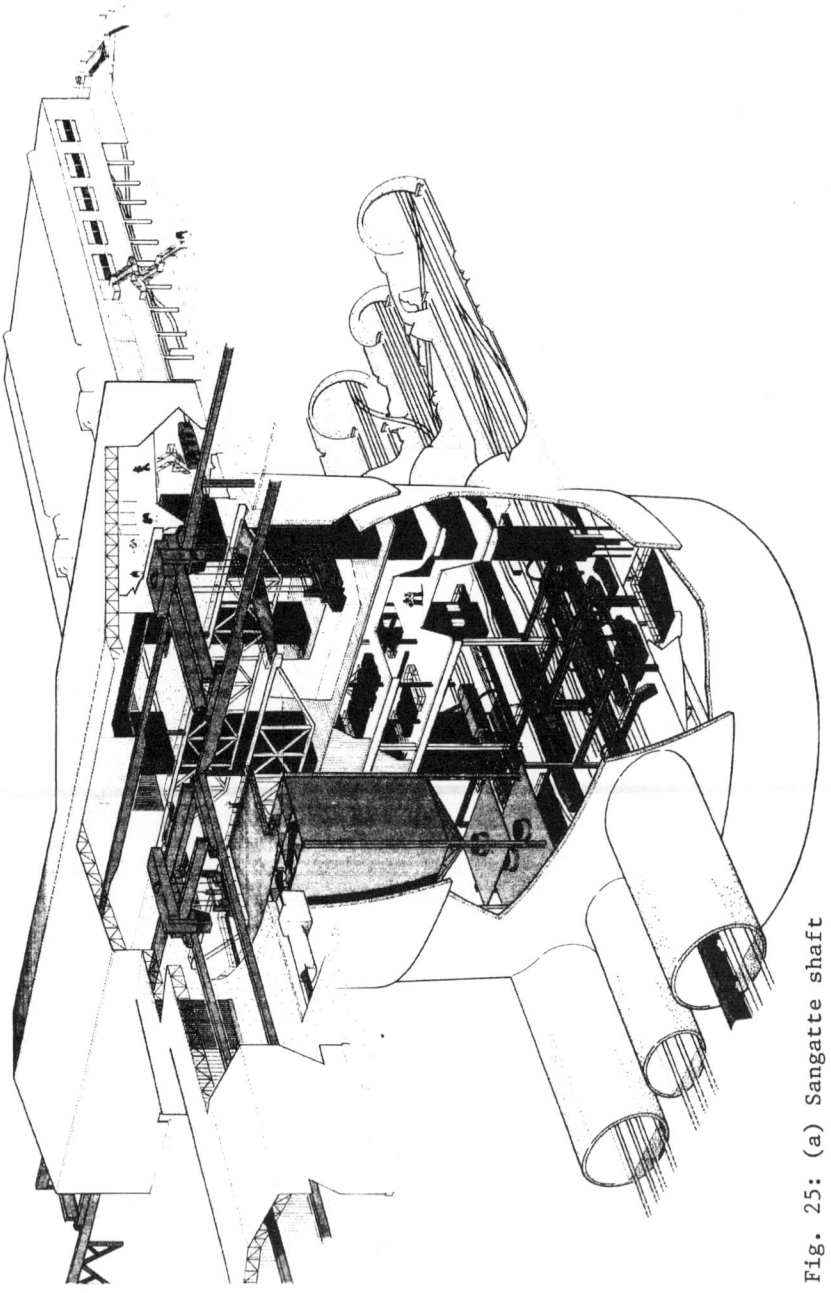

Fig. 25: (a) Sangatte shaft

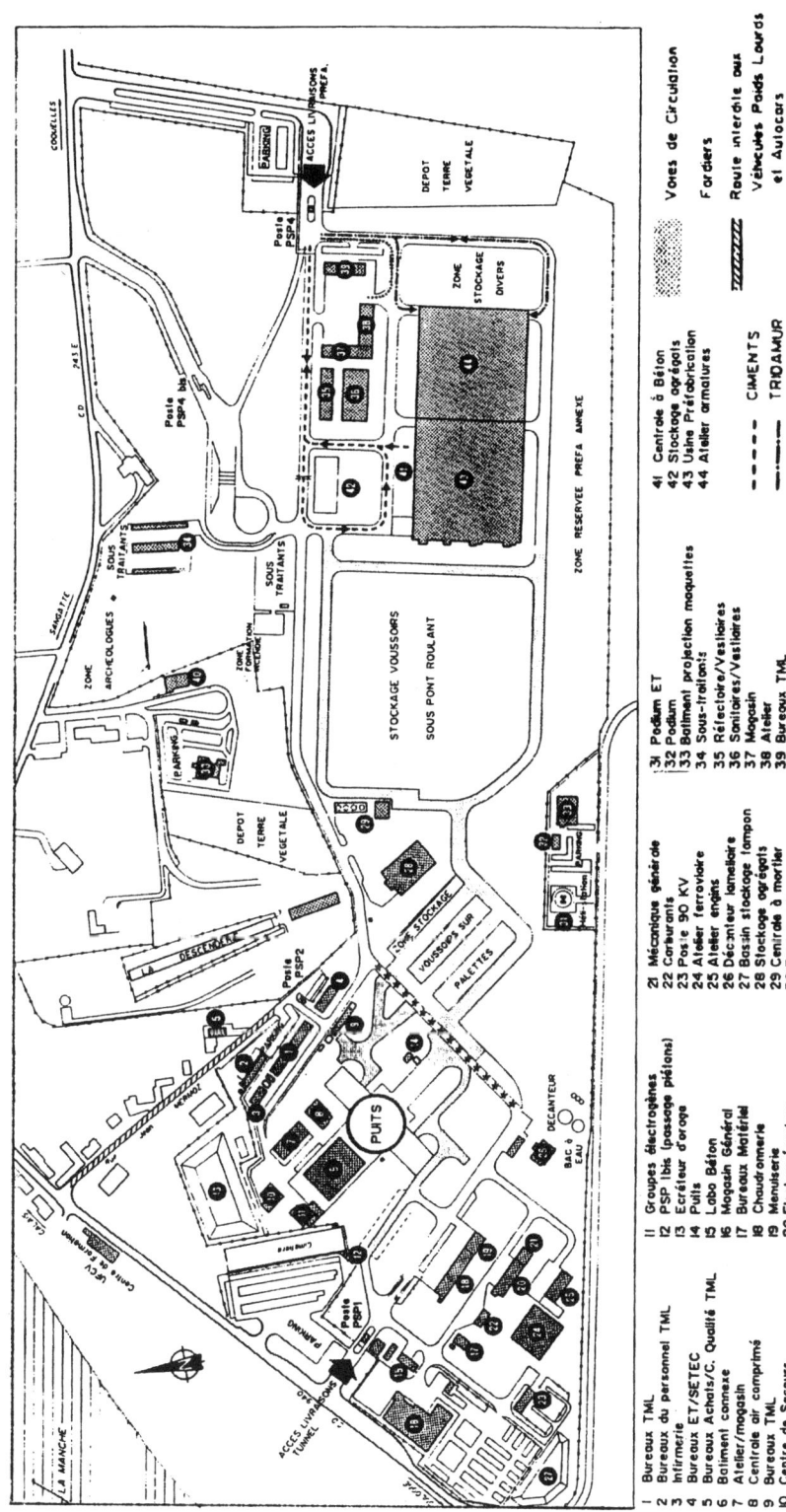

Fig. 25: (b) Layout of Sangatte shaft and works

1 Bureaux TML
2 Bureaux du personnel TML
3 Infirmerie
4 Bureaux ET/SETEC
5 Bureaux Achats/C. Qualité TML
6 Bâtiment connexe
7 Atelier/magasin
8 Centrale air comprimé
9 Bureaux TML
10 Centre de Secours

11 Groupes électrogènes
12 PSP 1bis (passage piétons)
13 Ecréteur d'orage
14 Puits
15 Labo Béton
16 Magasin Général
17 Bureaux Matériel
18 Chaudronnerie
19 Menuiserie
20 Electromécanique

21 Mécanique générale
22 Concurrents
23 Poste 90 KV
24 Atelier ferroviaire
25 Atelier engins
26 Décanteur lamellaire
27 Bassin stockage tampon
28 Stockage agrégats
29 Centrale à mortier
30 Restaurant

31 Podium ET
32 Podium
33 Bâtiment projection maquettes
34 Sous-traitants
35 Réfectoire/Vestiaires
36 Sanitaires/Vestiaires
37 Magasin
38 Atelier
39 Bureaux TML
40 Locaux Syndicaux-Salle Polyvalente

41 Centrale à Béton
42 Stockage agrégats
43 Usine Préfabrication
44 Atelier armatures

Voies de Circulation

Fardiers

Route interdite aux Véhicules Poids Lourds et Autocars

Zone de circulation interdite

– – – – CIMENTS
————   TRIDAMUR
————   MAGASIN

131

SANGATTE
Sangatte Shaft
121. The Sangatte Shaft is situated near Sangatte about
3.2km from the Beussingue Portal. The shaft is used during
construction for the movement of tunnelling equipment to and
from the tunnels and for the evacuation of tunnel spoil.
The shaft will house permanent mechanical and electrical
facilities during operation.

Sangatte Shaft Structure
122. The shaft is in effect a vertical cylinder, 66 m deep,
with an internal diameter of 55 m. The thickness of the
concrete wall varies from 1 m to 1.1 m.
123. At tunnel level, the tunnels are enlarged to form the
marshalling area.
124. Under tunnel level, the shaft houses the structures
and equipment for spoil treatment and pumping to Fond Pignon
Dam.
125. The shaft itself and the tunnel marshalling areas are
protected against water pressure and inflows by a
surrounding in situ concrete, flexible wall keyed into the
chalk marl.
126. The shaft is designed to withstand the loads from the
ground and water pressure.

Sangatte shaft during construction
127. During construction Sangatte Shaft is used for the
following functions:

   (a) treatment of tunnel spoil from TBM's and spoil pumping
       to Fond Pignon Dam
   (b) precast segments are supplied down the shaft to the
       tunnel faces from the precast plant which is located
       near the top of the shaft
   (c) erection of the TBM's
   (d) pumping of seepage water to the water treatment plant
       located near the top of the shaft
   (e) storage and handling of batteries for locos
   (f) access lifts for personnel
   (g) housing of the control centre for construction

Sangatte Shaft during operation
128. During operation, Sangatte Shaft will be used to
house permanent installations as follows:-

   (a) a pumping station for the seepage water coming from
       the underland portion of the tunnel
   (b) an electrical substation for power supply
   (c) rooms for signalling equipment
   (d) a passage for passenger evacuation
   (e) pressure relief ducts

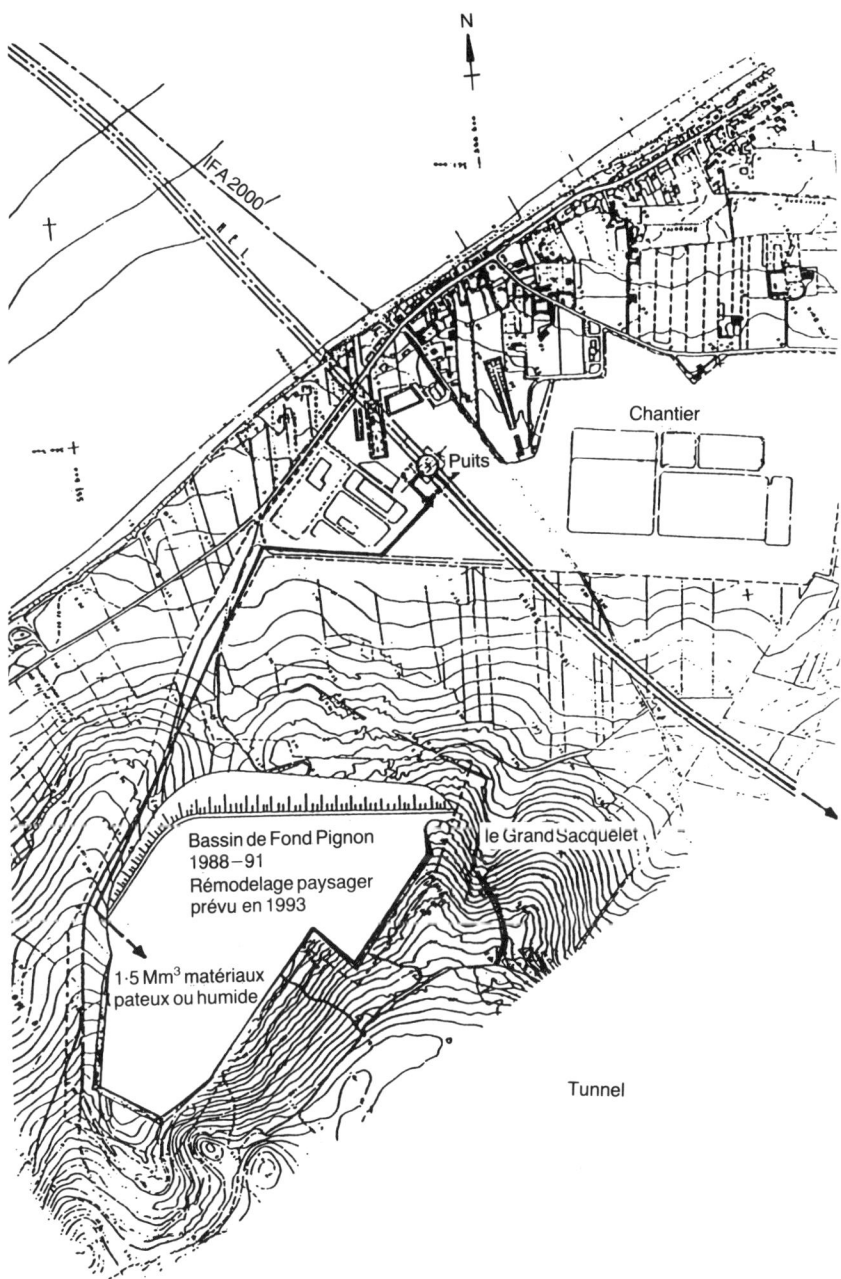

Fig. 26. Fond Pignon Dam site

129. At ground level the permanent installations will be:

(a) a cooling plant including low cooling towers
(b) a ventilation plant for service and emergency use
(c) fire fighting facilities, including pumps, tanks and associated electrical equipment
(d) a water treatment plant and associated sea discharge

130. The layout of Sangatte Shaft during construction is shown on Figure 25.

Fond Pignon Dam

131. 'Fond Pignon' Dam is located in 'Mt Saint Hubert' vale, about 1km from Sangatte Shaft site. It covers approximately 30 ha and is included in the French government concession area.

132. The dam stores the excavated tunnel spoil. Spoil from the TBMs is crushed and mixed with water so that the mixture can be easily pumped out to the storage area to be decanted.

133. The site is designed to be constructed in three phases. Volumes in each phase are:-

(a) Phase 1 - 750,000 m^3
(b) Phase 2 - 575,000 m^3
(c) Phase 3 - 50,000 m^3

134. The general layout of the site is shown on Figure 26 and includes:-

(a) a watertight upstream zone
(b) a drainage pit connecting to drainage trenches
(c) a stability downstream zone
(d) a water evacuation system
(e) a settlement area

The overall storage capacity of the dam is about 4.5 Mm^3 allowing the storage of 3 Mm^3 of in situ materials.

ACKNOWLEDGEMENTS

The Authors gratefully acknowledge the help and work carried out by their colleagues in Transmanche-Link and their Consultants on which this paper is based.

RESUME

Cet article présente les principaux aspects de la conception des ouvrages de génie civil souterrains du Tunnel sous la Manche :

- la description de la disposition générale des ouvrages situés entre le portail de Castle Hill et le Portail de Beussingue ;
- une brève description des conditions géologiques et des caractéristiques des terrains est donnée, ainsi que les principaux critères de conception du tracé.

Ensuite, la description des ouvrages les plus caractér-istiques est développée :
- les stations de pompage ;
- les ouvrages nécessaires aux traversées jonctions, qui permettent la circulation des trains et navettes en voie unique, lors des opérations de maintenance ;
- les rameaux d'évacuation des passagers, qui relient les tunnels ferroviaires au tunnel de service ;
- les rameaux de pistonnement, qui permettent de réduire la puissance nécessaire à la traction des navettes par la réduction des effets aérodynamiques.

Un chapitre particulier traite plus en détail de la conception des revêtements des tunnels :
- les revêtements en béton préfabriqué ;
- les revêtements en fonte, et rappelle les principales hypothèses de calcul.

Un chapitre est consacré à la description détaillée des ouvrages réalisés à Shakespeare Cliff pour le départ des tunneliers et la logistique nécessaire au percement des tunnels à partir de la Grande-Bretagne. Les fonctions de ces ouvrages, après la mise en service du tunnel, sont indiquées.

Le dernier chapitre traite des ouvrages nécessaires pour réaliser les tunnels à partir de la France, le puits de Sangatte et le barrage de Fond Pignon utilisé pour le stockage des déblais. Une brève description des ouvrages à réaliser pour la phase d'exploitation du tunnel est indiquée.

8. Conception des terminaux

Y. MISSOFFE, Ecole Supérieure des Travaux Publics, Engineering
Manager—Terminaux et Sécurité (France)

SYNOPSIS. Jonction entre le Système de Transport
Eurotunnel, et les réseaux routiers et ferroviaires
européens des deux côtés de la Manche (Figure 1), les
Terminaux accueillent l'ensemble des installations
nécessaires tant au service de l'usager qu'au
fonctionnement, à la gestion, à l'entretien et à la sécurité
du Système. La conception des Terminaux, tout en respectant
le caractère spécifique de chaque site et la nécessité d'une
bonne intégration dans l'environnement, doit assurer une
perception aisée par l'usager de la continuité du
cheminement qui sera le sien entre sa sortie des réseaux
routiers nationaux, sa prise en charge par Eurotunnel et sa
libre sortie vers le réseau routier de destination.

FONCTIONS DES TERMINAUX
1. Les fonctions principales d'un Terminal sont
d'assurer, pour les usagers :

(a) au départ, la prise en charge de tous les types de
véhicules routiers admissibles dans le système,
depuis le réseau routier extérieur jusqu'à
l'embarquement sur les navettes, incluant les
opérations suivantes : péage, contrôles frontaliers
et de sûreté et affectation aux navettes,
(b) à l'arrivée, les opérations depuis le débarquement
des véhicules jusqu'à leur insertion sur le réseau
routier extérieur,
(c) en outre, l'implantation d'un certain nombre de
services et commerces répond à la fonction d'accueil
des usagers, tant au départ qu'à l'arrivée.

2. A ces fonctions propres à l'usager, s'ajoutent les
fonctions à remplir par l'exploitant en vue de la bonne
marche du système de transport (Direction, administration,
exploitation, entretien, sécurité, sûreté, etc...).
3. Les qualités attendues de ce système de transport,
auxquelles une bonne conception des terminaux doit
contribuer sont :
(a) La permanence du transport (24h/24, tous les jours de
l'année), et sa rapidité (temps de transit dans les

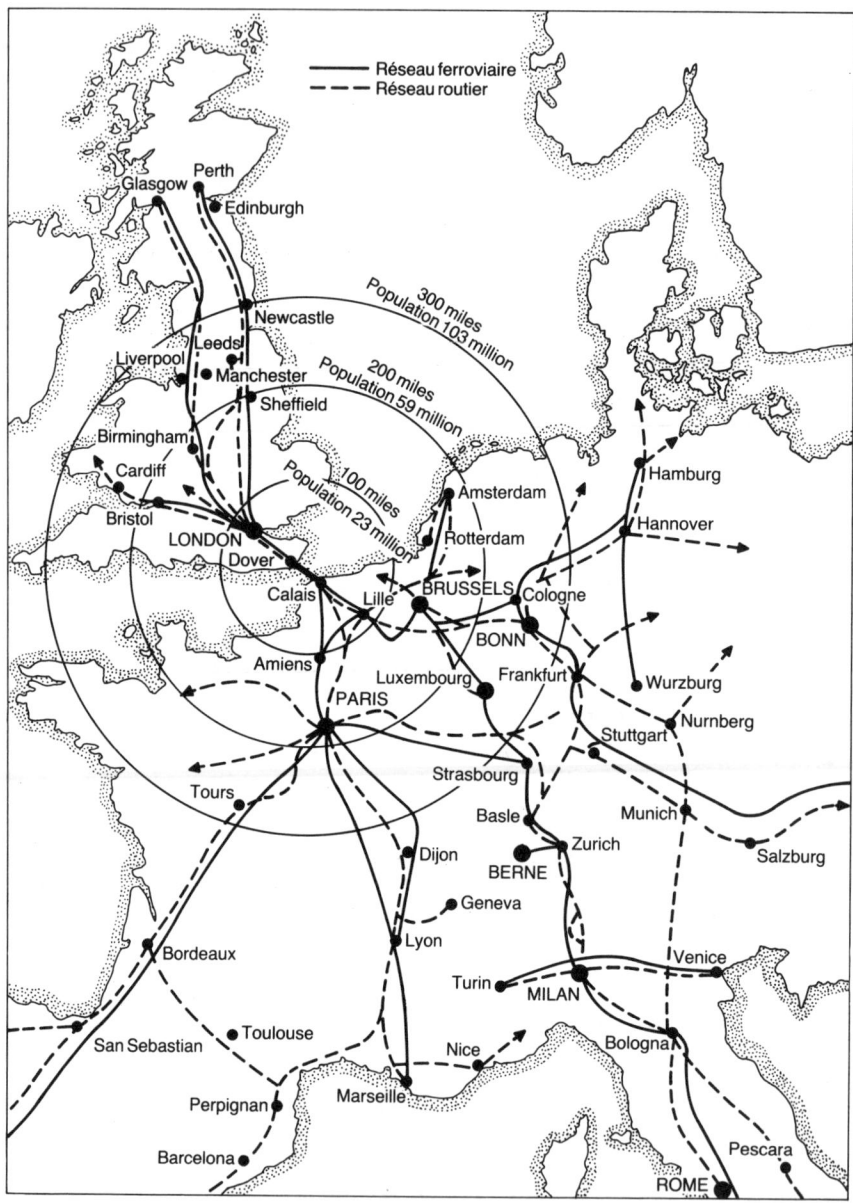

Fig. 1. Le tunnel sous la Manche au sein des réseaux
ferroviaires et routiers européens

terminaux aussi faible que possible),
(b) La facilité d'embarquement et de débarquement par un cheminement aisé et des manoeuvres simples,
(c) L'adaptabilité aux variations de trafic,
(d) La flexibilité : réaction à tout évènement anormal ou à tout incident routier ou ferroviaire sans dégradation excessive du service,
(e) La sécurité des usagers et du personnel
(f) La sûreté des installations,
(g) Une exploitation facile et efficace.

TRAFIC DE DIMENSIONNEMNT DES TERMINAUX

4. Les infrastructures des terminaux sont dimensionnées pour assurer normalement l'acheminement d'un trafic routier correspondant à la 30ème heure de pointe (sur la courbe des débits classés) de l'année 2003.

5. Ce trafic unidirectionnel, évalué pour les périodes de pointe tourisme et fret, est donné dans le tableau suivant :

Tableau 1. Trafic de 30ème heure de pointe pour l'horizon 2003

Période	Pointe tourisme	Pointe fret
Véhicules légers	785	260
Autocars	74	25
Poids lourds	55	130

6. Par ailleurs, la faisabilité de l'extension des installations au fur et à mesure de l'accroissement du trafic jusqu'à la saturation du système a été prise en compte dans la conception des plans de masse.

7. Les Terminaux sont également organisés pour supporter les importantes variations saisonnières, hebdomadaires et journalières que subit le trafic routier.

PLAN DE MASSE ET ORGANISATION GENERALE DES TERMINAUX.

8. Les études du plan de masse définissent la localisation et les liaisons des infrastructures implantées sur les sites des Terminaux, soit directement liées à l'exploitation du système de transport Eurotunnel (zone des quais, zone d'entretien...), soit directement induites par la présence des Terminaux (activités de service et de développement).

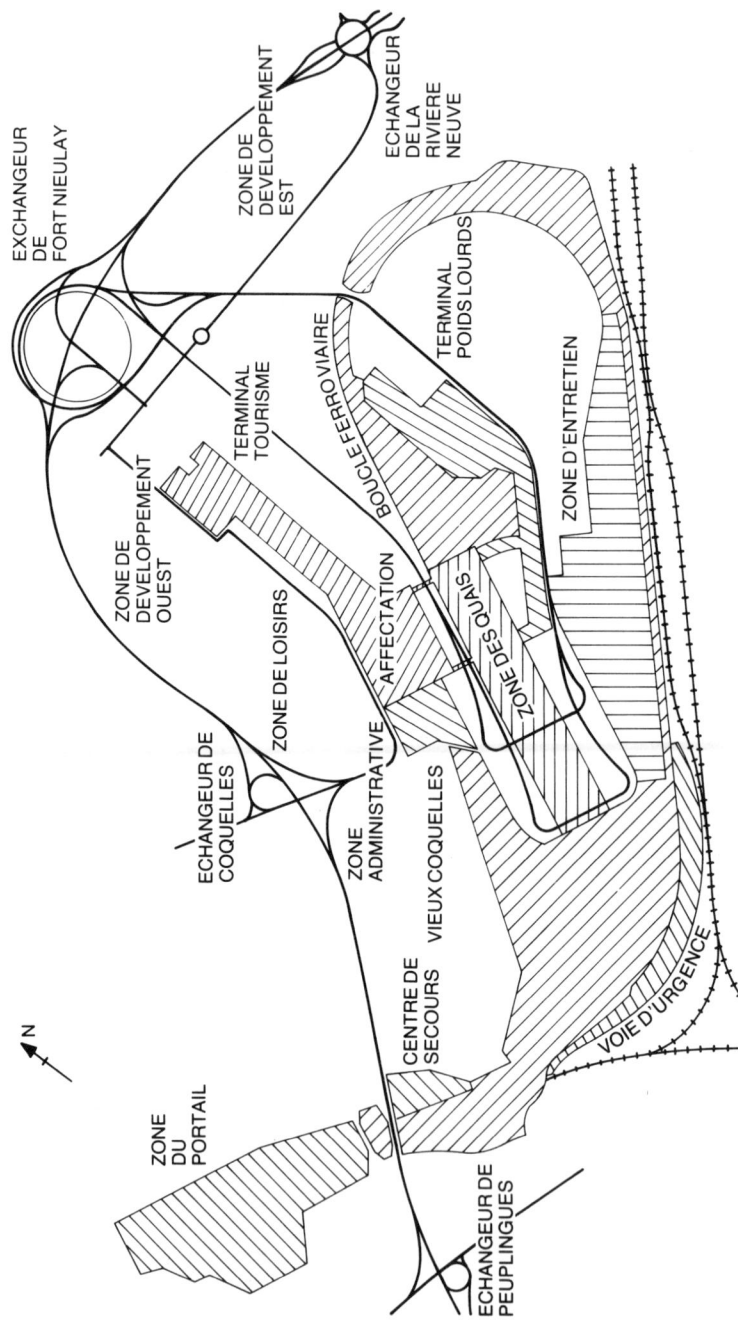

Fig. 2. Terminal de Coquelles - plan de zones

Les différentes zones des Terminaux

9. Un Terminal est composé de plusieurs zones qui, compte tenu des fonctions définies aux paragraphes 1. et 2., peuvent être classées comme suit :

(a) Zones accessibles aux usagers du lien fixe
  . le terminal tourisme
  . le terminal commercial
  . la zone des quais (embarquement et débarquement)
  . les zones de service et de développement
(b) Zones principales requises par l'exploitation
  . la zone administrative
  . la zone d'entretien
(c) Autres zones (plus spécifiquement liées à la sécurité)
  . le centre de secours
  . la voie d'urgence
  . les installations situées près du portail

Les spécificités du Terminal de Coquelles (France)(Figure 2)

10. Le Terminal de Coquelles occupe une superficie de 650 ha environ, dont 470 ha pour les installations directes du Lien Fixe. Il est situé sur 4 communes : Calais, Coquelles, Fréthun, Calais-Boulogne, à l'Ouest par la voie SNCF Paris-Londres et la tranchée d'accès au tunnel, au Sud par la voie SNCF Paris Calais et à l'Est par le Canal de Rivière Neuve et la future route pénétrante Ouest du Calaisis.

11. Le Terminal français présente, en outre, un caractère spécifique, celui de permettre l'implantation au voisinage des installations du Lien Fixe, de zones d'activités de différentes natures s'intégrant dans le développement économique du Calaisis et de la Région Nord/Pas-de-Calais. Ces zones seront définies dans le cadre de la procédure de ZAC (Zone d'Aménagement Concerté).

12. La superficie importante du Terminal de Coquelles lui permet d'accueillir le dépôt d'entretien lourd de l'ensemble du matériel roulant ferroviaire Eurotunnel.

13. Les contraintes géotechniques et l'existence de terrains tourbeux compressibles ont déterminé la configuration du site Français et l'implantation de la boucle ferroviaire située entre les tracés SNCF et la voie rapide autoroutière Calais-Boulogne. Celà a conduit à une séparation franche entre les installations du Terminal et les grandes liaisons routières et ferroviaires européennes.

14. Le choix du tracé ferroviaire, le niveau élevé de la nappe aquifère et la nature des terrains rencontrés ont guidé l'organisation générale du Terminal, dans le but de limiter les quantités de matériaux d'apport (rare dans cette région) et les travaux de consolidation des terrains compressibles. Cette organisation se traduit par les implantations suivantes :

(a) Le Terminal tourisme situé entre la voie rapide et la

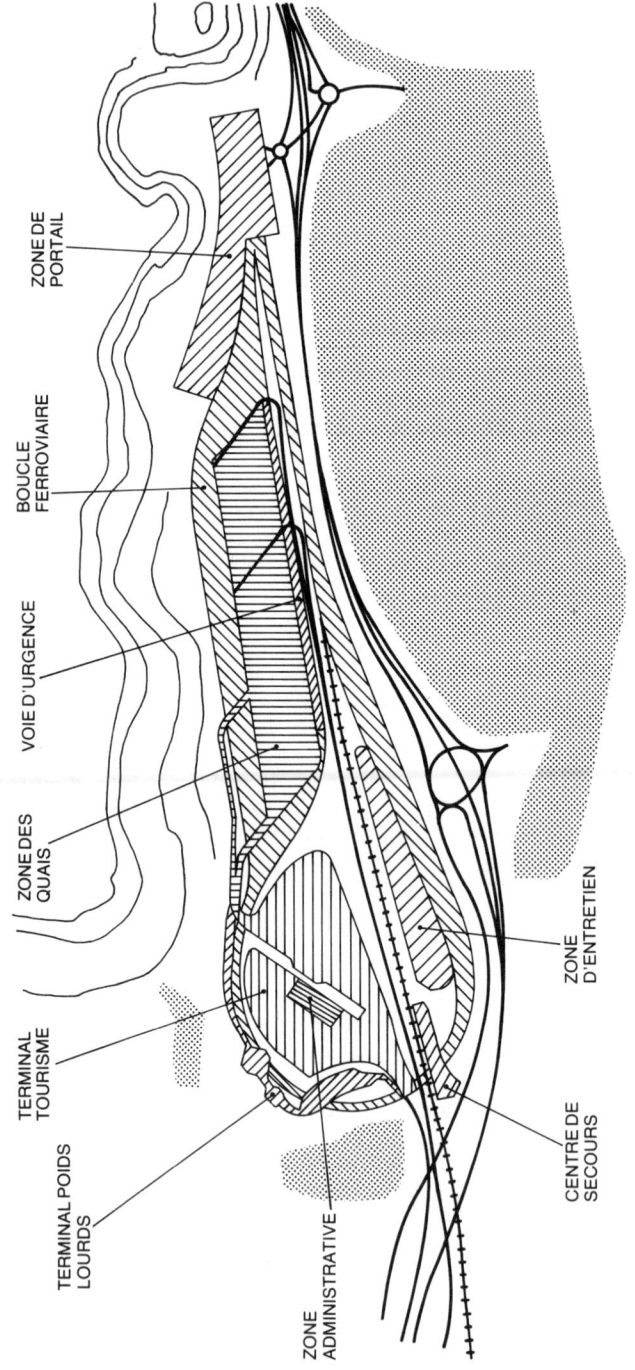

ZONE DE
PORTAIL

BOUCLE
FERROVIAIRE

VOIE D'URGENCE

ZONE DES
QUAIS

TERMINAL
TOURISME

TERMINAL POIDS
LOURDS

ZONE
ADMINISTRATIVE

CENTRE DE
SECOURS

ZONE
D'ENTRETIEN

Fig. 3. Terminal de Folkestone – plan de zones

boucle ferroviaire sur des terrains sains en léger
surplomb par rapport au reste du site,

(b) Le Terminal commercial, pour le fret, installé dans
la boucle ferroviaire sur des terrains tourbeux,

(c) La zone des quais d'embarquement et de débarquement
située le long du terminal tourisme le plus possible
hors de la zone tourbeuse,

(d) La zone d'entretien du matériel ferroviaire roulant
située au Sud du Terminal, sur des terrains tourbeux,
entre la zone des quais et les voies SNCF.

Les spécificités du Terminal de Folkestone (UK) (Figure 3)

15. Le Terminal de Folkestone s'étend sur une zone
d'environ 170 ha, située en longueur entre l'autoroute M20
vers Londres et l'escarpement de North Downs qui surplombe
le site de 60m. Les villages de Newington et de Peene
bordent à l'Ouest et au Nord les installations principales
et le portail situé sur la face Ouest de Castle Hill limite
le Terminal à l'Est.

16. L'organisation du Terminal a été grandement
influencée par la taille du site, sa configuration, la
protection de l'environnement, de même que par le tracé et
les profils ferroviaires à réaliser. Celà a conduit à
relever le niveau général du site en utilisant de grandes
quantités de matériaux d'apport.

17. Le site du Terminal, très compact compte tenu du peu
d'espace disponible, est divisé longitudinalement en deux
d'Ouest en Est par la voie ferrée principale continentale
CML. La zone administrative et des services, le terminal de
tourisme, le terminal commercial et la zone des quais sont
situés au Nord de la CML avec accès par l'échangeur
principal routier, et la zone d'entretien au Sud entre la
CML et l'autoroute M20, avec l'accès par la route de service
à Longport.

18. La proximité des villages de Newington et de Peene a
conduit à recouvrir la boucle ferroviaire Est pour limiter
les nuisances possibles.

LE LIEN FIXE ET LES USAGERS

19. La conception des Terminaux repose sur le principe de
la séparation maximale des trafics tourisme et poids lourds
depuis l'accès routier principal jusqu'à la zone des quais.

20. Le principe de cheminement de l'usager depuis le
réseau routier national est, malgré les contraintes propres
à chaque site, le même pour les deux Terminaux. Il se
décompose de la façon suivante (Figure 4) :

(a) l'échangeur principal routier qui sépare les flots et
dirige les véhicules de tourisme et les poids lourds
vers leurs installations terminales respectives,

(b) une zone d'entrée dans laquelle les usagers peuvent
opter pour une visite dans la zone d'activités
commerciales ou pour un accès direct au système,

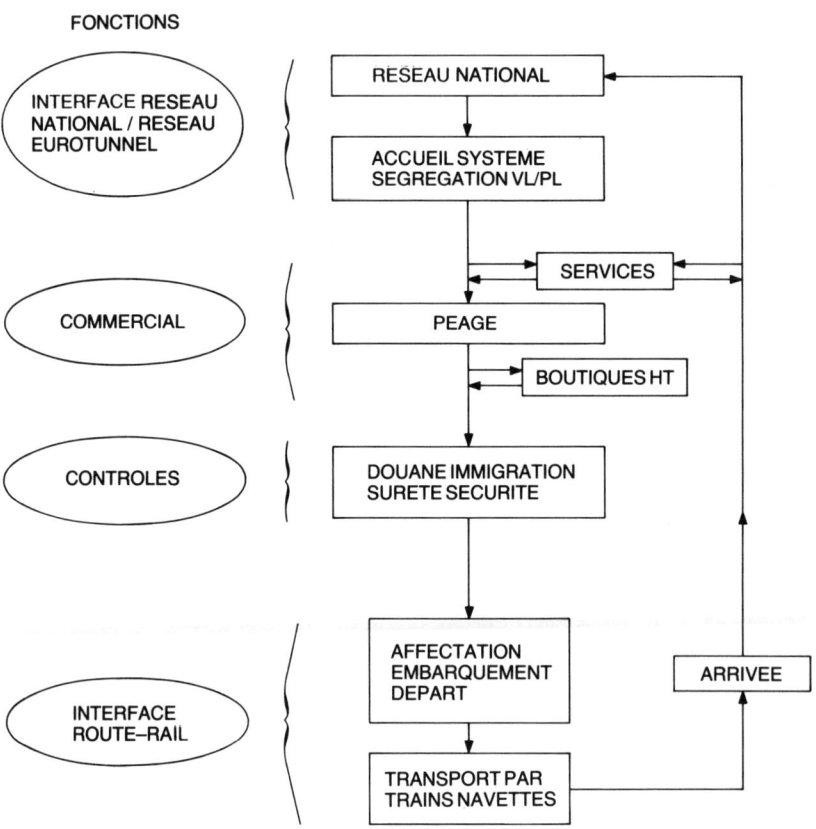

Fig. 4. Organisation fonctionelle des terminaux

(c) la barrière d'entrée couplée avec le péage (12 voies équipées de cabines doubles desservant deux voies contigües),

(d) les contrôles frontaliers et de sûreté de sortie du territoire,

(e) les contrôles frontaliers et de sûreté d'entrée dans l'autre territoire,

(f) la zone d'affectation permettant le tri des véhicules selon leurs caractéristiques, leur ordonnancement et l'attente dans quatre unités de stockage correspondant à la capacité de 4 navettes de type Simple Pont/Double Pont.

(g) la zone des quais qui comprend :
. deux ponts d'embarquement et rampes associées.
. deux ponts de débarquement et rampes associées,
. 10 voies banalisées, à l'ouverture du Lien Fixe
. des quais situés entre chaque voie
. une voie de service continue le long d'un côté de chaque navette (soit un quai sur deux), des rampes auxiliaires permettant le passage depuis les quais jusqu'au niveau de la voie (recyclage des véhicules)
cette configuration permettant :
. le chargement des deux niveaux d'une rame à double pont à partir de deux rampes voisines, évitant le mélange des véhicules de files différentes,
. le chargement de deux rames à double pont sur deux voies adjacentes (situation exceptionnelle)

21. L'organisation des zones de contrôles frontaliers et de sûreté a fait l'objet de nombreuses réunions de travail entre Eurotunnel, TML et les Administrations concernées françaises et britanniques. Les principes retenus sont les suivants :

(a) séparation des flux tourismes et poids lourds,
(b) installations suffisamment flexibles pour permettre :
. une action indépendante de chaque Administration
. des niveaux de contrôle renforcés en cas de nécessité
. une évolution aisée vers la situation Marché Unique Européen (MUE)
(c) exécutions des contrôles français et britanniques dans deux modules séparés,
(d) conception des modules pour éviter au trafic général la gêne des véhicules subissant un contrôle prolongé
(e) libre sortie du système Eurotunnel

22. Après son débarquement des navettes et en vertu du principe de Libre Sortie l'usager peut se diriger vers la destination de son choix sans contrôle spécifique.

23. Des zones de stationnement, d'une capacité globale d'environ 3 h de stockage du trafic de dimensionnement, sont réparties tout le long du cheminement de l'usager et

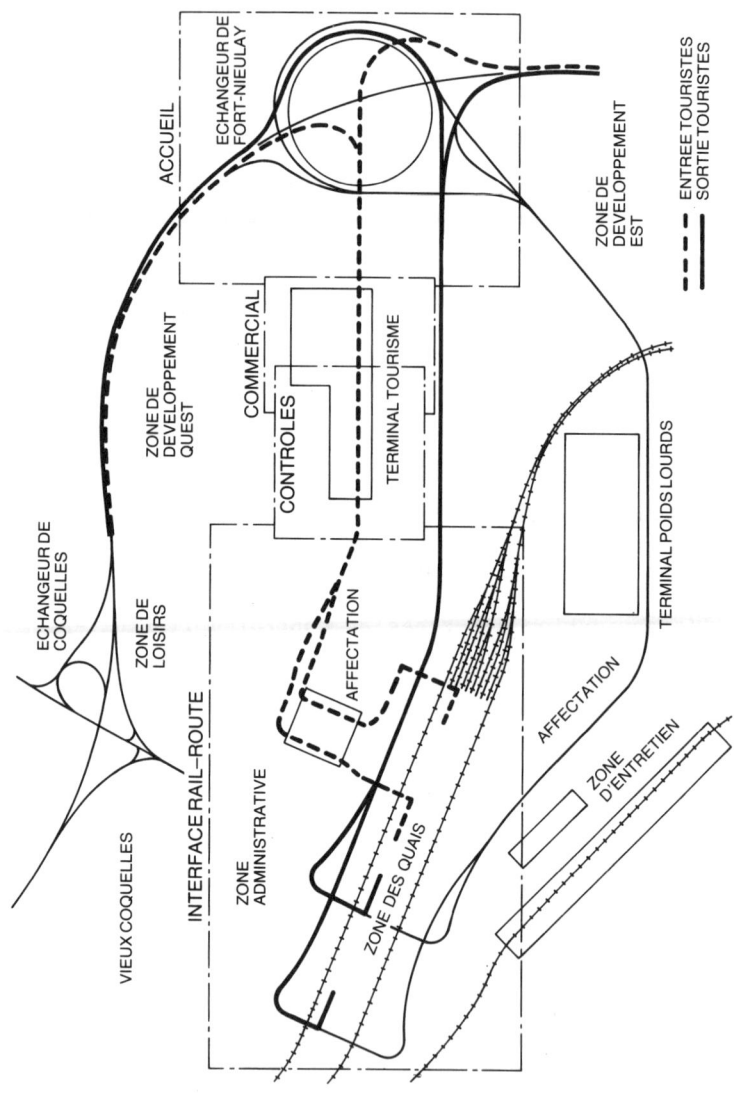

Fig. 5. Terminal de Coquelles - traitements des touristes

permettent de faire face aux super-pointes ou à l'accumulation de trafic par suite d'un incident :

(a) Stationnement organisé, capacité 1 h, en amont du péage,
(b) Stationnement des zones commerciales et de service,
(c) Stockage correspondant à la zone d'affectation.

Les spécificités du Terminal de Coquelles (FR) (Figure 5)

24. A partir de l'échangeur de Fort Nieulay, point fort de l'aménagement architectural avec son plan d'eau de 400 m de diamètre et son arche monumentale, l'ensemble des installations s'alignent le long d'un axe unique. Grâce à un profil en long judicieusement étudié, les usagers pourront percevoir ces installations dés l'entrée dans le système au niveau de l'arche.

25. Compte tenu de l'implantation des terminaux tourisme et commercial de part et d'autre des voies ferrées la zone des quais présente les caractéristiques suivantes :

(a) deux ponts d'embarquement accessibles d'un côté aux véhicules de tourisme, de l'autre aux poids lourds,
(b) deux ponts de débarquement avec sortie côté Nord-Ouest pour les véhicules de tourisme et côté Sud-Est pour les poids lourds,
(c) les quatre ponts situés perpendiculairement à un réseau de quais parrallèles et de même longueur.

26. Cette configuration permet de séparer totalement les circulations de véhicules de tourisme et de poids lourds, tant à l'entrée qu'à la sortie du système tout en garantissant un grand niveau de flexibilité.

Les spécificités du Terminal de Folkestone (UK) (Figure 6)

27. Compte-tenu des contraintes d'espace sur le Terminal britannique, les installations pour les usagers sont toutes siutées dans la boucle ferroviaire, au Nord de la CML.

28. Le Terminal tourisme, implanté au Sud-Est de cette zone, est séparé du terminal commercial, situé au Nord-Ouest, par la zone administrative et l'ensemble des activités commerciales et de service.

29. L'exiguité du site a conduit à implanter le parking d'affectation perpendiculairement au cheminement de l'usager et à donner à la zone des quais les caractéristiques suivantes :

(a) deux ponts d'embarquement accessibles du côté Nord-Ouest aux véhicules de tourisme et aux poids lourds,
(b) deux ponts de débarquement avec sortie côté Sud-Est pour les véhicules de tourisme et pour les poids-lourds,

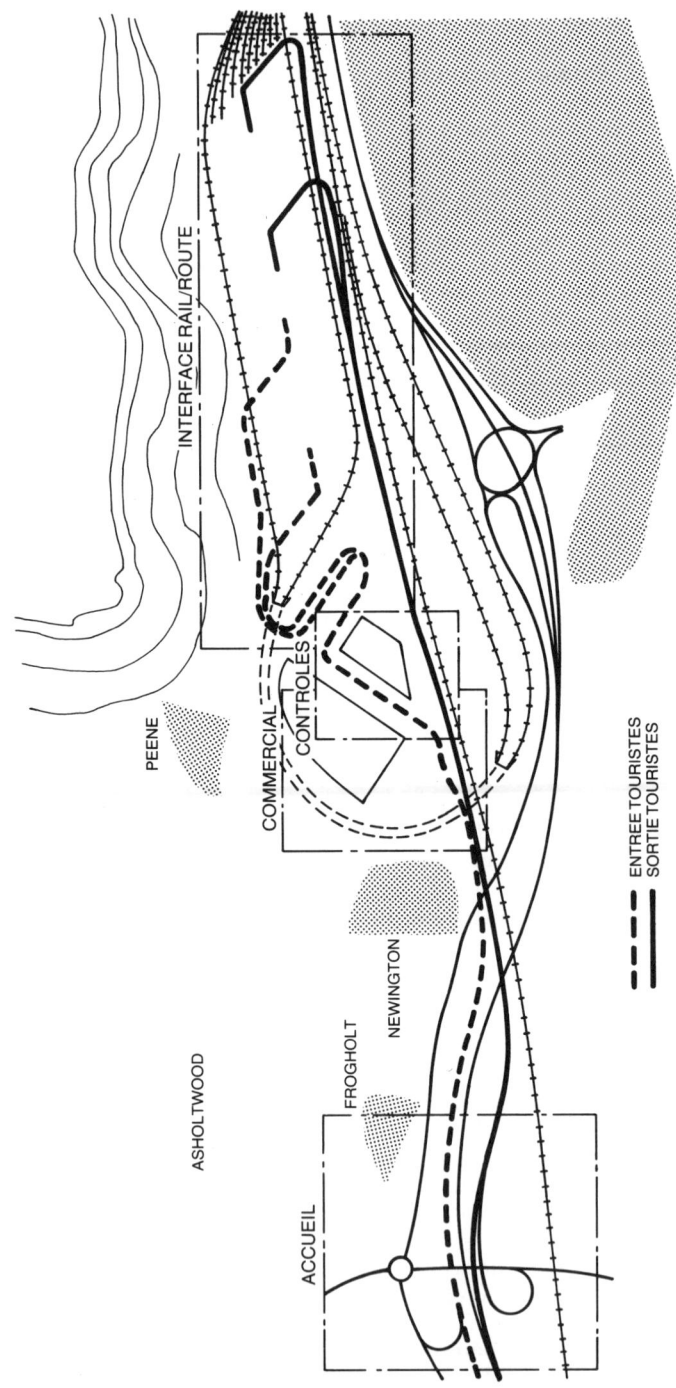

Fig. 6. Terminal de Folkestone – traitements des touristes

(c) les quatre ponts situés en biais par rapport à un réseau de quais parrallèles et de même longueur.

30. Cette configuration ne permet pas la séparation des flots véhicules tourisme et poids-lourds à l'entrée et à la sortie des quais et impose une disposition particulière des voies routières sur les ponts tout en facilitant les liaisons ponts-rampes.

LES TRACES FERROVIAIRES SUR LES SITES DES TERMINAUX

31. Dés 1986, des études très poussées d'optimisation du tracé ferroviaire dans les Terminaux ont été entreprises pour chercher à améliorer, de façon importante, les conditions d'exploitation tant des trains des réseaux nationaux que des trains navettes.

32. Différentes configurations ont été examinées dont celle dite en cul-de-sac mais c'est en définitive la solution en boucle qui a été retenue compte tenu de ses performances.

Les spécificités du Terminal de Coquelles (FR) (Figure 7)

33. Le tracé ferroviaire est imposé par les contraintes suivantes :

(a) l'axe des tunnels sous-terrestres entre Sangatte et Beussingue,
(b) la géométrie de la boucle fixée par des contraintes d'exploitation des navettes (grande vitesse d'approche)
(c) les raccordements au réseau ferré national et à la zone d'entretien,
(d) l'implantation des voies et quais au maximum sur des terrains non compressibles.

34. Le schéma ferroviaire proposé présente d'importantes améliorations par rapport à celui d'Octobre 1985 :

(a) les raccordements au réseau SNCF ont été modifiés pour permettre aux trains des réseaux de pénétrer dans le tunnel à plus grande vitesse et les points de convergence ont été repoussés vers le portail,
(b) le dispositif de communication croisée, près du portail, diminue fortement les perturbations en cas d'entretien ou d'incident sur une aiguille (pour les deux types de trains),
(c) le doublement des voies départ et arrivée des trains navettes améliore la fiabilité du système de transport Eurotunnel,
(d) la symétrisation maximale des faisceaux d'entrée et de sortie de la zone des quais contribue à une grande flexibilité de l'exploitation du système.
(e) le tracé a été entièrement repris pour limiter la

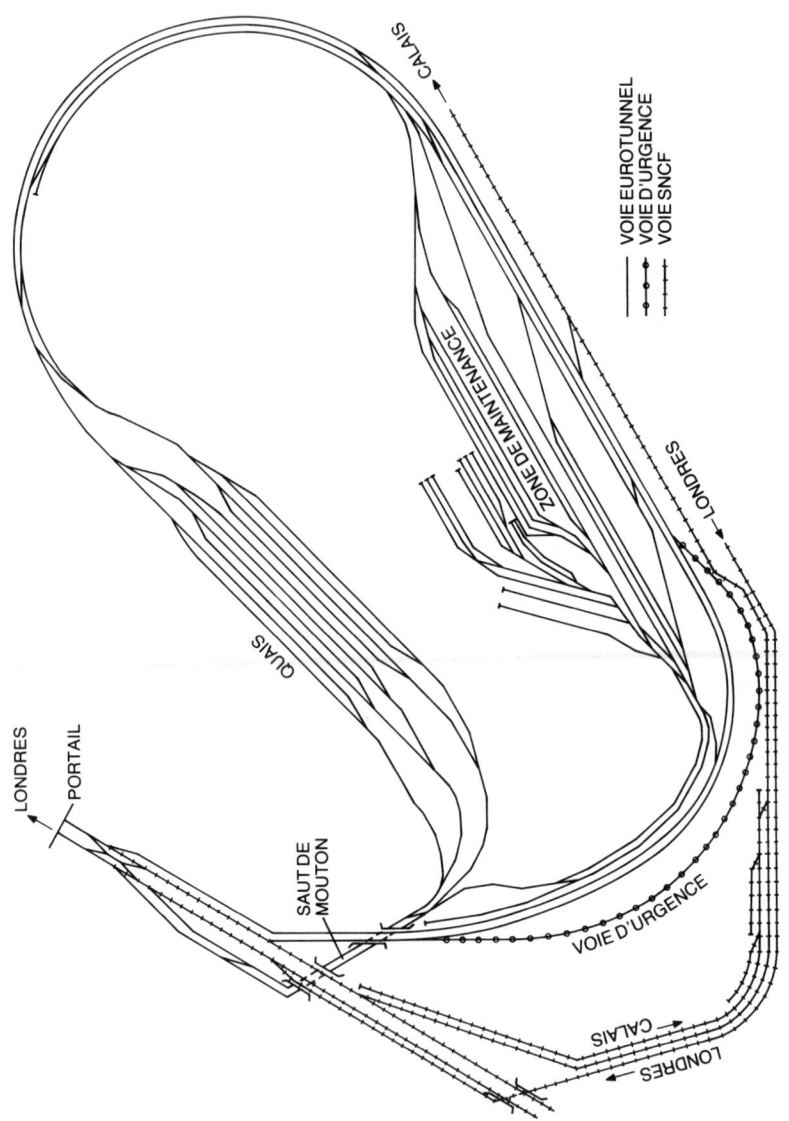

Fig. 7. Terminal de Coquelles – synoptique général des voies ferrées

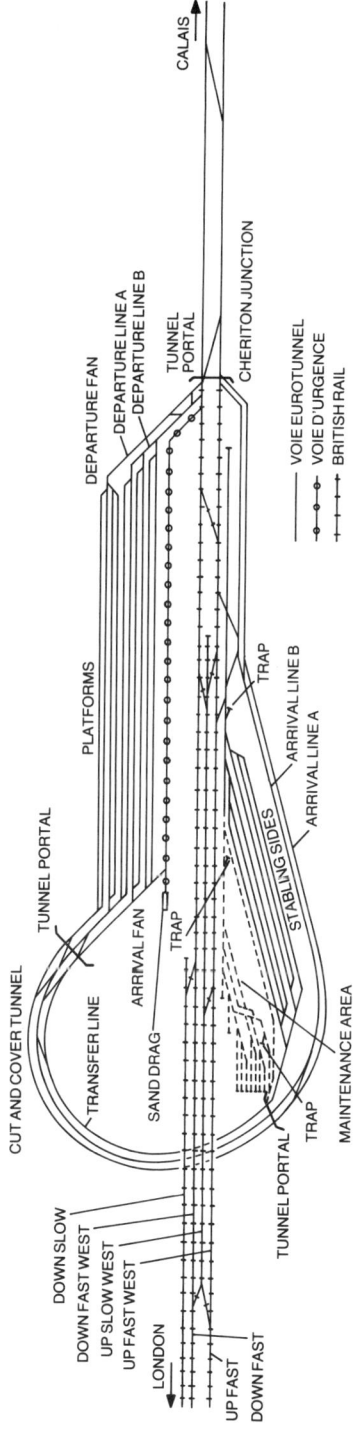

Fig. 8. Terminal de Folkestone – synoptique général des voies ferrées

gauche de voie (variation de devers en sortie courbe) à 1/1000 ce qui améliore considérablement la sécurité et le confort des passagers.

35. D'autre part, les conditions topographiques et de tracé ont conduit, côté français, à croiser la boucle au moyen d'un ouvrage dit en saut de mouton.

Les spécificités du Terminal de Folkestone (UK) (Figure 8)

36. Le tracé ferroviaire sur le site britannique est en boucle non croisée, traversé longitudinalement par la CML à partir de la sortie du portail.

37. La géométrie du tracé est conditionnée par les facteurs suivants :

(a) en plan :
. par les limites du site qui définissent les rayons minimum des courbes en relation avec les vitesses des navettes,
. par la longueur de la zone des quais,
. par la position du portail côté Est et par la jonction de la CML et des voies du réseau Sud-Est des BR côté Ouest,
(b) en élévation :
. par le niveau des voies ferrées au portail et par le croisement de l'autoroute M20 par la CML.

38. Comme pour le Terminal français, d'importantes améliorations ont été apportées tant en ce qui concerne la zone des quais (symétrisation des faisceaux) que pour la limitation du gauche de voie.

ARCHITECTURE GENERALE ET ENVIRONNEMENT

39. Une part importante de la réussite du Lien Fixe, projet majeur qui suscite l'intérêt manifeste du public, se joue autour des installations terminales, qu'il s'agisse, bien sûr, de la simplicité et de la lisibilité des accès routiers, de la fluidité du trafic jusqu'à l'embarquement, de l'efficacité des services offerts, mais aussi de la perception du traitement architectural et paysager.

40. Les études d'architecture générale ont suivi quatre grandes orientations :

(a) l'intégration harmonieuse dans les plans de développement locaux et régionaux,
(b) la recherche d'un plan d'ensemble cohérent, en dépit de la multiplicité des fonctions à assurer, et ordonné pour permettre et maîtriser le développement progressif des installations,
(c) l'intégration dans l'environnement paysager,
(d) la perception par l'usager d'une identité du système de transport Eurotunnel indépendamment du lieu de départ

41. A partir de ces axes de travail les architectes français Paul Andreu et Pierre-Michel Delpeuch (Aéroports de Paris) et britannique Bob Smart (Building Design Partnership) ont oeuvré en parfaite coordination pour établir les projets propres aux deux Terminaux.

42. En conformité avec les réglementations propres aux deux Nations, des études d'impact ont été entreprises, tant du point de vue de la définition de l'état initial que de l'impact des ouvrages et des mesures appropriées de compensation et ce sur les thèmes majeurs tels que :

(a) cadre physique (hydrogéologie, détermination des vents dominants, qualité de l'air),
(b) cadre naturel biologique (faune et flore),
(c) cadre agricole
(d) cadre paysager,
(e) cadre urbain,
(f) effet des infrastructures de transport (nuisances sonores des trafics routiers et ferroviaires, effet des circulations routières sur la qualité de l'air),
(g) patrimoine historique et culturel,
(h) cadre socio-économique.

Les spécificités du Terminal de Coquelles (FR) (Figure 9)

43. Compte tenu de l'absence de mouvement de terrain prononcé et de la grande superficie du site, il a paru nécessaire d'organiser l'espace autour de trois points forts architecturaux :

(a) l'échangeur de Fort Nieulay, articulé autour d'un lac central de 400 m de diamètre, associé à une arche monumentale de 50 m de hauteur signifiant l'entrée dans le système Eurotunnel,
(b) dans l'axe majeur de composition du terminal tourisme, la succession rythmée de série d'auvents métalliques haubannés (18m x 18m), couvrant les péages et les contrôles frontaliers,
(c) la zone des quais d'embarquement surmontée par de grandes structures métalliques haubannées supportant les caténaires, l'éclairage, la sonorisation.

44. A partir des points forts architecturaux, l'aménagement paysager s'articule comme suit :

(a) à partir de l'échangeur, deux rangées d'arbres en alignement encadrent les zones de péage et de contrôle,
(b) les parkings d'affectation recevront un traitement particulier tant interne que périphérique,
(c) le long de la voie principale de sortie tourisme s'étend un jardin ordonné par une série de digues plantées à espacement régulier entre lesquelles s'intercalent des pièces d'eau et des jardins bas,

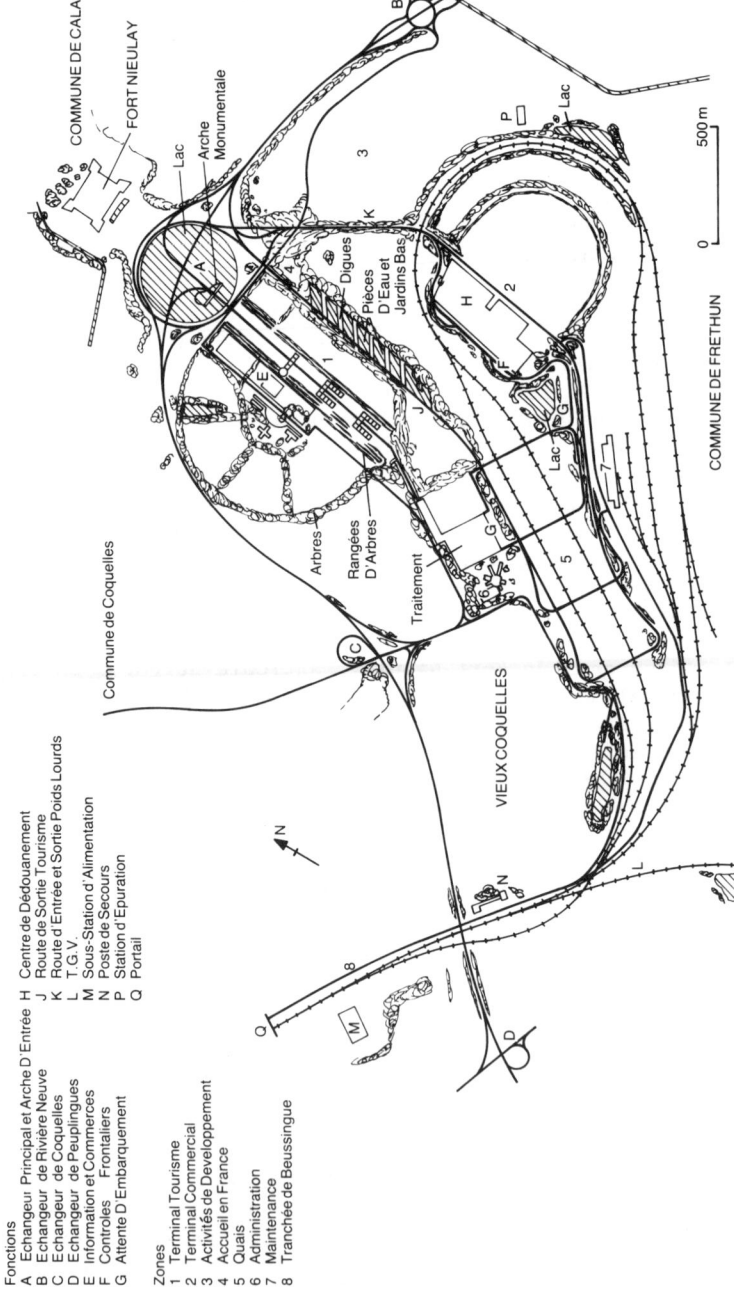

Fonctions

A Echangeur Principal et Arche D'Entrée   H Centre de Dédouanement
B Echangeur de Rivière Neuve   J Route de Sortie Tourisme
C Echangeur de Coquelles   K Route d'Entrée et Sortie Poids Lourds
D Echangeur de Peuplingues   L T.G.V.
E Information et Commerces   M Sous-Station d'Alimentation
F Controles   Frontaliers   N Poste de Secours
G Attente D'Embarquement   P Station d'Epuration
                                           Q Portail

Zones

1 Terminal Tourisme
2 Terminal Commercial
3 Activités de Developpement
4 Accueil en France
5 Quais
6 Administration
7 Maintenance
8 Tranchée de Beussingue

Fig. 9. Terminal de Coquelles - aménagement paysager

154

(d) des aménagements spécifiques sont prévus pour intégrer des sites ou des ouvrages à caractère technique.

Les spécificités du Terminal de Folkestone (UK) (Figure 10)

45. La localisation en vallée aux pieds de l'escarpement des North Downs, l'étalement le long de l'autoroute M20 et de la voie ferrée CML, la proximité des villages de Newington, de Peene et de Longport House sont autant d'éléments qui guident le développement architectural et l'aménagement paysager du Terminal britannique :

    (a) pas d'élément architectural monumental,
    (b) la réalisation de la boucle Ouest en tranchée couverte,
    (c) des mouvements de terre et plantations extensifs à proximité de Newington et de Peene,
    (d) un traitement particulier du terminal tourisme et des parkings d'affectation.

46. D'autre part, l'existence sur ou à proximité immédiate du site de cinq zones protégées, d'un grand intérêt pour la nature, impose une parfaite intégration des installations dans l'environnement.

47. En conséquence, la stratégie d'aménagement paysager est fondée sur le respect des principes suivants :

    (a) assimilation des installations
    (b) atténuation des perturbations
    (c) défense de l'environnement
    (d) la qualité de la conception.

LE TERMINAL ET LA SECURITE

48. Le fonctionnement des installations Eurotunnel doit satisfaire à des conditions bien définies de sécurité.

49. Des mesures spécifiques doivent être prises sur les Terminaux, tant pour assurer la sécurité dans les tunnels (c'est l'un des rôles du tunnel de service qui aboutit aux portails) que sur les Terminaux proprement dits.

50. Les critères de conception des Terminaux en matière de sécurité sont :

    (a) prévention des risques naturels et artificiels,
    (b) centralisation des alertes et organisation de secours,
    (c) lutte contre les sinistres : incendies, explosion,... tant dans les Tunnels qu'à bord des navettes et au sein des Terminaux,
    (d) protection des personnes en cas d'accident majeur ou d'incident.

51. A partir de ces critères des dispositions

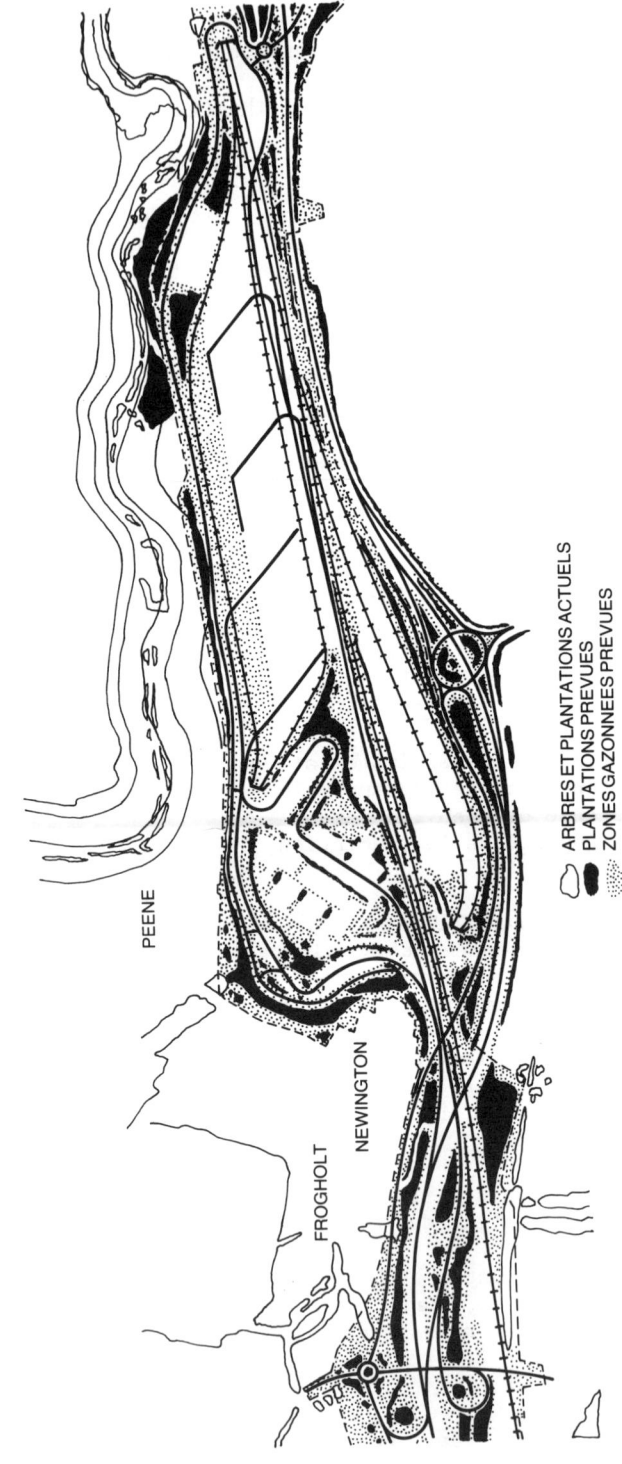

ARBRES ET PLANTATIONS ACTUELS
PLANTATIONS PREVUES
ZONES GAZONNEES PREVUES

PEENE

FROGHOLT

NEWINGTON

Fig. 10. Terminal de Folkestone – aménagement paysager

constructives particulières ont été prévues sur les sites des Terminaux :

(a) un réseau d'eau, enterré, maillé à partir de deux sources indépendantes,

(b) un centre de secours comportant les locaux abritant les personnels de secours et le garage des véhicules du Système de Transport du Tunnel de Service affectés à la sécurité,

(c) une voie d'urgence destinée à accueillir les trains en difficulté,

(d) des accès de secours depuis l'extérieur et à l'intérieur des Terminaux,

(e) un train de secours (en zone d'entretien) pour acheminer des moyens lourds dans le tunnel.

52. Ces dispositifs et un ensemble adéquat de procédures d'exploitation font du Système de Transport Eurotunnel un moyen de transport particulièrement sûr.

SUMMARY
GENERAL CHARACTERISTICS OF THE TERMINALS

1. Connecting points between the Eurotunnel Transport System and the road and rail networks both sides of the Channel, the Terminal installations are designed to ensure both the services to the users and the operation, management, maintenance and safety of the System.

2. The traffic flows at the 30th peak hour occuring in 2003 have been adopted as the design traffic flow for the Terminal design at opening.

3. The Terminals are split up into several major functional zones as follows :

(a) accessible zones to the users
. main road exchange
. tolls
. terminal amenities
. frontier controls
. allocation holding area
. platform area

(b) zones required for the operation of the system
. administrative area
. maintenance area

(c) other zones (linked to safety aspects)
. emergency centre
. emergency sidings
. portal installations

4. The platform area is designed to ensure the maximum flexibility and consists of two access overbridges, two egress overbridges and 10 common tracks.

5. From 1986, the rail layouts have been re-examined and optimised, improving greatly the operating conditions for both the national trains and the shuttle trains. The loop design has been finalised.

6. Depending on the characteristics of the two sites, particular attention has been given to the architectural aspects, the environmental integration and the safety of the terminal installations.

SPECIFIC ASPECTS OF THE COQUELLES TERMINAL (FR)

7. The main parameters of the French Terminal are :

(a) availability of space (650ha on which 470ha are directly dedicated to the fixed link installations)
(b) existence of compressible soil guiding the organisation of the general layout and track layout
(c) flat configuration

8. Starting from those parameters the specific aspects for the design are :

(a) installation of all the rolling stock maintenance facilities on the French site
(b) a clear and complete partition between the tourist and freight facilities and user roads
(c) strong architectural points such as the main exchange of Fort Nieulay, canopies on frontier control areas etc...
(d) a landscape strategy concentrated on the users areas (access road, allocation area, egress road

SPECIFIC ASPECTS OF THE FOLKESTONE TERMINAL (UK)
9. The main parameters of the English Terminal are

(a) restriction of space (170ha)
(b) location in length in a valley, back to the North Downs escarpment
(c) vicinity of Newington and Peene villages
(d) the M20 motorway at the South border
(e) the crossing by the CML from East to West

10. Starting from those parameters the specific aspects of the design are :

(a) realisation of part of the rail loop in a cut and cover tunnel
(b) partition of tourist and freight facilities
(c) a landscape strategy based on four principles :
. assimilation of the development within its setting
. minimisation of disturbance and visual intrusion
. conservation
. enhancement of the terminal design

9. The role of the Maître d'Oeuvre

H. GRIMOND, Chairman, SETEC and P. MIDDLETON,
Executive Maître d'Oeuvre

1. INTRODUCTION

The Maître d'Oeuvre (MdO) role is a well understood concept in France and is used for Public Works, such as the Channel Tunnel Project.

The management concept is illustrated in Figure 1.

Maître d'Ouvrage (Owner)

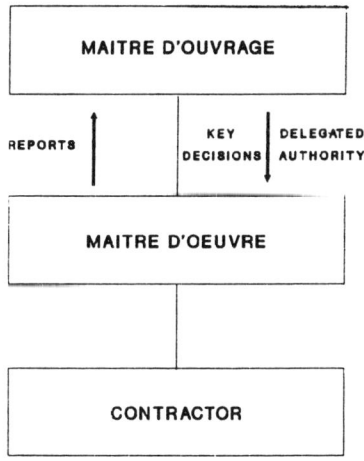

Figure 1

Maître d'Oeuvre (Project Manager)

Contractor

As is normal the owner outlines the project, obtains finance, and prepares to utilise the project once built.

The Maître d'Oeuvre prepares the preliminary design, elaborates the final design (drawings, specifications and tender documents), analyses the bids, helps the Owner to prepare the contract documents and supervises the construction. This includes the detailed management of the construction contract, quality assurance and quality control, cost and time management. He maintains a considerable independence, and is thus able to advise the owner as to his strategy, although responsibility for this strategy rests with the owner.

Broadly speaking, his role is close to a Project Management role in the UK although these are some aspects of his position which are closer to that of The Engineer.

Although this was clearly the starting point of the concept on the Channel Tunnel project, today the MdO on the Channel Tunnel performs a quite different role, which to be understood needs some historical development.

2. THE CONCESSION

In 1985 the Governments issued a competitive invitation to various promotional groups to apply for a concession to build a Fixed Link across the Channel. The role of MdO in the project was a requirement of the Governments.

A group of banks and contractors in France and the UK formed a promotional consortium, consisting of France Manche and the Channel Tunnel Group, to respond to this invitation. An offer was prepared and submitted on 31 October 1985.
SETEC, a major French engineering consultancy group, had been involved in the Channel Tunnel project since 1957, particularly on the design concept which utilised two railway tunnels carrying vehicles through in shuttles. During the phase leading to the submittal of the proposal to the Governments, SETEC was employed by the French Banks to prepare a traffic forecast and to carry out a technical audit of the project.

WS ATKINS, a major UK engineering consultancy group, was employed by the Channel Tunnel Group to provide an independent technical audit used to support the submission of the proposal to the Governments through the U.K. route.

The MdO role was supported by the embryo concessionaire and the banks and in 1986 ATKINS and SETEC formed a joint venture, ATKINS-SETEC. This joint venture took on the role of MdO when EUROTUNNEL, formed by France Manche and Channel Tunnel Group, became the concessionnaire for the Fixed Link at the end of 1986.

The organisation is shown in Figure 2.

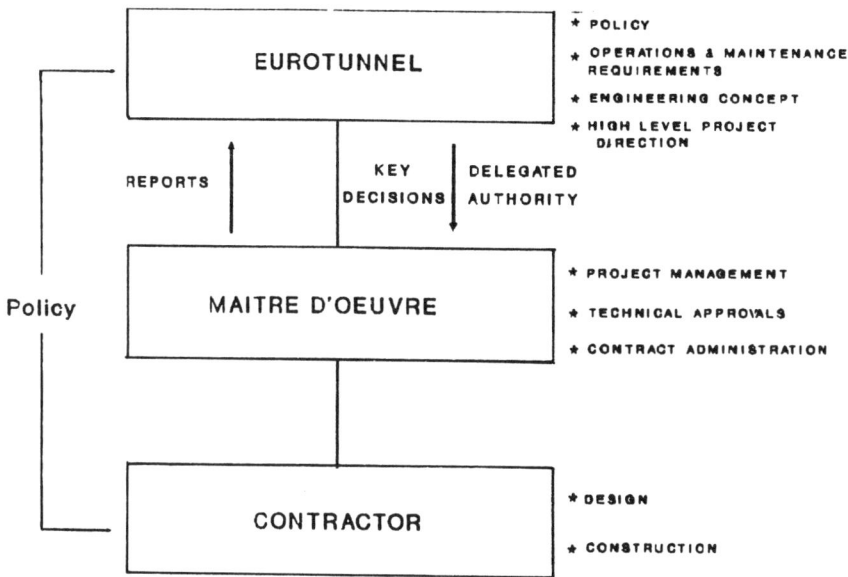

Figure 2

The MdO role is described in several major documents :

(i) The CONCESSION granted by the two Governments to CTG/ France Manche, and signed on 14 March 1986,

(ii) The MAITRE D'OEUVRE contract signed between CTG France Manche and ATKINS-SETEC on 13 August 1986,

(iii) The LOAN AGREEMENT between Eurotunnel and the BANKS signed on 4 November 1987.

(iv) The CONSTRUCTION ("design and construct") contract between EUROTUNNEL AND TML - the main contractor for the project - signed on 13 August 1986.

Towards the end of 1987 Eurotunnel decided to integrate the commercial, operations and engineering forces of the MdO into its own team.

The effects of this reorganisation were :-

- to strengthen of the Owner's team by the transfer into his organisation of those members of the MdO team who had, up to this date, been carrying out the traditional MdO role between the owner and the contractor,

- and emergence of a new team, the "INDEPENDENT MDO", which was responsible for the overall monitoring and audit of the Project on behalf of the Intergovernmental Commission and the Banks, and as expert advisers to Eurotunnel.

The following diagram (Figure 3) shows the organisation which became effective early in 1988.

FIGURE 3

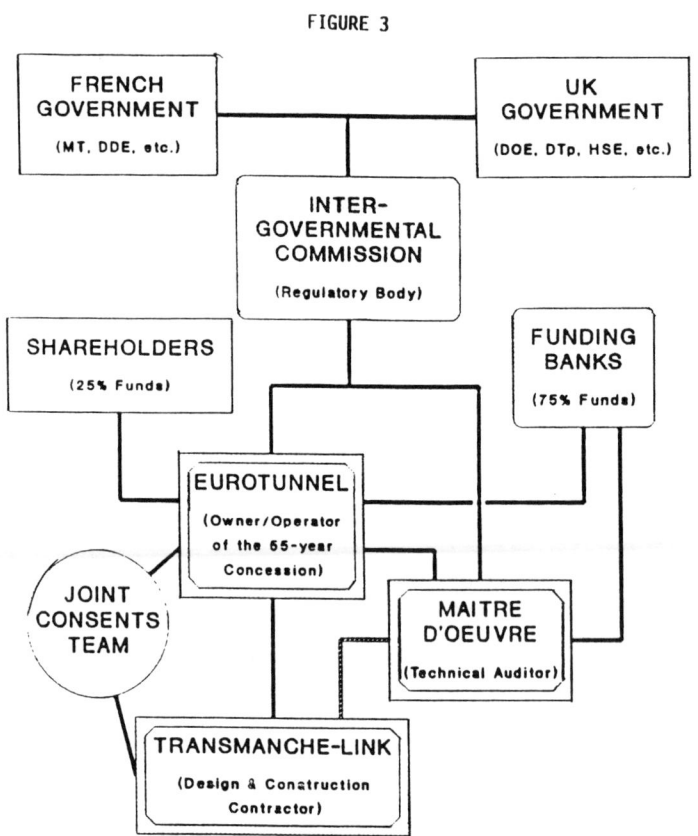

3. PRESENT ROLE OF THE MDO

Three executive aspects of the MdO's work are considered to be independent roles :

- The Quarterly Progress Report, produced in two languages,
- The Quality Assurance role,
- Safety of both construction and operations.

The main parties to the project and their relationship are shown in the diagram (Figure 4).

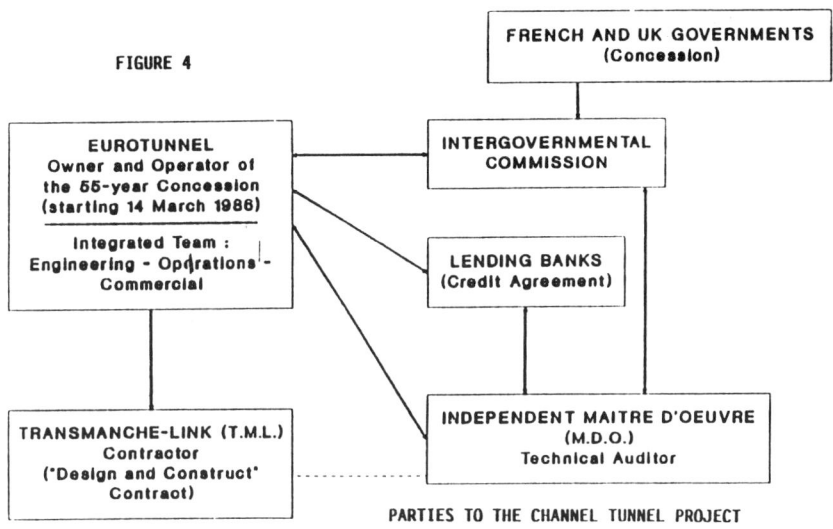

FIGURE 4

PARTIES TO THE CHANNEL TUNNEL PROJECT

The MdO has a strong relationship with the four major entities :

- The owner - Eurotunnel,
- The InterGovernmental Commission,
- The Banks,
- The Contractor - TML.

The main functions of the MdO are described below.

3.1 Relationship with the IGC
Engineering and environment
The MdO has the duty to prepare an independent view on all engineering submissions made to the IGC. This, in general terms, is almost a power of veto. This role continues through the detailed certification procedures leading up to the final operating certificate.

Safety

The MdO has a responsibility to perform a similar function with regard to the safety of the operating transportation system. The monitoring of the operating transportation system safety development will culminate in a full analysis and safety report prior to the issue of the final operating certificate.

The MdO has a specific responsibility with regard to construction safety. Because the law regarding construction safety in the United Kingdom and France are different, the MdO approach to this matter is also different. The approach may need to change, however, after the breakthrough of the service tunnel next year.

165

Project Progress

The MdO quarterly report (see below) is submitted to the IGC for information and we have certain specific duties in reporting the progress of the project, particularly with regard to the availability of funds.

Procedures

The IGC rely upon us to produce and negotiate appropriate procedures with all the parties leading up to specific key events eg. engineering submissions and operating certificate.

Quality Assurance

The MdO is required to assure the IGC that the work executed is in line with the engineering submissions made. This has led to our leading role in QA.

3.2 Relationship with the Banks

Quarterly Report

We prepare an independent detailed quarterly report on the project, which is aimed specifically at the Banks. The Banks require a particular knowledge of progress against cost and time, and although the report covers the complete project, a special emphasis is given to this aspect.

Much of the report is factual, but one section is devoted to the MdO's opinion of the state of the Project.

Appendix II describes this in more detail.

Other Duties

The MdO is required to issue periodic reports on programme and cost to complete (forecast reports), and to respond to requests from the Banks for independent assessments of project matters.

3.3 Relationship with ET

The MdO is required to report to ET on all aspects of the construction contract, including engineering, construction, cost and progress. To this end the MdO monitors the contractor's activities and submissions, liaises with ET and reports as necessary.

Summary and Conclusions

The MdO has a key role on the Channel Tunnel Project, in which it represents the interests of both the British and French Governments and the Banks, provides an independent overview of the total Project, identifies potential problem areas, and advises on remedial measures where the MdO considers them to be necessary in the interests of the Project.

RESUME

Le rôle traditionnel du Maître d'Oeuvre (MdO) tel qu'il existe en France est décrit. Le MdO assume la direction du projet pour le compte du Maître d'Ouvrage tout en gardant une certaine indépendance et peut, pour certains aspects de sa tâche, agir en qualité d'Ingénieur.

L'évolution du rôle du MdO au sein du projet du tunnel sous la Manche est ensuite expliquée. Le rôle actuel du MdO s'apparente plus étroitement à celui d'un auditeur technique indépendant qu'au rôle traditionnel décrit précédemment.

Les tâches exécutées par le MdO sont décrites et les diverses relations avec les principaux intervenants sont explorées.

10. The role of Eurotunnel

Dr T. M. RIDLEY, Managing Director — Project, Eurotunnel

SYNOPSIS. As owner and future operator of the Channel Tunnel, Eurotunnel must manage a whole series of interfaces. These include the main construction contract, contacts with the Governments, principally through the Intergovernmental Commission, and the interface with the lending banks providing the bulk of the financing. At the same time Eurotunnel must plan and provide for its future role as operator of the cross-Channel transportation system. This paper describes the role of Eurotunnel in the project as at the time of writing.

INTRODUCTION

1. The Channel Tunnel project has a long history but, before entering into the specifics of Eurotunnel's role as owner, it is useful to look at it from a recent historical perspective. The next paper, by TML, provides overall historical detail and therefore the references here are to Eurotunnel itself insofar as possible.

2. Figure 1 illustrates major Eurotunnel legal and contractual relationships. Eurotunnel, as a newly-created organisation in 1986, was faced with major legal, financial and technical challenges - challenges which established world-class industrial companies would find daunting.

3. As shown in Figure 2 Eurotunnel grew out of the original promoting group of French and British contractors and their banks. During the preparation of the proposal by the promoting group, it had become evident to the contractors and the banks that a strong, independent owner was required. The Concession from the two Governments would be granted to a single owner/operator and the owner would also have to deal with the project financing. Thus, Eurotunnel assumed its role as owner/operator in August 1986, when the Construction Contract was signed, and its role was confirmed in October 1986, when the second - and private - placement of Equity was accomplished. Overnight the UK Channel Tunnel Group and

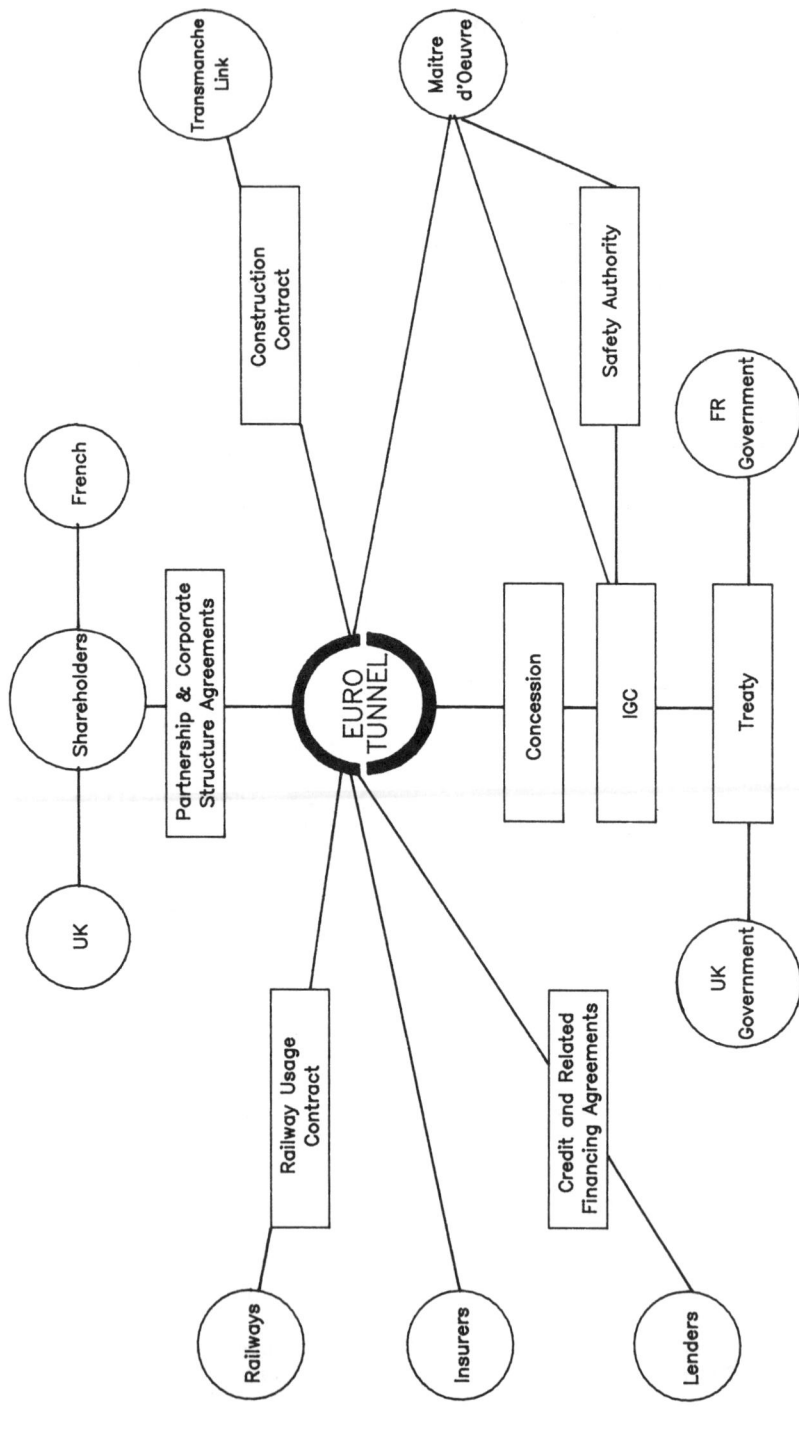

Fig. 1. Project relations structure

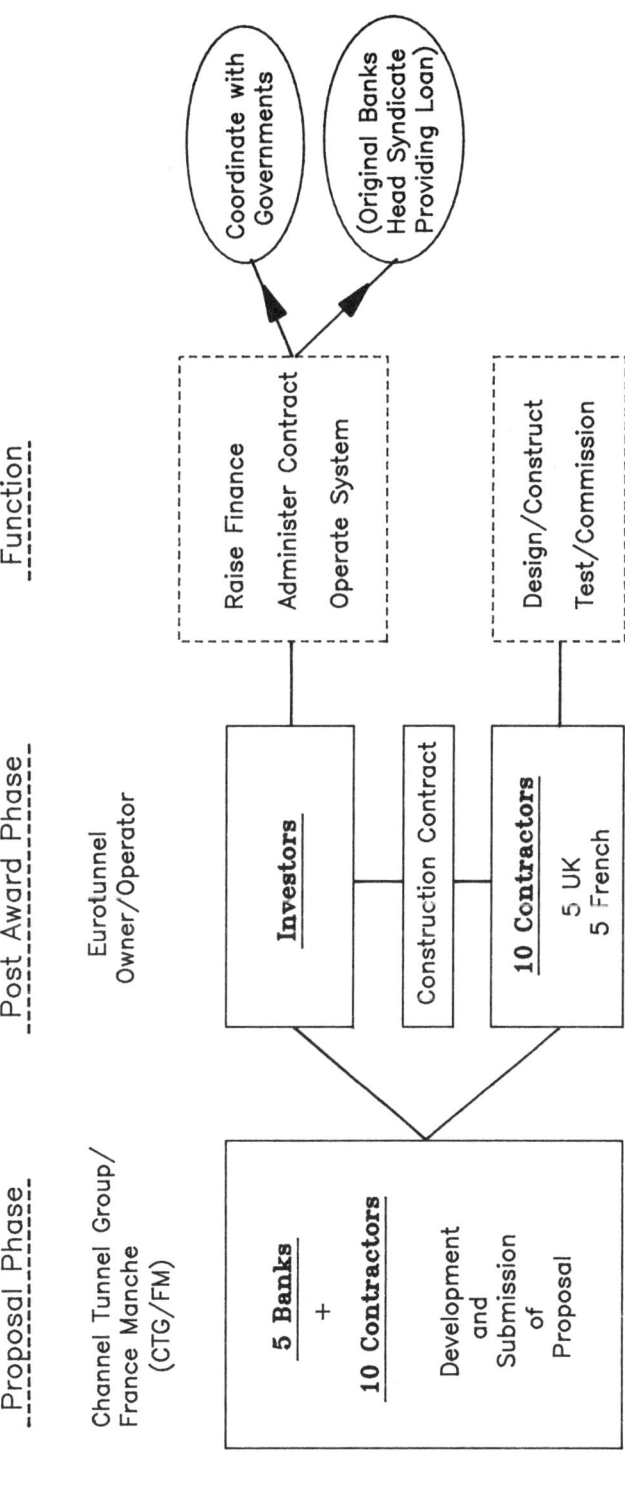

Fig. 2. Channel Tunnel organisation evolution

their French counterpart France Manche were translated into Eurotunnel and Transmanche-Link, their designated contractor.

4.   It is instructive to examine some of the complexities which had to be (and still have to be) dealt with. These complexities are at once interesting and challenging and often unique:

        (a) Project Origin
        (b) Requirements of the Concession
        (c) Cultural Differences
        (d) Geographic Dispersion
        (e) Financial Arrangements
        (f) Contractor Organisation
        (g) Owner Organisation

Project Origin
5.   The Channel Tunnel is one of the very few major international projects where the promoters have been (at different times) both the owner/operator and the contractor. This came about through the evolution of the capital structure which left the initial shareholders (the ten promoting contractors and five banks) in a minority position after the ownership base was expanded and the Eurotunnel corporate structure was set up and staff put in place. When the Concession was first granted, the Contractors' share of the capital was over 50 per cent whereas today it is less than 7 per cent. Once established, Eurotunnel negotiated the design/build Construction Contract with the original promoters, thus completing their transition from owner to contractor, responsible for the design and construction of the entire project. This has significance for the relationship between the two contracting parties.

Requirements of the Concession
6.   A fifty-five year Concession was granted by the two Governments to the Channel Tunnel Group/France Manche in March 1986, to build, own and operate the tunnel and transportation system. The Concession itself is unusual in that it specifies financial, design and operational requirements for the project and it stipulates the establishment of the Intergovernmental Commission (IGC) whose role is to ensure compliance with these requirements. The IGC is supported by a Safety Authority which reviews all aspects of safety and security of the total project and its future operations. The Concession also stipulates the establishment of an independent project manager - the Maître d'Oeuvre (MdO) - to ensure that the works are carried out to the required specifications and construction codes and that the project budget and timetable be respected. The Concession also requires that the Outline Design documents for all aspects of the project be submitted to the IGC for comment prior to construction. Each Outline Design

submission must be accompanied by a report on the design by the MdO. This requirement is important to Eurotunnel in that it adds another loop to the design approval process. The IGC can only object for reasons of safety, defence, security, environment, or non-conformity with the performance and facilities specified within the Concession itself.

Cultural Differences

7.   There are both national (British vs French) and industrial (civil construction vs international engineering - construction) cultural differences at play on the project. These differences can be a source of conflict between the owner and the contractor. National differences include the obvious language barrier, attitudes toward work, perspectives on time, degrees of formality in analysing problems and degrees of confidence in the chosen solutions, to name just a few. Industrial differences are found in varying degrees of sophistication in project management techniques, varying levels of appreciation of the benefits of modern management techniques and the inverse willingness to devote resources to them, as well as the expected differences in codes and standards.

Geographic Dispersion

8.   While the distances involved here are not great by international standards there is a great dispersion of Eurotunnel and Transmanche-Link staff throughout south-east England and north-west France. Eurotunnel has its headquarters in London but the Project Implementation Division (PID) is 15 miles south in Sutton and there are British construction offices in four other locations near Folkestone 75 miles to the east, a precast facility 65 miles north-east at the Isle of Grain, as well as two French construction offices at the Sangatte site near Calais. TML's engineering design is also dispersed with the UK work being done by two design sub-consultants whose offices are in South London with the French work being done at two other design offices in/near Paris.

Financial Arrangements

9.   The Construction Contract itself specifies three different payment arrangements:

(a)   Reimbursible with a Target Cost (and sharing of cost over/under runs) for the Tunnels and all associated civil works.

(b)   Lump Sum for the Terminal Civil Works and all the mechanical and electrical systems in the Terminal areas and through the tunnels.

(c)   Cost Reimbursible with fee for the Rolling Stock.

10.   These arrangements have necessitated development of strict rules for allocation of the actual costs and segregation of resources so that Eurotunnel is assured of correct accounting for each item of cost.

11.   The funding of the project which, is entirely through private capital, is largely in Pounds Sterling and Francs so that currency exchange rate risks are minimised. Twenty per cent of the capital is in equity and eighty per cent in debt which was obtained through a single credit agreement arranged by some 40 banks with a total syndication involving over 200 banks world-wide. The arranging of this credit is one of the major accomplishments of Eurotunnel to-date. The Credit Agreement itself contains many provisions and conditions which must be met in addition to the normally expected commercial terms.

Contractor Organisation
12.   Moulding individuals from ten independent engineering/contracting companies into one cohesive project organisation while simultaneously carrying out the design and construction of one of the world's greatest engineering undertakings is without doubt the most complex of all the complexities of the Project. National and industrial differences are magnified and it is essential to provide firm direction and leadership to counteract any tendency to pull apart.

Owner Organisation
13.   Similar divisions and differences also exist within the Owner. Executives, managers, engineers and other employees have been drawn from numerous British, French, other European and American companies. These include not only major engineering companies and consultants, but also companies in the French steel industry, the British and French National Railways, private and public transportation organisations, public utilities, private petroleum groups and many more. This has brought together people from varying backgrounds who have for the most part not worked together previously.

14.   Eurotunnel is an evolving organisation. It has completed its role as promoter, having successfully obtained all necessary governmental permissions and having obtained financing for the project. It is now fully engaged as project manager in the overall sense - recognising direct management of the engineering, procurement and construction rests with the Contractor. Eurotunnel is already preparing for its ultimate role as a major transportation company, and by 1993 its evolution from promoter through builder to operator will be complete. It will then operate a unique railway system consisting of "shuttle" trains carrying cars, coaches and lorries together with through-trains from British Rail and SNCF.

PREPARING FOR THE PROJECT

15. The technical design of the project has been treated
in detail in London earlier in the programme. Early in
Eurotunnel's life it was involved in basic studies defining
the project and, as its operating and commercial management
has been formed, its involvement has continued.

16. A more critical role for Eurotunnel as a fledgeling
company was piloting the legislative efforts necessary to
authorise the construction of the project in the United
Kingdom and France.

17. In February 1986, France and the UK signed the Treaty
authorising a fixed link between the two countries. The
Concession giving Eurotunnel the right to develop, finance,
construct and operate the tunnel as that fixed link was
signed in March 1986, subject to a number of conditions.
Neither the Treaty nor the Concession could be put in effect,
however, until after enabling legislation was passed by the
legislative branches of the two Governments.

18. In France this proved to be a relatively
straightforward process as the legislative machinery did not
require a long process. The Déclaration d'Utilité Publique
was signed on 6 May 1986. L'Assemblée Nationale and le Sénat
unanimously approved laws permitting ratification of the
Treaty and approving the Concession.

19. In the UK, the approval of the Government required the
introduction of a Hybrid Bill in Parliament and its passage
through both houses, with hearings before a Select Committee
in each. A Hybrid Bill as a combination of a private bill
and a public bill allowing a reasonably efficient and timely
flow through the parliamentary process while at the same time
giving public access to contribute to the debate. This
process mobilised a great part of Eurotunnel's total
resources over a period of many month beginning in the middle
of 1986 and concluding in July 1987. UK parliamentary
elections caused delay in May 1987 bu the Channel Tunnel Bill
was finally given Royal assent on 23 July 1987.

20. With legislation completed in both countries the
Treaty was ratified at a ceremony in Paris on 29 July 1987.

21. The passage of the Hybrid Bill and the Déclaration
d'Utilité Publique were important not only to obtain
necessary Government authorisation but because the
authorisation was key to proceeding with the financing plan
for the project i.e. permitting Eurotunnel to proceed at full
speed with finalisation of the Credit Agreement with the
Banks and the sale of equity (Equity 3) to the public.

KEY STEPS IN FINANCING

1987:

JULY	●	Completion of Parliamentary processes in UK and France/ Treaty Ratified/Concession in force
JULY	●	Railway Usage Contract signed
AUGUST	●	50 Banks Underwrite £5 Billion Credit Agreement
SEPTEMBER	●	EIB £1 Billion Credit Agreement signed
OCTOBER	●	Loan Syndication completed
OCTOBER	●	French High Speed Railway Line announced
NOVEMBER	●	198 Banks Sign Credit Agreement
NOVEMBER	●	Equity 3 (£770 Million Public Issue) Underwritten
DECEMBER	●	Credit National FF 4000 Million Credit Agreement Signed

Fig. 3. Key steps in financing

22. In addition, the obtaining of the Government authorisations permitted the Contractor access to the UK site, on which initial critical preparatory work could not be undertaken beforehand.

23. The original programme for the project required the legislative and financial hurdles to be overcome by July 1987 and the delays encountered, though relatively short, nevertheless impacted the Contractor's intended work plan. After the Government authorisations were obtained, the obtaining of the money to proceed with the work became the number one priority.

24. Let us recall the amounts in question. The Credit Agreement covers a loan of the equivalent of £4 billion (or FF 40 billion at 10FF to £1) plus a stand-by credit of another £1 billion (or FF 10 billion at 10FF to £1). It was arranged by the UK and French banks which originally backed the project plus a number of major international banks.

25. In the second half of 1987 it was necessary to complete the sale of equity (Equity 3) of £7770 million equivalent, and to complete the credit (loan) agreement. Figure 3 shows the main steps - all of which had to be achieved in the order shown. It should be noted that Equity 3 was placed in the month after the "crash" of October 1987.

26. Thus, at the end of 1987 the financing was in place. It should be noted that there were (and are) no comparable projects as precedents. The major international oil and gas pipelines built in Europe during the 1960s have some similarities, but their owners were established major companies with ample financial resources from their basic businesses.

RESPONSIBILITY AS OWNER/PROJECT MANAGER
27. Figure 1 has shown Eurotunnel as it relates to many other organisations. For the project to succeed Eurotunnel must manage all the functions represented by this chart.

28. All the interfaces are important and interdependent. The most obvious and high-profile interface is that of Eurotunnel with TML. This is handled primarily by Eurotunnel's Project Implementation Division (PID) - based in Sutton, but with important teams on the sites in both the UK and France. PID is a project management team acting for and on behalf of Eurotunnel and comprises managers and engineers from Eurotunnel and the engineering firms of Bechtel, W.S. Atkins, Sir William Halcrow, SETEC and Tractebel. PID will be supplemented by personnel from Eurotunnel's Operations Division, who will become the future engineering department of the operating company.

29. Through PID, Eurotunnel administers the Construction Contract, assures that the future Operations department participates adequately in the design process, monitors and approves as necessary design and construction work and monitors the cost and schedules. It assures that intervention of outside agencies such as the Intergovernmental Commission and the Railways occurs in an orderly, programmed manner so as to be compatible with TML's duties as designer/builder.

30. PID prepares monthly reports for Eurotunnel project and general management. These reports also serve to keep the lending banks and their advisers informed. Monitoring costs and progress and reporting thereon takes on an extra dimension in Eurotunnel compared to most major projects, because Eurotunnel has no cash flow - nor will it have until it opens for business in 1993. Thus the Banks' reporting requirements are very stringent.

31. The Credit Agreement with the Banks contains many covenants which Eurotunnel must satisfy prior to each drawdown of the loan. Many covenants involve financial ratios which are based on forecasts of future revenues and on forecasts of costs to complete the project. The traffic studies on which the project's financial viability is based are updated twice yearly.

32. In addition to the reports and advice from the MdO, the Banks are supported by a Technical Adviser comprising specialised personnel from the firms of Parsons-De Leuw Cather from the USA and Lahmeyer International from Germany. The Technical Adviser acts on behalf of the Banks to monitor compliance with the Credit Agreement.

33. As mentioned previously, Eurotunnel reports to the Banks and satisfies the many covenants on a regular basis. Eurotunnel meets with the four Agent Banks and the key members of the international banking syndicate periodically.

34. On major international projects it is normal for the owner companies to work closely with governments involved. In the case of the Channel Tunnel, the Concession stipulates in considerable detail the role of the two Governments in approving certain aspects of the design and in monitoring many other elements of the project. Thus Eurotunnel liaison with the Intergovernmental Commission (IGC) is of fundamental importance. The IGC, under terms of the Concession, examines and may raise objection to the design of the project as regards safety, security and the environment and monitors the cost and progress.

35. The IGC is served by the MdO, an independent technical and management team drawing on specialised personnel from

Atkins, Halcrow, SETEC and Tractebel, which provides the IGC with reports on all design submissions as well as providing cost and progress reports. Eurotunnel works closely with the MdO, which is represented on the various working groups established to assure that the IGC/Safety Authority is fully integrated into the process of design of the project.

36. While Eurotunnel has many interfaces to manage, it clearly must recognise at all times its primary concern for its shareholders. This responsibility is the everyday concern of Eurotunnel's corporate management and financial department and is a major consideration of overall management of the project.

37. Earlier in the conference, you heard from Alain Bertrand, Managing Director - Operations and Safety. The Eurotunnel project team works in close liaison with the Operations team and the latter will become increasingly involved as testing of the system, start up and commissioning approach. At the end of the day, the project team will evolve into the engineering and maintenance department of the operating company.

38. Eurotunnel is developing from a small spin-off from the original project promoters into a major corporation. In doing so, its role has changed from promoter to financing/project manager and is in the process of evolving further into its final role as a major transportation system operator. In its current role as overall manager of the project, it must assist the Contractor through facilitating financing, obtaining necessary approvals from the regulatory authorities and co-ordination with the future users - BR and SNCF. At the same time it must manage its contract with Transmanche-Link to assure that the project on completion provides a viable business for its shareholders.

RESUME

Ce rapport traite de l'évolution de la structure juridique,
financière et d'organisation d'Eurotunnel, ainsi que des
interfaces avec les Gouvernements, les banques, les
enterprises et les usagers.

Il retrace l'évolution d'Eurotunnel, à partir de la création
du groupe promoteur d'origine, composé d'enterprises de
construction et de banques françaises et britanniques. Cette
évolution s'est traduite par un élargissement de
l'actionnariat de base et par la transformation d'Eurotunnel
en une organisation capable d'assurer les fonctions de maître
d'ouvrage et d'exploitant chargés de réunir le financement et
passer les contrats de construction. En mars 1986, les
Gouvernements français et britanniques ont signé un contrat
de Concession pour une durée de 55 ans autorisant Eurotunnel
à construire, puis à exploiter le tunnel et le système de
transport. La Concession prevoit la mise en place d'une
Commission Intergouvernementale (CIG) destinée à assurer le
respect des divers règlements et d'un consultant indépendant
(le Maître d'Oeuvre), qui assume une mission de supervision
et assure que les travaux sont réalisés dans les limites du
budget et du calendrier, et dans le respect des normes de
construction. Le contrat de construction implique
l'intégration de dix enterprises de construction
indépendantes qui diffèrent aussi bien du point de vue de
leur nationalité que de leur expérience.

La mission d'Eurotunnel se modifie au fur et à mesure de
l'avancement du projet. Son action au titre de promoteur est
maintenance achevée et Eurotunnel agit à présent en qualité
de responsable du projet dans tous les sens du terme. Son
rôle ultime sera d'être une importante société de transport.

L'historique de la promulgation des lois nécessaires aussi
bien en France qu'au Royaume Uni coincide avec la manière
dont le financement et les garanties bancaires ont été mis en
place.

L'articulation avec d'autres organisations constitue une
partie importante de l'activité d 'Eurotunnel. Pendant la
phase de construction l'interlocuteur principal est
Transmanche-Link, l'organisation contractante. La
communication est assurée par la Division Réalisation du
Projet (PID) à Sutton dans le Surrey. Cette division est
responsable du travail de conception et de construction et
rend compte régulièrement des coûts et des progrès du projet
aux organismes de financement. Eurotunnel a aussi de
nombreaux contracts avec la CIG et le Comité de Sécurité avec
les actionnaires et le Maître d'Oeuvre. Par dessus tout,
Eurotunnel doit assurer le succès du projet et en faire une
société de transport viable.

11. Project management by Transmanche — Link joint venture

P. ESSIG, Diplomé: Ecole Polytechnique, Ecole Nationale Supérieure des Ponts et Chaussées, Chairman, and A. D. McDOWELL, FICE, FCIOB, Deputy Chairman, Transmanche—Link joint venture

SYNOPSIS. The Channel Tunnel Project must be one of the most challenging and difficult projects this century.
Challenging because of its sheer size and uniqueness regarding the entirely new transportation system. Difficult because of the financing arrangements which mean that the 200 funding banks involved have a very keen interest in progress and costs which undoubtedly inhibit the normal and expected contractual relationships between client and contractor. It is the first infrastructure project connecting France with the UK and this in itself provides a challenge and opportunity to learn from each other and improve our mutual understanding.

INTRODUCTION
1. Before explaining the Project Management of Transmanche-Link it is necessary to explain briefly the background leading to the contract.

2. In 1984 the Channel Tunnel Group (CTG) was formed by the amalgamation of Balfour Beatty, Costain, Tarmac, Taylor Woodrow and Wimpey, National Westminster Bank and Midland Bank.

3. Similarly a group was formed in France. This group consisted of Bouygues S.A., Dumez S.A., Société Auxiliaire d'Entreprises S.A., Société Générale d'Entreprises S.A., Spie Batignolles S.A., Banque Indosuez, Banque Nationale de Paris and Crédit Lyonnais. This group was known as Francemanche S.A. (FM).

4. In April 1985 the British and French Governments invited proposals for the construction of a totally privately financed Fixed Link between the UK and France. No guarantees would be given by the Governments except that compensation would be paid should the project be cancelled for political reasons.

5. The two groups, CTG and FM joined together to produce a proposal for the development, design, financing, construction and operation of a Fixed Link.

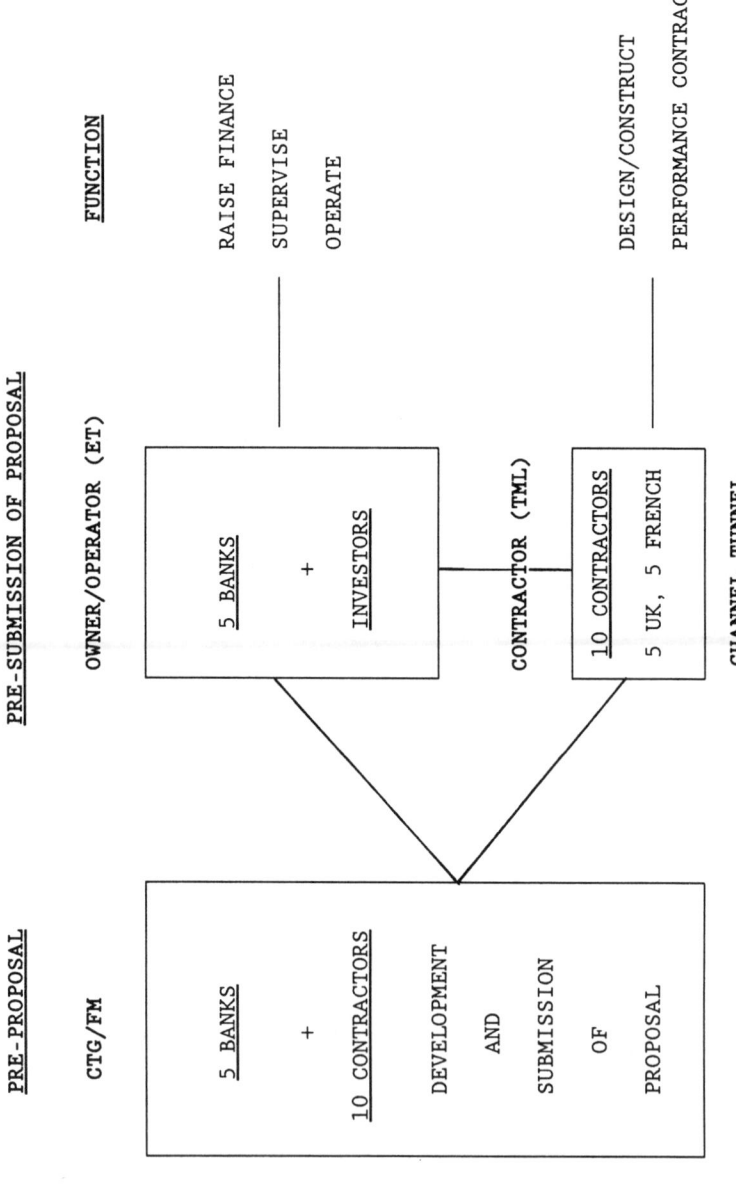

Fig. 1. Channel Tunnel financing and organisation

6.   CTG/FM decided after examining various options that the only scheme which was both technically feasible and financially viable was for a twin-bored rail tunnel to carry road and rail traffic, passengers and freight.

7.   The proposal was submitted to Governments on 31st October 1985. Four competing schemes were submitted

(a) The CTG/FM Twin-bored rail tunnel
(b) The Euro Route Combined bridge/road tunnel
(c) Channel Expressway - Combined road/rail tunnel
(d) Eurobridge - Road bridge

8.   The Governments decided in favour of the CTG/FM scheme and on the 20th January 1986 awarded the concession for the construction and operation of the Fixed Link.

9.   The conditions of the proposal made it necessary to develop two separate entities whose objectives would be

(a) To finance and operate the project)
(b) To design and construct the project.) Fig. 1.

10. During the preparation of the proposal it became apparent that one concession would be granted to a single owner operator company and that this owner would enter into a single construction contract. It was also obvious that the requirement for a strong owner, independent from the contractors, capable of commanding the respect of the City and Institutions, and able to raise the necessary monies to construct the channel Fixed Link, was of paramount importance.

11. The Channel Tunnel Group and Francemanche S.A formed themselves into a partnership which in due course became known as Eurotunnel, and the Employer under the contract. At this point the two construction Joint Ventures Translink (UK) and Transmanche (Groupement d'Interet Economique) joined together as Transmanche-Link. The ten shareholder companies are jointly and severally liable for the obligations of Transmanche-Link.   Fig. 2.

12. The ten contractors remain shareholders of Eurotunnel and were responsible for financing the proposal and development of the project with the founder banks up to Equity 2 (the placing of units to raise finance which took place during October 1986).

13. It also has to be remembered that the concession only came into force at the time the Channel Tunnel Treaty was ratified which was not until the summer of 1987 after the UK general election of May 1987. This situation did not only put founder members and other investors at considerable risk, but also did not encourage senior staff to leave secure positions to join the project which could have been so easily aborted during the summer of 1987.

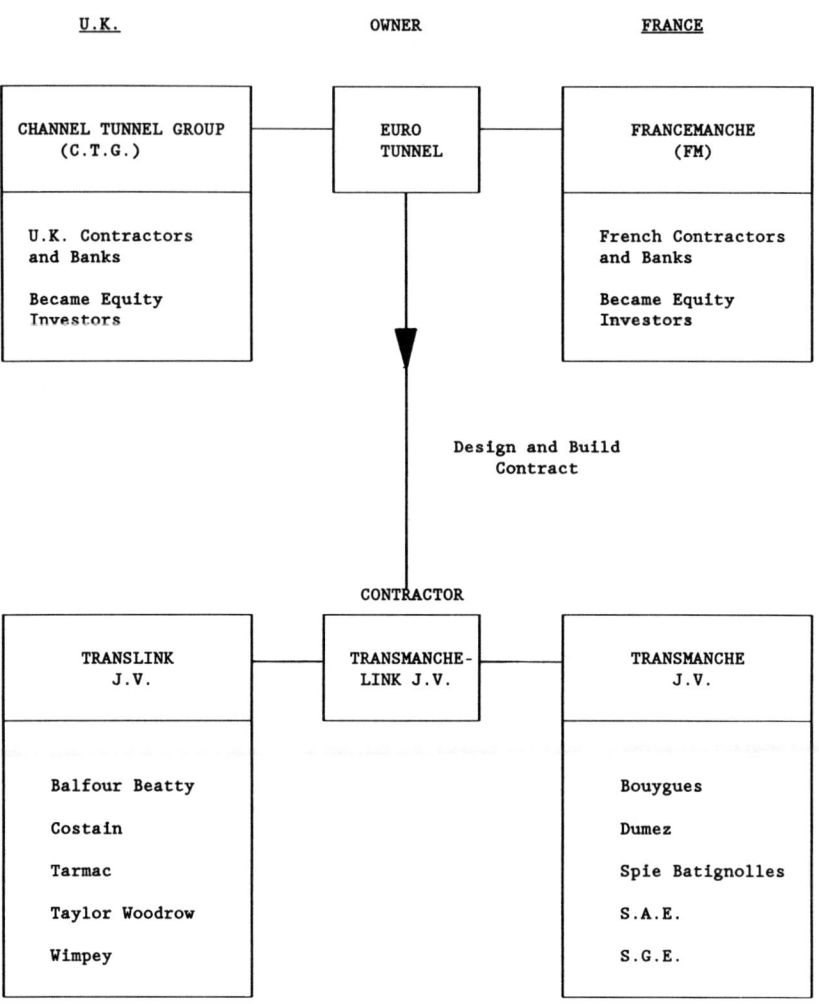

Fig. 2. Parties, relationship and contract involvement

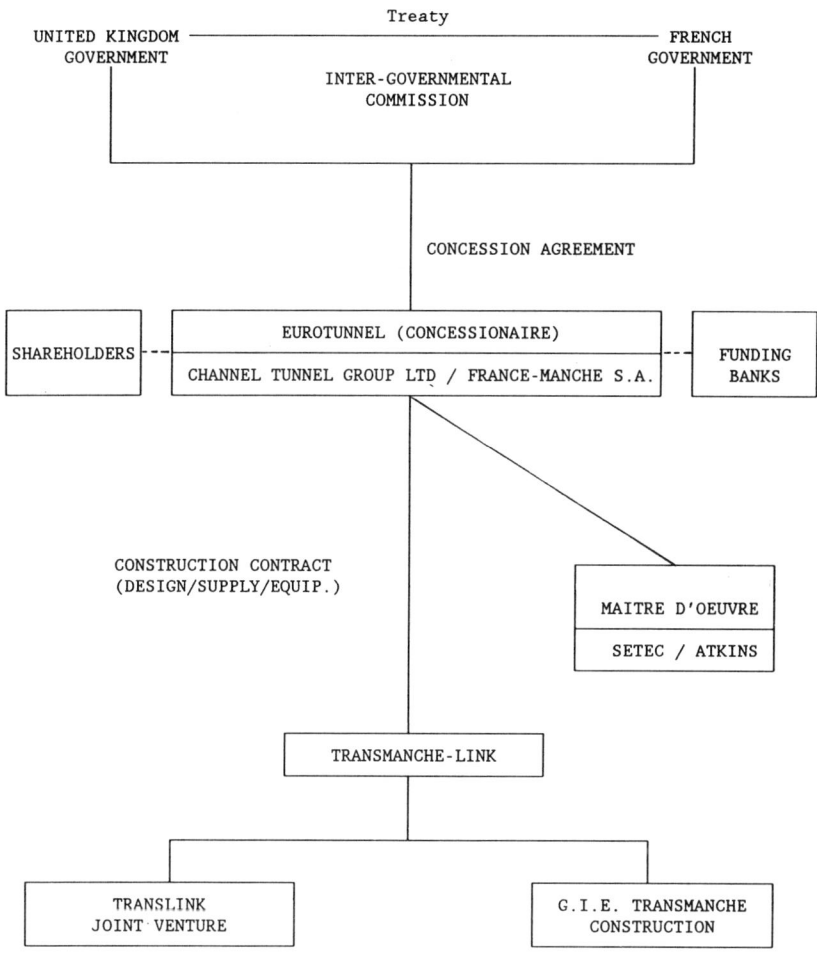

Fig. 3.   Governments - concessionaire - contractor relationship

14. The concession also required the concessionaires to appoint at their own expense "the Maitre d'Oeuvre" (MDO) whose responsibility is to review whether the works carried out conform to the relevant specifications, to the relevant construction codes, regulations and standards, to the relevant construction contract and to the timetable and relevant cost projections.

15. Under the treaty between the two Governments an Inter Governmental Commission (IGC) was set up to review all matters appertaining to safety, defence, security and environment.

16. The relationship of all parties connected with the project is shown on Fig. 3.

THE CONTRACT

17. After prolonged negotiations (mainly with the Shareholder Banks) the contract was signed on the 13th August 1986. The contract value was approximately £2,660M (September 1985 prices) with a construction period of seven years commencing in May 1986, with a completion date of May 1993.

18. The programme was extremely tight as it had been reduced at the proposal stage.

19. The contract consists of three specific sections:-

(a) <u>Target Works</u>- all tunnel works including fabrication of tunnel lining segments are paid on a basis of cost incurred, subject to a share of the saving or extra cost (difference between target cost and actual cost) between ET and TML plus a fee. Approximate value £1286M. The target can be adjusted by means of variation orders

(b) <u>Lump Sum Works</u> - the whole of the Terminal Works and the Railway Fixed Equipment Mechanical and Electrical Works of Terminals and Tunnels are paid on lump sum. Any changes required by the owner are subject to variation orders. Approximate value £1070M

(c) <u>Procurement Items</u> - the supply of Rolling Stock (shuttles and locomotives) is paid on a cost reimbursement basis with in addition a procurement fee. Approximate value £245M.

CONDITIONS OF CONTRACT

20. The conditions of contract are long and complicated but the more important ones related to finance and payments are as follows:-

(a) <u>Payment</u> - monthly basis including one months forward funding which includes off site materials

(b) <u>Retention</u> - maximum 5% (bond provided in lieu)

(c) <u>Escalation</u> - according to the provision of the contract (agreed formulae) covering all work elements

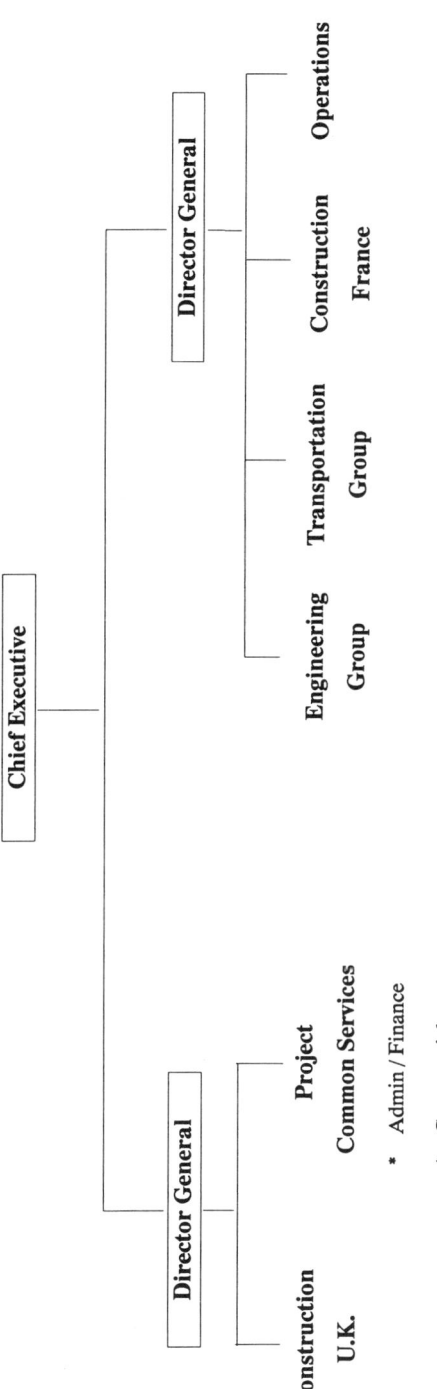

Fig. 4. Director Generals' responsibilities

(d) <u>Liquidated Damages</u> - the contract imposes heavy liquidated damages for failure to achieve completion of the works within the period for completion. Advanced liquidated damages are imposed for failure to achieve completion of the works required within the various milestones (key) dates identified within the project programme

(e) <u>Warranty Periods</u> - Building & Civil Works - 12 months
Electrical & Mechanical Works - 24 months
Procurement Items - 12 months

(f) <u>Bonds</u> - performance bond - 10% of total contract value.

21. <u>Administrative Conditions</u>

(a) <u>Insurance</u> - Eurotunnel is responsible for the provision of the "All Risks" insurance for the works (both temporary and permanent)

(b) <u>Labour</u> - Legislation British or French as applicable. The contract allows for 7 days working 24 hours per day, subject to the applicable laws and regulations of England and France

(c) <u>Funding</u> - TML have to provide cash flow forecasts to Eurotunnel and to liaise on co-ordination of executed work and forecasts

(d) <u>Taxes</u> - French and UK tax rules applicable

(e) <u>Audits</u> - the target works nature of the contract requires fully detailed records and accounts to be maintained throughout the duration of the cost-plus-fee situation

(f) <u>Sub contractors and suppliers</u> -the contract sets out procedures regarding sub-contracting. The procedures differ according to whether the subcontract relates to the Target Works, the Lump Sum Works or the Procurement Items.
The contractor will ensure that tenders are invited from third parties for not less than 30% of the works in accordance with EEC Procedures

(g) <u>Limitation of funds</u> - TML on Eurotunnel instructions will adjust the programme according to Eurotunnel's financial resources. If the works are delayed as a result TML is entitled to an extension of time

(h) <u>Language</u> - contractually the English and French languages have equal validity and either language may be used

(i) <u>Legal Conditions</u> - the contract is subject to the common principles of French law and English law and where necessary to the general principles of international trade law. Disputes shall be referred in the first instance to a panel of three persons acting as independent experts not arbitrators. Arbitration under the rules of the International Chamber of Commerce shall be provided for thereafter.

PROGRAMME

22. The programme called for a site start at Sangatte October 1986 and at Shakespeare Cliff 1st April 1987. The initial work in the UK was to prepare the existing adit and 4.5m diameter marine service tunnel (constructed in 1974) for the arrival of the first tunnel boring machine by August 1987 with a start up milestone date of December 1st 1987.

23. The first milestone date for the start up of the French tunnel boring machine (service tunnel marine) was 1st December 1987. Other details of the programme are the subject of other papers.

24. The placing of orders for these two machines had to be by 15 October 1986 (UK) and 1st October 1986 (France).

25. These machines had to be designed and fabricated extremely quickly as did the concrete segment linings, fabrication works for the segments, the sea wall on the UK works and the Fond Pignon reservoir on the French side to accept the spoil from the borings. All this and other works called for the rapid mobilisation of staff not only for construction, but also engineering and design staff.

26. The forecast turnover was as follows:-

	1986	1987	1988	1989	1990	1991	1992	1993
£M's	16	217	403	607	640	523	214	24

The UK works account for £M1403 and the French works FFM12400 at September 1985 prices.

27. By any standards this project was to be "Fast Track" with over £M100 of turnover on each side of the channel, to be completed in the second year of operation.

ORGANISATION
Legal Structures

28. The contractor consists of ten construction companies (5 French & 5 UK). The five French companies are grouped together in an organisation known as GIE Transmanche Construction. GIE stands for Groupement d'Intérêt Economique and under French law is regarded as having a separate legal personality for certain purposes.

29. The five UK companies are grouped together as Translink Joint Venture which, as the name implies, is the usual UK form of bringing together more than one construction company within a project and does not have a separate legal personality.

30. Transmanche and Translink are in turn grouped together under the umbrella organisation of Transmanche-Link (TML) which is regarded as a joint venture under UK law and a Société de Participation under French law. TML, however, has no trading function and is merely a co-ordinating and liaison organisation between the French and UK halves of the

project. This is because the contract has been unable to
secure satisfactory exemptions from the fiscal authorities
in either the UK or France, and therefore Translink Joint
Venture can only trade within the UK and Transmanche within
France, without risking unfortunate tax consequences for
member companies. The most practical result of this is that
sub-contracts are always entered into either by Transmanche
GIE or Translink JV and never by TML.

31. Whilst Eurotunnel accepts this position it regards
TML as the contractor and as previously mentioned all ten
member companies are jointly and severally liable. It should
be noted that the parent companies do not participate
directly in the physical construction work, except through
specialist subsidiaries who might win work competitively.
The parent companies have seconded many staff into TML.
This was crucial for the initial organisation.

Management Structures

32. Reflecting the Anglo-French nature of the project,
and the necessity for a co-ordinating and project management
organisation, it was decided to set up a TML office firstly
at Croydon, which subsequently moved to its present office
at Sutton.

33. The Translink and Transmanche organisations which had
been formed during the early stages of the project were to
be moved to site as quickly as possible i.e. Folkestone and
Calais. The acquisition of suitable office accommodation,
particularly in the UK was a continuous problem.
There are now 400 staff in TML offices at Sutton and 254
at Shearway House Folkestone and 109 at TMC Calais. These
are of course in addition to the site organisations at
Sangatte and Coquelles, Shakespeare Cliff, Cheriton and Isle
of Grain. All staff and labour are employed by Translink
or Transmanche.

34. The management structure for TML is shown in Fig. 4.
This was changed early 1989.

35. The Translink and Transmanche organisations at
Folkestone and Calais are totally autonomous as is described
elsewhere.

36. The two Director Generals are responsible for the two
national joint ventures which are the two profit centres of
TML. They report to the Chief Executive who is responsible
for the overall success of the project, and profitability of
TML.

37. The Chief Executive and Director Generals report to
the TML Supervisory Board. In addition the Director
Generals report to their respective national joint venture
boards.

38. The Director Generals have other responsibilities as
shown on Fig. 4 but basically the French Director General is
responsible for Engineering & Transportation System and the
UK Director General is responsible for Commercial, Project

Services and Finance, in addition to their national
construction responsibilities.

39. The Boards up to the end of 1988 are shown in Fig. 5.
The 10 chairmen of the parent group companies form "The
Board" and meet twice yearly. The Managing Directors of the
shareholder companies form a Supervisory Board which meets
monthly. It should be noted that the shareholder companies
in the UK are subsidiaries of the parent companies, i.e.
Wimpey Major Projects, Taylor Woodrow Construction, Balfour
Beatty Construction, Costain Civil Engineering and Tarmac
Construction.

40. All members of these two boards are part-time
including the chairman of the Supervisory Board. They are
non executive directors of TML.

41. The Executive Board of TML meets weekly and is
chaired by the Chief Executive.

42. The structure worked well and remarkable progress has
been achieved particularly on the civil construction work.

43. Those readers who have been involved in large joint
ventures overseas will be aware that different nationalities
put together in a foreign country soon work together and
abide by a common set of rules, not only those implemented
by the joint venture but also those of the host country.
The Channel Tunnel is very different in so far as the
British stay in the UK and the French in France working
under their own national legal, tax and labour laws. This
has made it extremely difficult for TML to implement a set
of procedures by which both national joint ventures will
work.

44. At the end of 1988 discussions took place between ET
and TML when it was agreed that:-

(a) It was necessary for Eurotunnel and TML to
    become more compatible for the second half
    of the project. This required a restructure of both
    organisations

(b) It was necessary to organise TML more on a "Company"
    basis

(c) It was necessary to have a more unified approach to
    the Transportation System (all orders to be placed
    during 1989).

45. The result of these deliberations was that two very
senior Railway Executives of international reputation were
recruited. One as Managing Director of Eurotunnel and the
other Executive Chairman of TML.

46. From the beginning of 1989 TML would only have one
Board. This would replace the original Board and
Supervisory Board.

47. The chairmen of the parent companies would continue
to meet twice a year but as a "Members Assembly".

48. The Board whose members would remain basically the

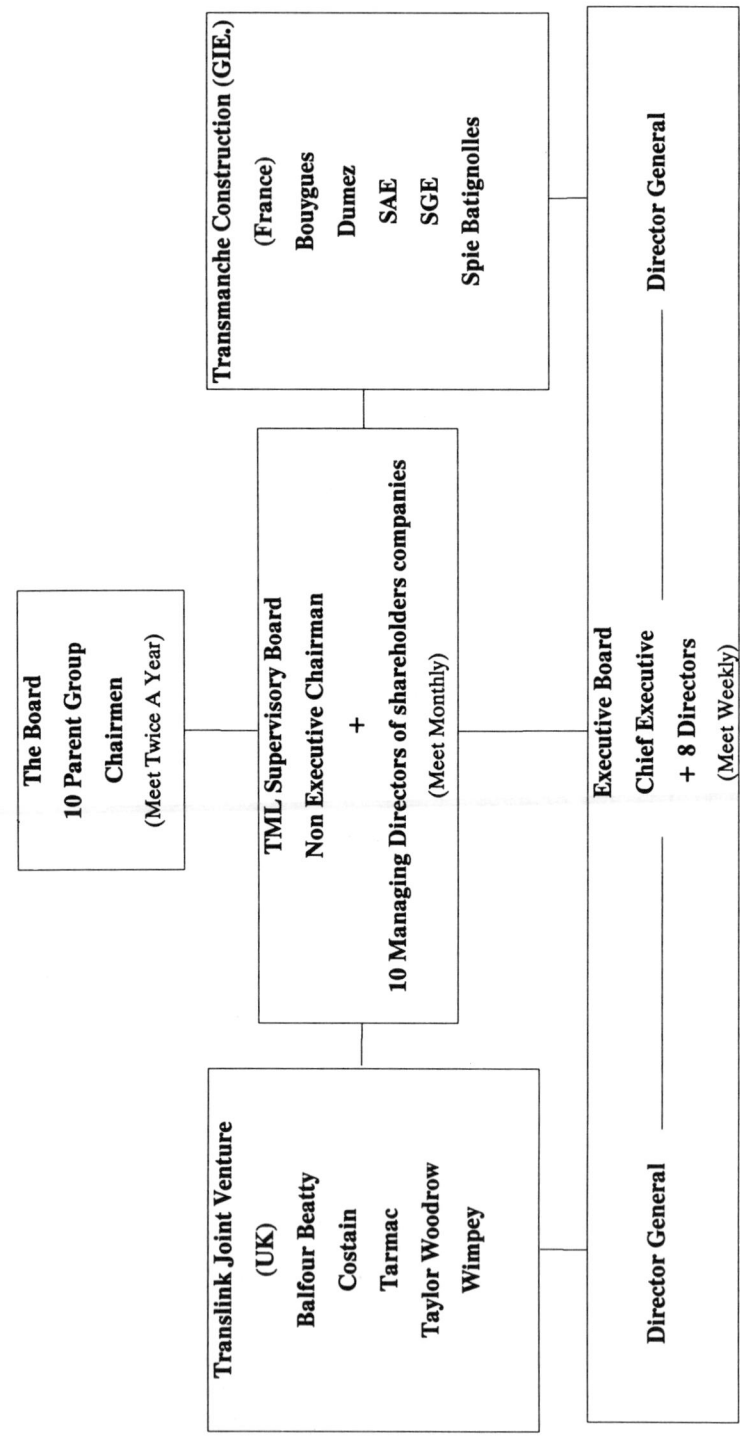

Fig. 5. Transmanche Link Boards 1986–88

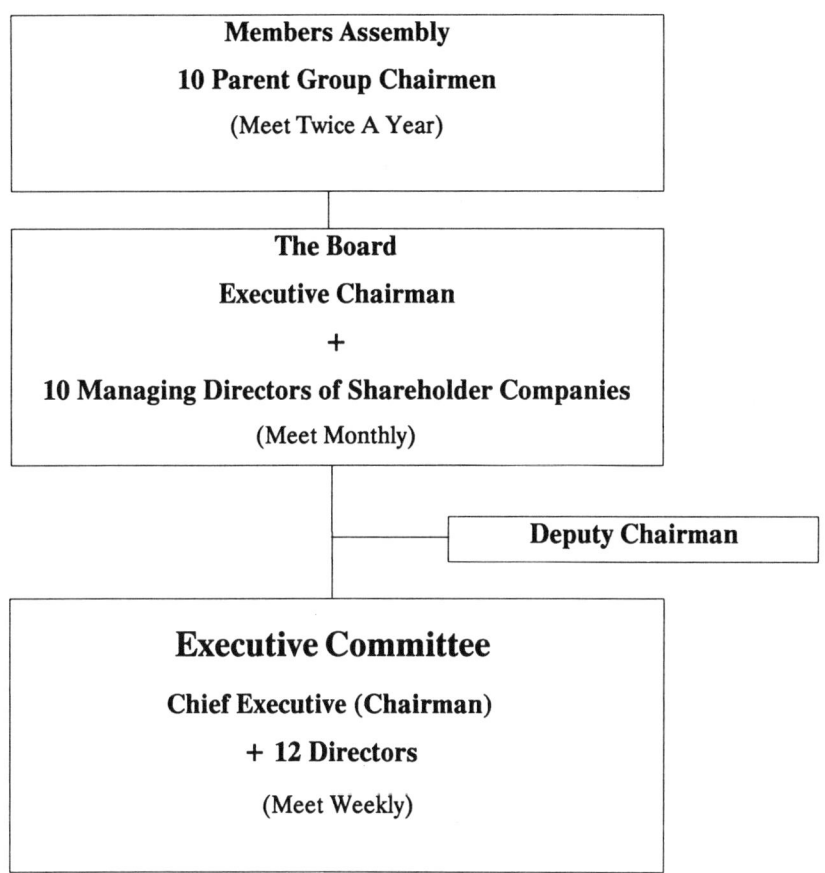

Fig. 6. Transmanche Link Board post 1988

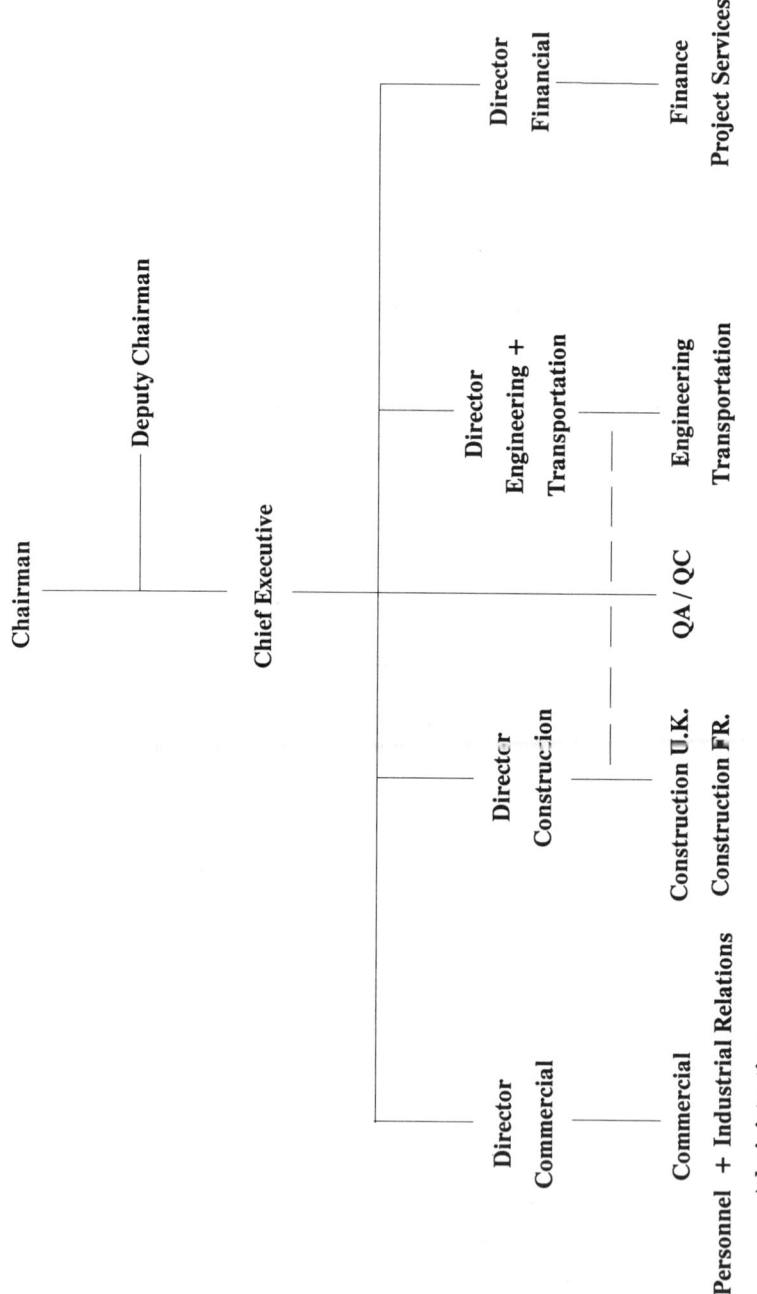

Fig. 7. Executive structure

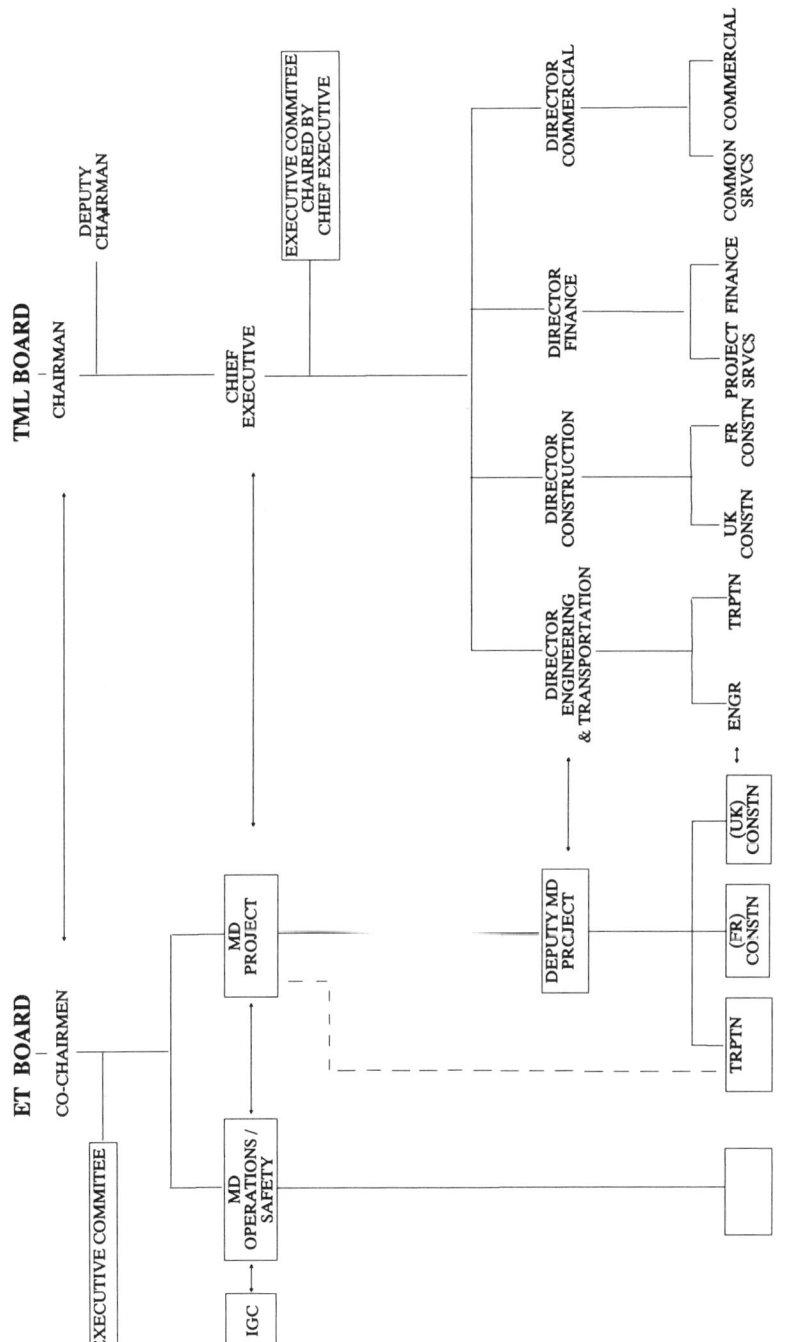

Fig. 8. Interrelations

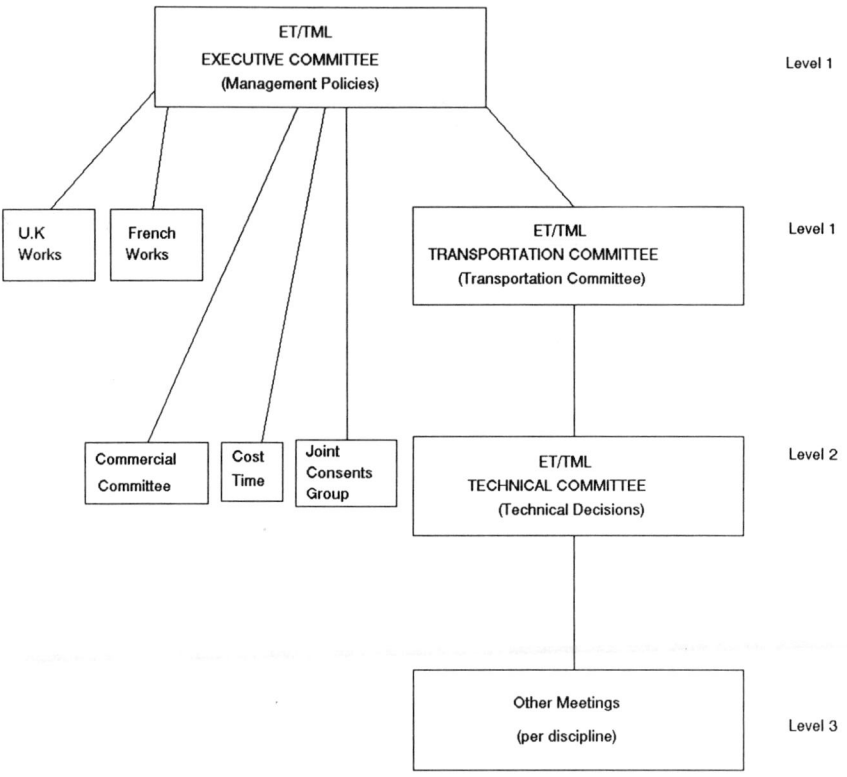

Fig. 9. Eurotunnel and Transmanche Link joint committees and meetings

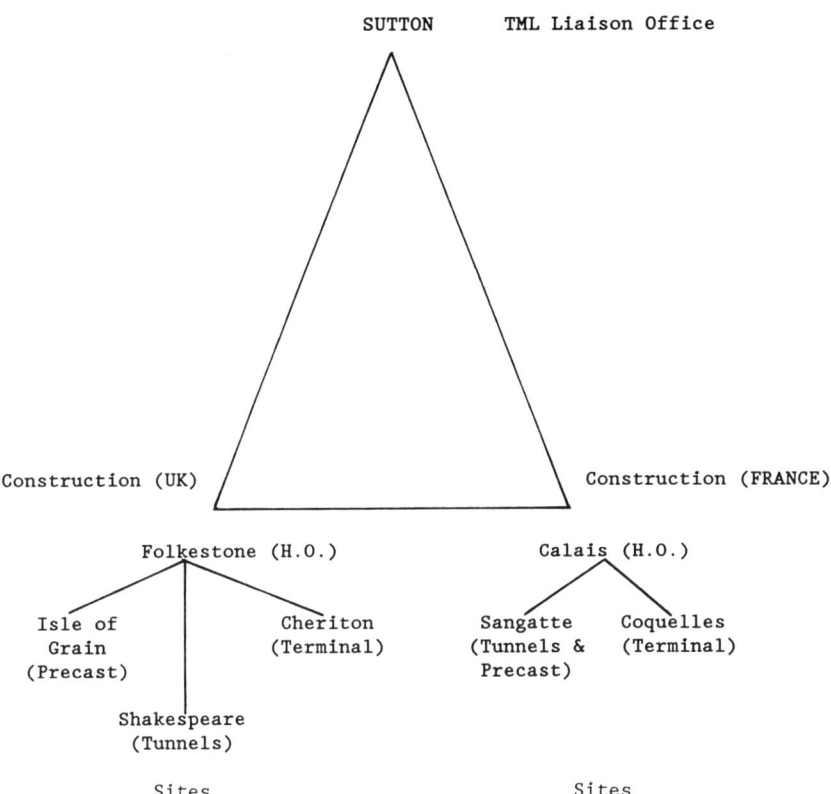

Fig. 10. Project management location

same as the Supervisory Board would have an Executive Chairman (full time).

49. There would be a Deputy Chairman (non executive), and a Chief Executive who would chair the Executive Committee. Fig. 6.

50. The Executive Structure of TML was revised as shown in Fig. 7. It will be seen that this organisation is more similar to that of a company and introduces new top executive positions.

51. The interrelations of the Eurotunnel and TML organisations from early 1989 is shown on Fig. 8.

52. Various joint committees have been established and these are shown on Fig. 9. The decision making joint committees are:-

(a) Executive Committee
(b) Transportation Committee
(c) Technical Committee

TML PROJECT MANAGEMENT
53. Location

(a) All project management staff are employed and located either in project site offices in France or the UK or in Head Offices at Folkestone or Calais. Fig. 10.
(b) All major co-ordination is carried out from the TML office at Sutton.

Other papers are dealing in detail with operational management of the project, particularly construction, engineering and transportation. However it is necessary to briefly outline the responsibilities of the various central departments of TML.

54. Transportation System Group (TSG) has overall responsibility for the budget, programme and technical performance of the packages of works for all the fixed equipment (mechanical and electrical) for the Railway System and rolling stock. This includes activities from the completion of definitive design (when applicable), detailed design through manufacturing and installation to completion and commissioning. The TSG departmental strength is 94 approximately.

55. Engineering Department. The Engineering department is responsible for the whole of the engineering for the project including the Transportation System. The function commences with development studies to optimise design concept through to construction drawings. Because of the design construct nature of the project close coordination is maintained between the engineering and construction departments. Strength at Sutton is 188. Much of the design is carried out through External Consulting Organisations. In the UK alone this has peaked at 550 staff.

56. Quality Assurance Department. At the outset it was

decided that the project would be completed within the framework of an efficient quality system supported by a Quality Assurance Programme.

57. The Director of the Central Quality Assurance Department is responsible for the overall implementation of TML Quality Assurance programme and common Quality Assurance procedures throughout the TML organisation and for ensuring that the quality of the "Works" will fully comply with all applicable codes, regulations and contractual requirements.

58. Until the recent organisational changes the Quality Assurance Director was responsible to the UK Director General. It is now considered necessary to increase his authority and organisational freedom so he will in future report directly to the Chief Executive. The department organisation is shown Fig. 11.

59. Commercial Department. The central TML Commercial Department provides general commercial policies, and common procedures for management of the construction contract, procurement and sub-contracts and legal services throughout the organisation. It ensures

- that specific procedures are complying with the general policies and common procedures
- the co-ordination of the relevant activities and the consolidation of information and results at TML level
- the contractual interfaces with Eurotunnel, including payment.

60. Financial Department. The TML Financial Directorate oversees all financial and accounting functions throughout the organisation in terms of policy and co-ordination. This includes all aspects of Accounting

(a) Financial Accounting (balance sheet, profit and loss account, fixed assets and depreciation, bought ledger, sub-contract ledgers, debtors, payroll, banking and cash)

(b) Management Accounting - analysis by cost centres and by cost elements

(c) Finance - treasury, distribution, provisions, bonds etc

(d) Cost Control - in terms of cost coding, accruals etc

(e) Administration - corporate affairs (financial), insurances, stores and equipment, office management etc.

61. Management Services Department. A central Management Services Department leads and co-ordinates all UK aspects of Human Resources (recruitment, personnel, training, industrial relations) external relations and security.

62. Project Services In order to serve the needs for information of the TML Executive, the Shareholders and also the Client, TML have emphasised the role of the Project Services Group covering specifically planning, cost management and computer services. Fig. 12.

63. The planning of the project is managed on the basis

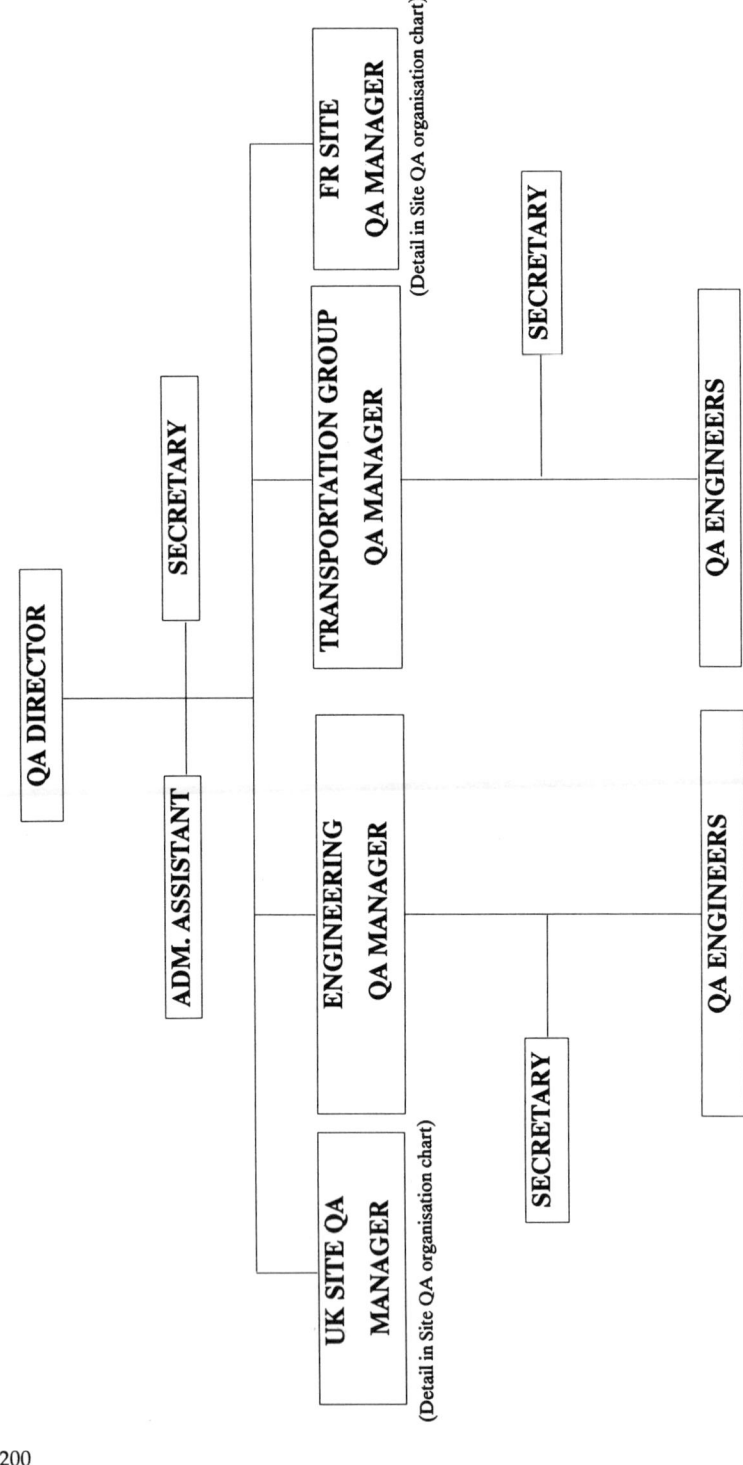

Fig. 11. TML QA organisation

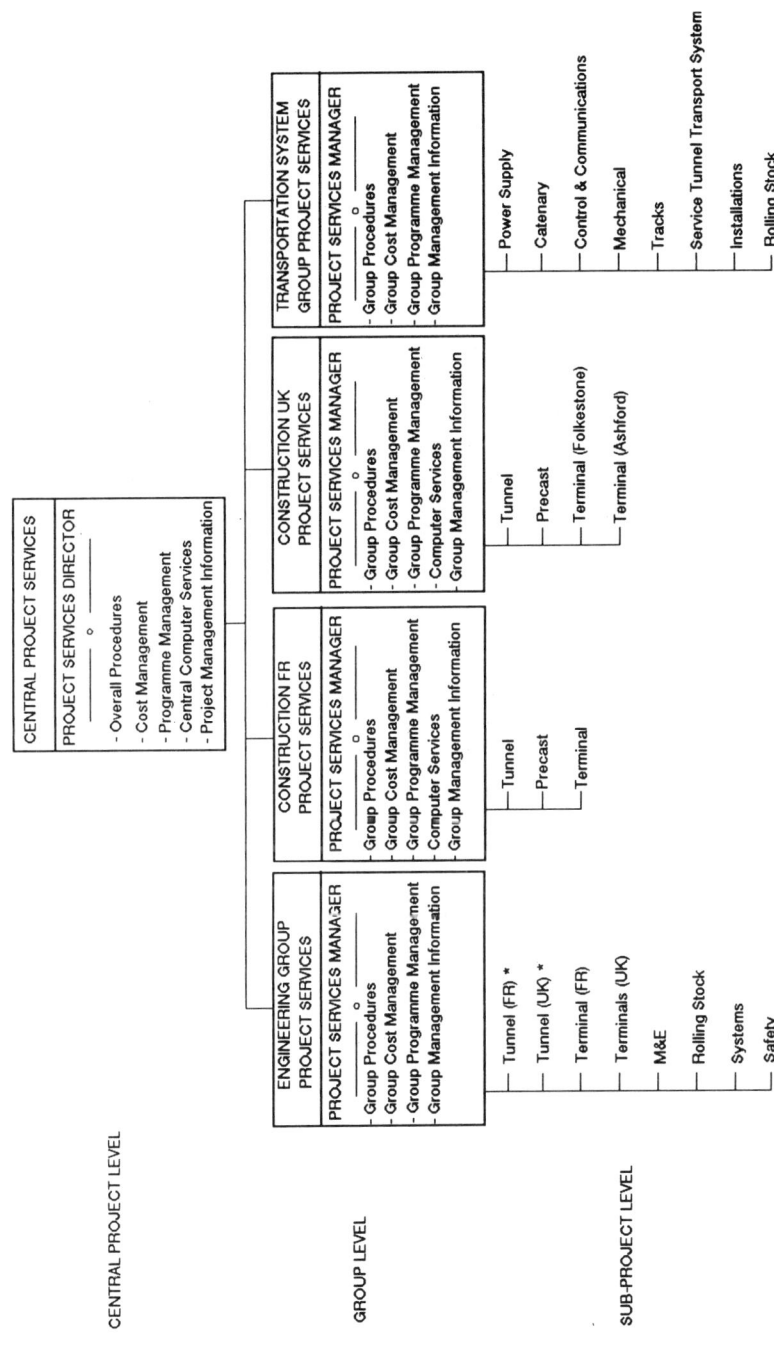

Fig. 12. TML project services organisation

of 4 levels, giving visibility to the schedule at executive, co-ordination, sub-project working and sub-project detail levels.

64. The objectives of the project schedules are set by milestones at the co-ordination level (level 2). Also at this level the major activity interfaces between separate groups are established and controlled so as to give each group and sub-project managers clear definitions of their time boundaries and the appropriate degree of freedom within which to manage their own work.

65. Extensive use is made of the critical path method with central reporting and the management of interfaces and float.

66. The Cost Management approach has been to allocate the total project budget to each individual sub-project thereby establishing the control budget for all work. Constant monitoring of design, scope and performance changes takes place in order to ensure control. Major re-estimates of the total costs are made on a six monthly cycle in order to establish formal positions of profitability. Contingency is managed at levels above that of the sub-project manager.

67. The project has established 3 significant computer centres located in Paris, Folkestone and Sutton making use of IBM and Prime machines all linked to the in-house data communications network serving voice, fax and data transmissions. Part of this network is shared with the Client.

68. Planning and Cost Management make extensive use of TML's data communications network linking all sub-projects to the data centres so as to access the local and common management systems.

69. Extensive use is made of micro computers most of which are attached to the communications network.

RESUME

<u>Synthese</u> Le Projet du Tunnel sous la Manche est probablement
l'un des projets qui présente les défis les plus importants
et les plus ambitieux de notre siècle, défi de par sa taille
et de par l'originalité de son Système de Transport.

Le montage de cette opération en concession privée et son
financement, impliquant plus de 200 Banques dans le monde,
représentent une expérience exceptionnelle et unique.
C'est le premier projet d'infrastructure liant la France et
le Royaume Uni et ceci constitue en même temps, un défi et
une opportunité, celle de se connaître et d'améliorer la
compréhension réciproque entre ces deux nations.

<u>Historique</u> Les Gouvernements français et britannique se
décidèrent en faveur de la proposition de Channel Tunnel
Group/France Manche S.A (Eurotunnel) et en janvier 1986, en
accordèrent la concession pour sa construction et son
exploitation.
En août 1986, le Channel Tunnel Group/France Manche S.A
(Eurotunnel) attribuent a Transmanche-Link Joint Venture, le
contrat de la réalisation du tunnel, dont la valeur est de 26
milliards de francs (valeur 1985).
Le délai de construction de sept ans commençait en mai 1986.

<u>Contrat</u> Le programme du contrat est extrêmement serré, avec
une date de début de travaux en mai 1986 et une date
d'achèvement des travaux prévue en mai 1993. La croissance
de Projet est extrêmement rapide avec un chiffre d'affaires
allant de zéro en 1986 à plus de 800m£ en 1989.
Les travaux sont découpés en trois sections distinctes:

  a) Travaux en dépenses contrôlées - couvrant tous les
      travaux de Tunnel y compris la fabrication des
      voussoirs. Ces travaux sont payés sur les bases d'un
      remboursement de dépenses exposées et d'un partage des
      économies et des surcoûts réalisés par rapport au prix
      objectif (valeur approximative de 1286m£ en 1985)

  b) Travaux au forfait - le totalité des travaux des
      Terminaux, les équipements fixes électromécaniques des
      Tunnels et des Terminaux sont payés sur la base d'un
      forfait (valeur approximative 1070m£ en 1985).

  c) Fournitures - TML est payé de la valeur du matériel
      roulant (locomotives et navettes) par Eurotunnel, sur
      la base d'un remboursement du coût plus un fee (valeur
      approximative 245m£ en 1985).

Le programme du contrat prévoit une série de dates jalons.
De fortes pénalités sont imposées si ces dates ne sont pas
respectées ou si les travaux ne sont pas achevés dans le
delais prévus.

Organisation et management de TML La coordination franco-britannique est effectuée à partir de Sutton où Eurotunnel a également basé son equipe de projet.
Les deux organisations de construction et de gestion des activités nationales sont basées à Calais (Transmanche) et à Folkestone (Translink), l'ensemble du personnel de production étant réparti sur les différents sites (voir Fig 10

Pendant le première phase du Projet, les structures d'organisation étaient articulées autour de deux Directeurs Généraux (France et Grande Bretagne), responsables des activités de Transmanche et de Translink; ils partageaient la direction et la responsabilité des fonctions centrales (Ingénierie, Etudes, Commercial, Finances).

Les Directeurs Généraux rendaient compte à un "Chief Executive" qui était responsable de mener à bonne fin le Projet et qui, à son tour, rendait compte au Conseil d'Administration de TML comprenant un représentant de chacune des dix Sociétés Mères de TML.

A la fin de l'année 1988, il a été décidé, pour tenir compte de l'évolution du Projet. de restructurer la direction générale de TML par une organisation plus classique et d'attacher plus d'importance au Système de Transport. Le management central de TML s'est transformé en une structure à quatre Directeurs, pleinement responsables de la Construction, de l'Ingénierie et du Système de Transport, des aspects commerciaux et financiers, chacun d'entre eux rendant compte au Chief Executive qui, lui, continue de rendre compte au Conseil d'Administration.

12A.   UK construction management

J. H. WINTON, MICE, Construction Director UK

SYNOPSIS

Under the 55 year period of the UK and French Governments concession for the channel tunnel link, there is *'The Construction Phase'* and *'The Operation Phase'*.

Transmanche Link contract with Eurotunnel (the concessionaire) is for the *Construction Phase* and its organisation includes the UK and French engineering design and construction groups and the UK and French transporation groups. The former provides for the civil engineering construction work while the latter provides rolling stock, equipment and commissions the fixed rail link up to the running of the first revenue train in 1993.

The *Operation Phase* of the concession is carried out by Eurotunnel until the year 2042.

U.K. ORGANISATION

Introduction

1.   The construction contract consists of target cost works, lump sum works and procurement items for rolling stock and locomotives.
The UK construction works are in the proportion of:

Target cost works	50.0%
Lump sum works	40.0%
Procurement items	10.0%

The total UK labour force is anticipated to be between 5000 and 6000 for management and operatives and there are some 50,000 man years for supply and sub-contract work from third parties.

UK Construction Management

2.   The UK construction organisation is simply two basic units 'construction' and 'construction support'.
Construction is divided into work areas dictated by type and location of the work. The U.K. construction work areas are fully described in the next section.

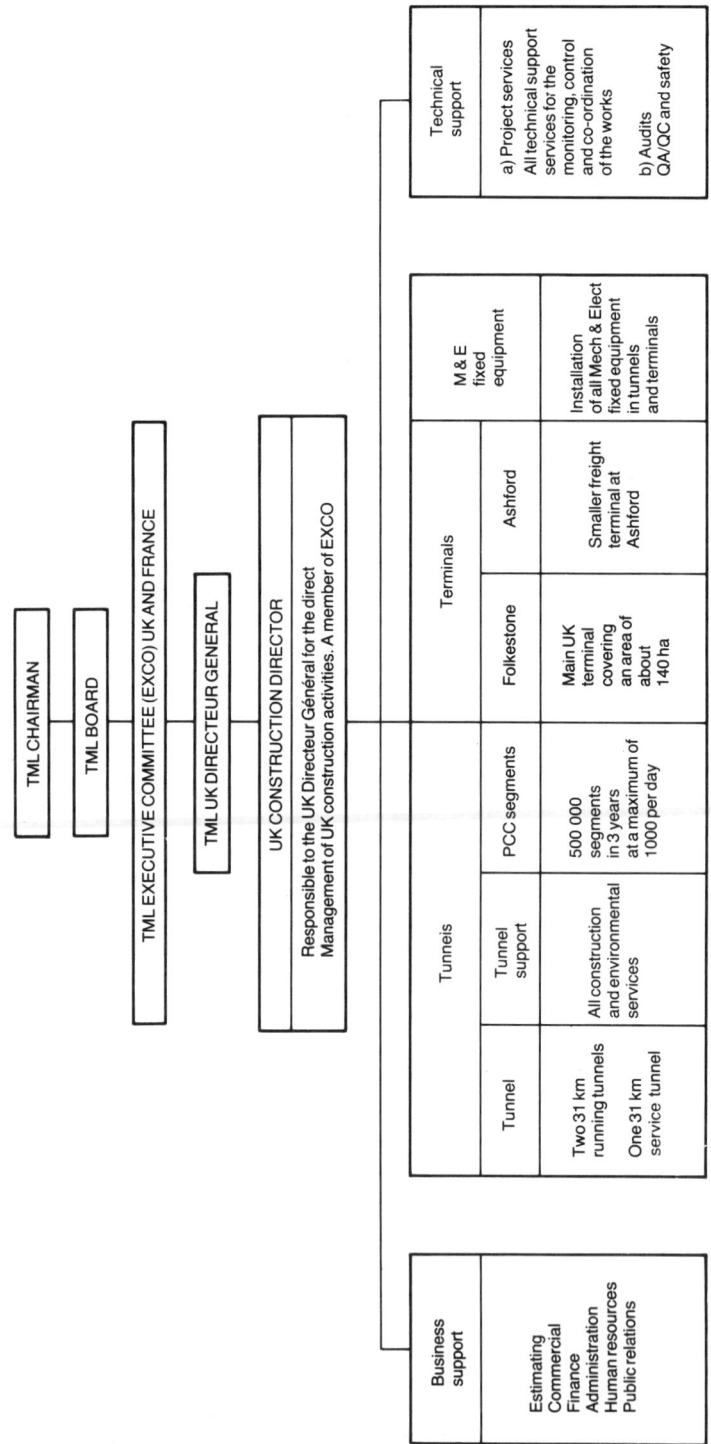

Fig. 1. UK Construction Organisation

Construction support is divided into business support and
technical support which are described under paragraph 20 and
paragraph 27 respectively. Business support includes those
services which are common to any business venture. Technical
support are those specialised services required by management
to co-ordinate the technical services for the various
disciplines of the work and provide an overall management
information service and control tool to achieve management
objectives in accordance with the contract.

The UK management organogram along these lines is set out in
chart form in Fig. 1.

UK CONSTRUCTION WORK AREAS

3.   The UK construction group activities are divided into
six sub-projects to provide manageable cost centres for the
target cost works and the lump sum works at the various
locations. The sub-projects are as follows:
   (1)   Tunnels
   (2)   Tunnel support services
   (3)   Precast concrete segment manufacture at Isle of Grain
   (4)   Folkestone terminal
   (5)   Freight terminal - Ashford
   (6)   M & E fixed equipment
Each sub-project is divided into sections under the control
of a section manager. It has its own plant organisation
and satellite administration and support services reporting
to the sub-project construction manager or managers and with
a functional link to the appropriate department heads at the
Folkestone head office, where overall management control is
exercised by the UK construction group directorate.

TUNNELS
4.   The sub-divided work areas for the tunnel sub-project
located at shakespeare cliff are as follows:
   (a)   Marshalling area
   (b)   Main tunnel drives
   (c)   Lower shakespeare cliff (spoil disposal)

Marshalling Area
5.   The marshalling area is the underground works at
shakespeare cliff where TBMs are erected and materials are
handled to and from the surface and personnel access to the
underground works is provided.

Main Tunnel Drives
6.   Main tunnel drives construction requirement is to
drive six tunnels from the marshalling area at shakespeare
cliff. Three seaward and three landward. The seaward
tunnels require a service tunnel and two running tunnels to
be driven some 22 kilometres to junction with the french
tunnels. The landward tunnels require a service tunnel and

two running tunnels to be driven some 9.0 kilometres to the Castle hill portal, leading into the terminal. The TBM drives end at holywell some 1000m short of the terminal portal which includes a 500m cut and cover section and 500m of construction through castle hill.

Lower Shakespeare Cliff (Spoil Disposal)

7. Construction of the UK tunnels and associated underground structures produce approximately 4.75 million m^3 of excavated material. It is intended on completion of the landward service tunnel to transport approx. 1.0 million m^3 to the terminal site for fill, leaving approx. 3.75 mllion m^3 for disposal in the reclamation area, behind the construction of a new seawall. This area is the key to the expansion of the lower shakespeare site to accommodate the services and facilities required to maintain the TBM drives and on completion certain permanent works services.

The tunnel sub-project works are target cost and is the largest of the UK sub-projects and carries out some 36% of the UK civil works. Its organisation consists of two principle areas, one technical and one for production and a total of some 4000 supervisory staff and operatives for 24 hours working 7 days a week. The sub-project has a major plant organisation and its own satellite administration, planning, costing, accounts, commercial and procurement departments with functional links to the Folkestone head office of UK construction

TUNNEL SUPPORT SERVICES

8. Tunnel support services cover the following construction requirements of the tunnel sub-project.
    (a) Electrical power supply and distribution
    (b) Tunnel ventilation
    (c) Water supply and distribution
    (d) Tunnel drainage
    (e) Tunnel services support brackets
    (f) Communications, monitoring and train signalling systems including radio and radio paging, a telephone system, and a public address system.

The tunnel support services are target cost and the sub-project accounts for 5% of the UK civil works. The sub-project has its own construction manager and 150-200 technical and supervisory staff. Its organisation utilises and is allied to the facilities of the tunnel sub-project departments. The sub-project operates similar to the tunnel sub-project on a 24 hour basis, 7 days a week.

PRECAST CONCRETE SEGMENT MANUFACTURE

9. The precast segment production site is on the Isle of Grain. The site has an area of 28 hectares with adjacent deep water jetty facilities for aggregate and existing rail and road connections. The production requirement is 1000 segments per day maximum and the manufacturing process is a continuous 24 hour operation nominally five days a week.

Segments are delivered to the tunnel site by British Rail.
The works are target cost and amount to approximately 10% of
the UK works. The organisation consists of a construction
manager and some 700-800 supervisory staff and operatives.
The sub-project has its own administration, planning,
costing, accounts, commercial and procurement departments
with functional links to the Folkestone head office of UK
construction.

FOLKESTONE TERMINAL
10. The sub-project work areas are defined as follows:
(a) Southern access road and structures
(b) Earthworks and drainage
(c) Lower structures
(d) Platform ramps and overbridges
(e) Terminal buildings

Southern Access Road and Structures
11. Construction includes the construction of a road system
to convey traffic onto and from the M20 and A20 and a rail
system. The work involves a new main line rail bridge, road
bridges and associated works across the M20 and A20 and the
new sidings at dollands moor.

Earthworks and Drainage
12. Construction includes some 3.7 million m^3 of earthworks
involved in reducing and raising the ground level of the
terminal to provide the correct formation for structures,
buildings and trackwork. It includes a pre-earthworks
drainage system diversion and a post earthworks drainage
system.

Lower Structures
13. Construction includes retaining walls, underpasses and
the arrival loop tunnel constructed in advance or in
conjunction with the earthworks.

Platforms, Ramps and Overbridges
14. Construction includes all civil works in the platform
area.

Terminal Buildings
15. Construction includes an administration building; an
amenities building; a maintenance building; UK and french
border control buildings; a security building; a sub-station
a tunnel services building and associated toll buildings.

16. The terminal sub-project works are lump sum and amount
to some 15% of the UK works. The organisation consists of a
construction manager and some 700-800 supervisory staff and
operatives. The sub-project has its own plant department and
satellite administration, planning, costing, accounts,

commercial and procurement departments with functional links to Folkestone head office of UK construction.

FREIGHT TERMINAL - ASHFORD

17. There is insufficient space at the Folkestone terminal for the checking by customs of road freight that has not been given prior clearance. The solution is the construction of an Inland Clearance Depot (ICD) SE of Ashford near the village of Sevington. The construction of the ICD has started and is due for completion in 1992.
The works are lump sum and amount to approximately 1.3% of the UK works.

M & E FIXED EQUIPMENT

18. The sub-project work areas covering tunnels and the terminal are as follows:

(a) Support steelwork, brackets and cable tray/ladder racks in tunnels
(b) Permanent tunnel pipework to fire and water mains
(c) Cross passage installation of M & E equipment
(d) Permanent high voltage cables
(e) Permanent low voltage cables
(f) Running tunnel rail introduction
(g) Running tunnel trackwork construction
(h) Service tunnel screed
(i) Trackside works for signalling system
(j) Overhead catenary system
(k) Installation of cooling for tunnel environment
(l) Ventilation fans to provide forced air draught in the tunnels.

The sub-project work is for an entire railway system and is managed as a whole for the UK and France. The work starts in 1990 with a 27 month overlap with civil work. The works are lump sum and the total UK and French proportion of the cost represents approximately 19% of the whole works. The total estimated work force for UK and France is 2500 supervisory staff and operatives.

CONSTRUCTION SUPPORT

19. Construction support is sub-divided into:
(a) Business support
(b) Technical support

BUSINESS SUPPORT

20. Business support includes the following departments:
(a) Estimating department
(b) Commercial department
(c) Finance department
(d) Administration department
(e) Human resources department
(f) Public relations

Estimating Department

21. The estimating department provides the following services:
- Analysis of the original offer and data to meet the requirements of the TML cost control system.
- Estimates for each section of the work at the various stages of its design development (APS and APD) with optimisation estimates of alternatives when necessary, all in accordance with the agreed contract procedures.
- Estimating advice to external consultants and 'in house' design teams for cost planning.
- Assists commercial department in the pricing of the preparation of the six monthly forecast of the end cost of the project, progressively taking into account actual cost and estimates based on the latest design.

Commercial Department

22. The commercial department effects a commercial service relating to the quantity surveying, contractural and insurance requirements, and procurement service activities of UK construction.

The contract is forward funded and the basis of the valuation calculation for the target cost works and the lump sum works are shown diagramatically below:

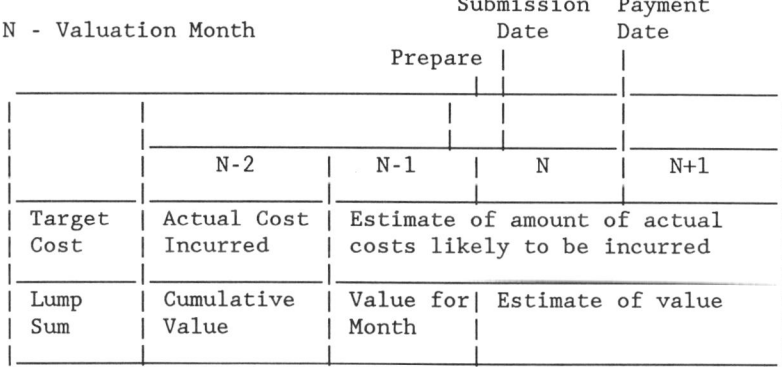

Valuation = (N-2)+(N-1)+(N)+(N+1) and is progressive for each subsequent valuation.

Comparison of actual against estimated is an ongoing reporting requirement under the contract.

Finance Department

23. The finance department activities includes the following:
- Control of financial and management accounts.
- Functional control of sub-project accounts department.
- Routine audits as per agreed programme.
- Ad hoc audits as required.

- Regular reporting of cost and budget values through the monthly financial report.
- Preparation of the target works application to Eurotunnel.
- Control of overheads and the financial accounting function including all ledgers and balance sheet items.
- Preparation of statutory accounts.
- Cashier and treasury departments.
- The operation of all 'Prime' computer hardware and operating systems including user support and micro support services for accounting.

Administration Department

24. The administration department activities include the following:
The Folkestone Head Office administration; all voice communication and word processing systems; staff and labour time and attendance and the hourly paid operatives payroll; stores and stock control; the construction village administration and the function control of sub-projects administration departments.

Human Resources

25. The project requires to be resourced by some 5000-6000 personnel to achieve the contract obligations. The efficient recruitment, administration and implementation of sound personnel policies and practices together with the training, motivation and retention of this resource is crucial to the final success of the project. The major requirement of the human resources function is to work closely with management at all levels to ensure the provision of a service which will respond to immediate needs and meet future planned requirements in what is currently a very active and aggressive labour market. The operation of this service has been organised on the basis of the personnel, industrial relations, recruitment, security and medical specialist functions to ensure an effective and responsive resource for management.

Public Relations

26. A public relation office is established in Folkestone head office and trained staff channel all complaints to construction and environmental personnel and advises complainent of action taken and ensures it is to his or hers satisfaction. The TML P.R. office also supports a number of requests for donations and sponsorships. It mounts displays at exhibitions; gives talks and slide presentations to clubs and local organisations; arranges site visits in response to requests from national and international engineering groups, the local councils and in particular from the people of East Kent. The P.R. office assists ET in organising and manning special open days at the channel tunnel exhibition centre at Folkestone.

Because of the national and international interest shown in
the project regular media days are held by the P.R. office
each month where press and media are kept informed of the
progress of the work and queries are answered.

TECHNICAL SUPPORT

27. Technical support is sub-divided into:
    (a) Project services
    (b) Audits

PROJECT SERVICES

28. Project services includes the following:
    (a) Project Services Department
    (b) Project Management
    (c) Programming
    (d) Cost Management
    (e) Planning Consents and the Environmental Specialist
    (f) Meetings
    (g) Procedures

Project Services Department

29. The project services department is responsible to the
UK construction director for the provision of technical
services to the UK sub-projects in accordance with TML policy
and procedures and includes:
- Programme planning and control.
- Cost management requiring setting up of budget and cost
  control procedures for UK construction and monitoring their
  application.
- The collation of all cost reports for onward transmission
  to the TML liaison office, Eurotunnel and the technical
  advisors to the banks.
- The implementation of computer systems to meet requirements
  of user departments.
- The control of all technical documents and drawings for
  temporary works.
- The monitoring of land survey observations in the UK.
- Co-ordination of construction activities with British
  Rail, in particular the delivery logistics of PC segments
  from the Isle of Grain to shakespeare cliff.
- Monitoring services provided by the planning consents
  team for the temporary works and environmental team for the
  measurement of data.

The project services department of UK construction has a
functional link with TML liaison office, Sutton and french
counterparts for the various disciplines.

Project Management

30. A project group has been set up to strengthen the
extended lines of communication between the TML offices at
Sutton, Folkestone and Calais. The group co-ordinates the
interfaces between the sub-projects and the engineering

group; the sub-projects and the support services departments; the construction sub-projects and the individual support services departments. The project group in no way takes away the authority and responsibility already vested in the sub-project managers; the support services directors and the support services managers.

Programming

31. The objective of the UK construction programme is to co-ordinate the planning and scheduling of the UK works to ensure the contract milestone dates and/or such other objectives set by EXCO are met and to collate and analyse progress data so that adverse trends are identified ahead of time and any corrective action is taken. The heirarchy of programmes is as follows:

Executive Summary
Level 1 Project   By TML Liaison Office Sutton
Level 2 - Sub-project }
Level 3 - Section   } By UK Construction Head Office
Level 4 - Sub-section } and Sub-Projects

The Level 2 Programme covers all the UK construction works and includes:- logic networks; bar charts; method statements; manpower histograms and activity listings identifying the interfaces with engineering (design), sub-project and transportation group activities.

Cost Management

32. The essential requirements of the cost management system is to provide and report at any given time the status of the following:

(a) Budget
(b) Variation orders
(c) Incurred cost
(d) Commitment on orders
(e) Earned values and performance
(f) Trends
(g) The last 6 monthly and latest forecast to end of contract
(h) Deviations from latest approved budget

The cost management reporting system has been developed incorporating the above factors to meet not only TML's internal requirements but also those of ET and the bank's technical advisors.

Given the above information, this allows management to exercise control over expenditure, commitments, physical progress and forecast cost to completion.

Planning Consents and the Environmental Specialist

33. The planning and environmental issues for the works are governed by the normal legislative framework plus the special provisions of the Channel Tunnel Act 1987.

The authorised development of planning permission and
technical approvals for both permanent and temporary works
are governed by the Channel Tunnel Act within the limits of
deviation. Outside the limits of deviation the normal
legislation applies.

At present responsibility for making the relevant planning
submissions and applications for technical approval is shared
by Eurotunnel and TML. The authorities responsible for
issuing the relevant consents and approvals include Kent
County Council, Dover District Council, Shepway District
Council, Ashford Borough Council, Southern Water Authority
and other relevant statutory undertakers. The total number
of consents and approvals is in excess of three hundred.

A special unit, the joint consents team, operating on behalf
of Eurotunnel and TML has the responsibility for preparing
all the submissions. The joint consents team issues regular
progress reports for the benefit of Eurotunnel and TML and
the authorities involved.

In addition, TML employ a team of environmental specialists
that maintains a continuous relationship with statutory
bodies such as the Nature Conservancy Council and local
authority environmental health officers to monitor progress
of the construction work and ensure the implementation of
appropriate measures to minimise the eventual environmental
impact of the tunnel system.

The requirement for consents and approvals under the Channel
Tunnel Act and other legislation forms the major element in
progressing the engineering design approvals for the overall
project programme.

Meetings

34. A *UK construction meeting* chaired by the
Construction Director and attended by departmental directors,
central managers and construction managers for precast yard,
terminals, tunnels, tunnel support services, M & E fixed
equipment, project services, finance and administration,
estimating, commercial, human resources, safety, QA, project
co-ordination and engineering is held on a regular monthly
basis at the Folkestone head office of UK construction.

A *Cost control meeting* chaired by the UK Directeur
General is also held monthly to review all cost management
reports, each sub-project activity is reviewed by cost centre
and corrective action is taken where necessary. The cost
control meetings are attended by the Construction Director,
construction and departmental managers with supporting
staff.

Each department and sub-project also holds monthly meetings

and minutes are circulated to the Construction Director and others.

Procedures

35. A system of procedures are established for all key project activities.

Procedures are classified into the following three categories:

    Category 1 - Common to all TML groups and liaison offices
    Category 2 - Specific to a group
    Category 3 - Specific to a sub-project or group

AUDITS

36. Audits include the following:
    (a) Quality Assurance
    (b) Quality Control
    (c) Safety

Quality Assurance

37. The TML executive committee has decreed that the construction of the channel tunnel project shall be carried out within the framework of an efficient quality system supported by a quality assurance programme. The principles of the TML quality system for the UK works are that construction management is responsible for constructing quality into the works by means of quality planning and initial quality control. The RE organisation is responsible for independent quality control to achieve the satisfactory execution of the works in accordance with the drawings, specifications and sub-project procedures.
The quality assurance department carries out a review audit and surveillance to ensure the effectiveness of the whole quality assurance system.

Quality Control

38. The engineering manager tunnels is responsible for all engineering matters relating to the construction of the permanent tunnel works and has on site an engineering design liaison team called the RE organisation reporting to him to assist and ensure early response to engineering problems as they arise. The engineering manager terminals has a similar engineering design liaison team for the terminal. In addition the RE organisation with inspectorate is directly responsible for independent quality control activities throughout UK construction.

Safety

39. TML is required to ensure his employees, agents and sub-contractors comply with all their duties and obligations under English law covered by the Factories Act 1961, the Occupiers Act 1957, the Control of Pollution Act 1974 and the Health and Safety at Work etc. Act 1974. The TML Safety

Manager reports to the Directeur General and the Construction Director and his duties are:- to establish and manage the safety function of the UK construction group; to supervise the UK construction safety organisation; liaise with external authorities and government bodies, and maintain an effective liaison with the French counterpart to ensure comparable standards.

CONSTRUCTION MANAGEMENT

RESUME
Introduction
   1.   Le Contrat de Construction comprend les Travaux en
Dépenses Controlées, les Travaux à Forfait et les
Fournitures pour le matériel roulant et les locomotives.
Les Travaux de Construction du côté britannique se
répartissent de la façon suivante :

   (a)   Travaux en dépenses controlées   50,0%
   (b)   Travaux à Forfait   40,0%
   (c)   Fournitures   10,0%

Le total des effectifs prévu pour l'encadrement et la main
d'oeuvre est de l'ordre de 5000 à 6000 ; les commandes et
les contrats de sous-traitance provenant de tiers
représentent l'équivalent de 50 000 années travaillées.

Direction Construction UK
   2.   La Construction UK est organisée à partir de deux
unités de base "Construction" et "Construction Support". La
Construction est divisée en zone d'activité selon le type et
l'implantation des travaux.
   3.   "Construction Support" est divisé en unités de
soutien "Affaires" (Business support) et "Technique"
(Technical support), cf paragraphes 20 et 27. "Business
support" comprend les services communs à toute entreprise.
"Technical support" comprend les services spécialisés
nécessaires à la gestion qui coordonnent les services
techniques des diverses disciplines des travaux et
fournissent un service d'informations général pour la
gestion ainsi qu'un moyen de contrôle afin d'atteindre les
objectifs de gestions prévus au Contrat.
L'organigramme correspondant est représenté en Fig.1.

Zones d'Activite de la Construction UK
   4.   Les activités du Groupe Construction UK sont divisées
en six sous-projets afin de faciliter la gestion des centres
d'imputation des travaux en dépenses controlées et des
travaux à forfait aux différentes implantations. Les sous
projets sont :

   (a)   Tunnels
   (b)   Services auxilliaires tunnel
   (c)   Usine de préfabrication des voussoirs en béton à
         l'Ile de Grain
   (d)   Terminal Folkestone
   (e)   Terminal Commercial - Ashford
   (f)   Equipements Fixes M&E

   5.   Chaque sous-projet est divisé en sections sous le
contrôle d'un chef de section. Le sous-projet possède sa

propre organisation des travaux et des services auxilliaires ainsi qu'une administration "satellite" qui sont placés sous la responsabilité du (des) Directeur(s) Sous-Projet Construction et en relations fonctionnelles avec les chefs de département concernés au siège de Folkestone, où le contrôle de gestion général est exercé par la Direction du Groupe Construction UK.

Construction Support
6.   Construction Support est sous-divisé en :

(a) Business Support
(b) Technical Support

7.   Business Support
Cette unité de soutien comprend :

(a) Département Etude de prix
(b) Département Commercial
(c) Département Finance
(d) Département Administration
(e) Département Ressources Humaines
(f) Relations Publiques

8.   Technical Support
Cette unité de soutien est sous-divisée en :

(a) Service Projet
(b) Audits

8.   Service Projet
Le service projet comprend :

(a) Département Service Projet
(b) Gestion Projet
(c) Planning
(d) Gestion des coûts
(e) Approbation des permis et spécialiste de l'environnement
(f) Réunions
(g) Procédures

9.   Audits
Les audits comprennent :

(a) Assurance Qualité
(b) Contrôle Qualité
(c) Sécurité

12B. Management de la construction France

P. MATHERON, Directeur de la Construction, France

OBJET. Pour faire face à la complexité et à l'ampleur des ouvrages à réaliser dans un délai très court (percement des tunnels, préfabrication des voussoirs, construction du terminal et installation des équipements définitifs), un concept d'organisation décentralisée a été mis en oeuvre autour de quatre directions opérationnelles, dénommées "sous-projets", responsables chacune d'un char. et cinq directions fonctionnelles, rattachées au Directeur de la Construction France, qui assistent les sous-projets et assurent la consolidation des activités et le contrôle de gestion. Les directeurs opérationnels sont responsables, devant le Directeur de la Construction, des objectifs généraux de leur sous-projet en matière de planning et budget. La cohérence de l'ensemble est assurée grâce à des procédures et un système informatique coordonné, mis en place par les directions fonctionnelles.

LE PROJET.
1. Rappelons brièvement les ouvrages à concevoir et à réaliser côté France, qui font l'objet d'autres exposés détaillés par ailleurs :

- 57 km de tunnels à forer à l'aide de 5 tunneliers sophistiqués, fabriqués "sur mesure", travaillant simultanément,
- une usine de préfabrication de voussoirs en béton destinés au revêtement des tunnels, spécialement créée sur le site et capable de produire 9 000 éléments par mois,
- un terminal, de la taille d'un aéroport international, dont la principale fonction consiste à assurer l'embarquement dans les navettes d'un flot de véhicules équivalent à celui d'une autoroute (3 450 véhicules/heure).

Outre le défi technique que constitue chacun de ces ouvrages pris séparément, il faut rappeler que le programme très tendu de l'opération, nécessite un recouvrement important des activités d'études, d'approvisionnement, de génie civil et de montage des équipements électromécaniques.

POLITIQUE DE MANAGEMENT

2. TML a donc mis en place un système de management franco-britannique unifié, basé sur des règles formalisées applicables à l'ensemble du projet, et sur une organisation décentralisée.

Objectifs généraux

3. Ce système de management, qui vous a été décrit par ailleurs, se traduit au niveau de la Construction France par les objectifs généraux suivants qui s'imposent à tous :

- l'adhésion des hommes au projet et le développement de leur capacité à travailler ensemble,
- une politique de sécurité rigoureuse,
- l'obtention de la qualité,
- le respect des délais d'exécution,
- la maîtrise des coûts.

Système de gestion

4. Pour atteindre ces objectifs, un système de gestion basé sur des règles du jeu clairement définies et formalisées dans les domaines suivants a été mis en place :

5. <u>Ressources humaines</u>. Les objectifs fondamentaux pour le personnel ouvrier sont le professionnalisme, la sécurité, l'emploi, la qualité, l'information et la communication. En ce qui concerne l'emploi, le GIE Transmanche s'était engagé à recruter 75 % du personnel ouvrier dans la région Nord-Pas-de-Calais. A ce jour, une politique volontariste et une collaboration avec l'Etat et la Région ont permis de dépasser largement cet objectif (proportion de l'ordre de 90 %). Ce recrutement local a demandé la mise en place d'un plan de formation de grande envergure avec toutes les parties prenantes. Pour le personnel d'encadrement, dont la moitié environ est originaire des 5 sociétés françaises constituant le GIE Transmanche, un mode de management participatif basé sur une large diffusion de l'information et l'établissement d'objectifs au niveau des secteurs a été mis en oeuvre.

6. <u>Sécurité</u>. Un service spécialisé en relation permanente avec les services techniques et opérationnels, a établi les procédures de sécurité. Une politique rigoureuse de prévention est mise en oeuvre. Elle passe par la formation de l'ensemble du personnel lors du stage d'accueil, complétée ultérieurement par des formations spécifiques : sauveteur, secouriste, brevet national de secourisme et réanimation, habilitations électriques, engins, appareils de levage. Une permanence de sécurité est assurée 24 h sur 24 par un pompier professionnel et un infirmier. Deux médecins du travail sont employés sur le site à temps plein.

7. <u>Qualité</u>. Un système de gestion de la qualité, placé sous la responsabilité de spécialistes qui ont acquis une

forte expérience de cette fonction sur de très grands
chantiers, a été mis en place. Il s'agit de créer un
consensus sur les avantages que l'on doit attendre d'une
gestion rigoureuse de la qualité tant en matière de
programme et de coûts qu'au plan des relations avec le
Client. Il s'agit aussi de former et de sensibiliser
l'ensemble du personnel : en définitive, chacun est
responsable de la qualité de sa production.

   8. <u>Délais d'exécution</u>. Le planning du projet respecte le
découpage de l'organisation et comprend les niveaux
suivants:

   - niveau 1 : planning général, géré par la structure de
     coordination,
   - niveau 2 : planning "sous-projet",
   - niveau 3 : planning "secteur".

Les délais sont gérés au moyen d'objectifs mis en place au
niveau des secteurs et consolidés au niveau des sous-
projets. Le planning de niveau 3 est établi sur support
informatique et identifie les tâches, les interfaces et les
ressources nécessaires. Il permet d'établir les courbes
d'avancement, les besoins en personnel et en trésorerie.
Les plannings de niveau 2 sont constitués par les plannings
d'études et de construction de chaque sous-projet. Enfin,
le planning général de niveau 3 intègre les activités des
différents sous-projets.

   9. <u>Coûts</u>. Chaque sous-projet est un centre de profit
autonome. Il maîtrise et gère ses propres coûts et recettes:

   - en mettant en place un système d'établissement et de
     suivi des objectifs financiers,
   - en procédant 2 fois par an à une prévision de coût final,
   - en révisant tous les 3 mois cette prévision de coût
     final.

La consolidation des activités de gestion des coûts et de
l'établissement des rapports sont assurés par la Direction
Suivi du Projet avec l'assistance de la Direction
Administrative et Financière.

<u>Etablissement et suivi des objectifs</u>
   10. Afin de pouvoir établir et suivre les objectifs mis en
place pour l'application des politiques de gestion définies
ci-dessus, un document de synthèse est établi et
périodiquement révisé. Ce document définit les objectifs à
court et moyen terme en matière de coûts, délais, choix
techniques et ressources humaines. Ce document est établi
et périodiquement révisé par chaque sous-projet ou direction
fonctionnelle (en principe tous les 3 mois) sous la
responsabilité du Directeur de sous-projet ou du Directeur
fonctionnel. Il est présenté et soumis à l'approbation du
Directeur de la Construction France et de la Direction
Générale.

ORGANISATION DE LA CONSTRUCTION FRANCE

L'organigramme du Groupe Construction France est représenté en Annexe 1.

Responsabilités de la Direction de Construction France

11. Le Directeur de la Construction France est responsable, devant la Direction Générale, de l'ensemble des activités de construction côté France.

Il est en particulier responsable du budget, du suivi des programmes des travaux et de la gestion de l'ensemble du personnel de la Construction France.

Il supervise l'activité de tous les sous-projets et directions fonctionnelles de la Construction France et il a entière autorité sur leurs directeurs.

Il délègue une partie de son autorité et de sa supervision directe aux directeurs adjoints.

12. Les deux directeurs adjoints de la Construction rendent compte au Directeur de la Construction France. Leurs responsabilités se répartissent de la façon suivante :

- autorité et supervision directe des activités des sous-projets de contruction des Tunnels et de la Préfabrication.
- suivi des grands problèmes de gestion des contrats et supervision des activités commerciales.

13. Les deux directeurs délégués ont les responsabilités suivantes :

- coordination entre Construction France et Construction UK, gestion des procédures de Construction France et consolidation des rapports d'activité de la Construction France.
- liaison avec les collectivités locales et régionales et avec la presse, visites de chantier et suivi de certains dossiers techniques communs aux sous-projets.

Responsabilité des Directions Opérationnelles

14. Elles sont au nombre de 4 :

- sous-projet Tunnels,
- sous-projet Préfabrication,
- sous-projet Terminal de Coquelles,
- département de montage Electromécanique.

Chacun des 3 sous-projets est placé sous la responsabilité conjointe d'un chef des études pour l'ingénierie et d'un directeur de sous-projet construction pour les travaux.

Le chef des études et le directeur du sous-projet construction sont conjointement responsables de la performance du sous-projet dans les domaines technique, délai et résultat.

La gestion des interfaces entre ingénierie et construction repose sur une similitude du découpage entre le Groupe Ingénierie et le Groupe Construction : à chaque directeur d'un sous-projet correspond un chef des études d'un département ingénierie, qui est responsable de la gestion des interfaces ingénierie-construction pour le sous-projet concerné.

15. Le Directeur d'un sous-projet Construction :

- définit la structure du sous-projet et contrôle la composition et la gestion de son personnel d'encadrement et ouvrier,
- est pleinement responsable de la sécurité du sous-projet,
- gère les relations avec le Client,
- gère la production du sous-projet : il est responsable des choix techniques concernant la construction, de la planification de la production et de l'établissement et du suivi des objectifs.
- décide du choix des sous-traitants et fournisseurs, signe les contrats de sous-traitance et les marchés et les gère.
- assure la gestion administrative et financière du sous-projet.

16. Le chef des études est responsable des études et de l'optimisation du sous-projet.

17. Les sous-projets opérationnels sont subdivisés en secteurs.
D'une manière générale, il y a un responsable opérationnel à la tête de chaque secteur travaux, chargé de la tenue des coûts, des délais et des choix techniques de construction, donc de la tenue des objectifs. A noter que le niveau de responsabilité d'un chef de secteur travaux est de l'ordre de 50 à 500 millions de francs, habituellement considéré dans les travaux publics comme un chantier important.
Les secteurs ou services fonctionnels assurent, pour le compte des secteurs travaux et sous la responsabilité d'un chef de secteur ou de service, des prestations communes au sous-projet de type administratif, financier, commercial ou technique.

18. Le directeur du département Montage Electromécanique dépend du Directeur de la Construction France. Il est responsable de :

- programmation du montage des installations électro-mécaniques définitives,
- supervision des montages avec gestion du budget de cette activité,
- gestions technique, financière et des programmes des travaux de montage, réalisés par TML ou sous-traités,
- mise en place des règles d'hygiène et de sécurité.

Cependant, il devra obtenir l'accord de la direction du Groupe Système de Transport concernant le manuel et les procédures du montage électromécanique, les budgets de supervision ou de travaux de montage, les avenants des contrats de sous-traitance.

Responsabilités des Directions Fonctionnelles

19. Les Directions Fonctionnelles sont au nombre de 5 :

- Direction Administrative et Financière,
- Direction Commerciale,
- Direction Suivi du Projet,
- Direction Ressources Humaines,
- Direction Assurance Qualité.

20. D'une manière générale, les responsabilités des directions fonctionnelles sont les suivantes :

- définition des procédures applicables à la Construction France, établies en conformité avec les procédures générales TML, et vérification de leur application par les sous-projets,
- assistance aux différents sous-projets, à leur demande,
- centralisation des éléments de gestion en provenance des sous-projets et consolidation au niveau Construction France à usage interne (gestion) ou externe (transmission au Client),
- gestion des fonctions non décentralisées dans les sous-projets.

21. Les directeurs fonctionnels ont un accès direct aux secteurs et services des sous-projets chargés des fonctions correspondantes. Les instructions sont émises par le canal des procédures ou du directeur du sous-projet concerné.

22. Les directions fonctionnelles sont découpées en secteurs ou services en fonction de leurs responsabilités respectives.

La communication au sein de Construction France

23. La politique de communication mise en oeuvre doit répondre aux contraintes spécifiques de ce grand projet :

- origines diverses des collaborateurs,
- rythme de travail (3 postes de travail, 7 jours sur 7),
- dimension du chantier,
- localisation des personnels (Paris, Londres, Calais, Sangatte, Coquelles),
- nécessité d'acquérir rapidement des réflexes communs, une bonne connaissance du projet et de son évolution constante.

ORGANISATION DE LA CONSTRUCTION FRANCE

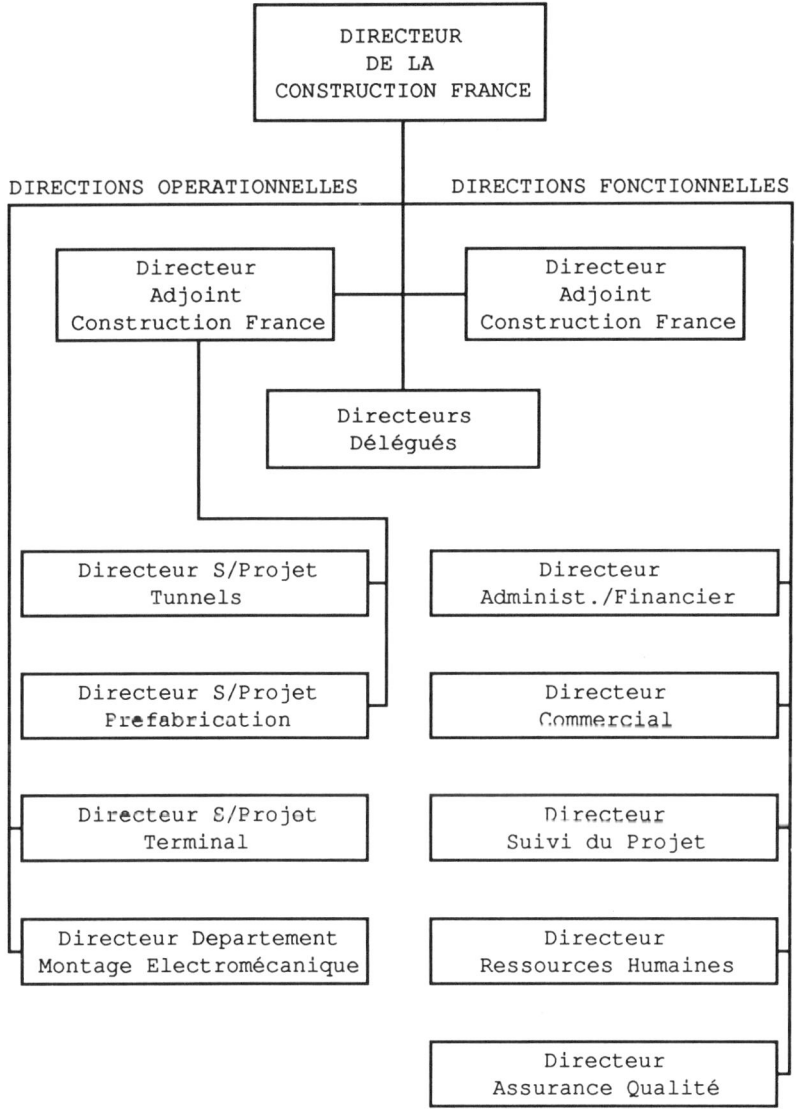

Annexe 1 - Organigramme du Groupe Construction France

Cette politique s'articule autour de deux grands axes :

- Les réunions :
    - le comité de direction et les comité de sous-projets se réunissent chaque semaine. Ils permettent une meilleure information et facilitent la prise de décision.
    - des réunions d'information périodiques rassemblant l'ensemble des employés et cadres.

- Les publications et l'audiovisuel :
    - deux publications, l'une mensuelle "Infos-Dernières", l'autre timestrielle et franco-britannique "Le Lien" apportent des nouvelles sur l'évolution du projet à l'ensemble du personnel.
    - un réseau de diffusion de 25 points de télévision interne répartis sur l'ensemble du chantier permet de diffuser à tout instant des vidéogrammes, des films ou des clips sur l'avancement des tunneliers, la sécurité, la formation, l'hygiène. En outre, un journal télévisé mensuel d'une durée de 20 mn, apporte à chacun les images de la vie de "son" chantier.

SUMMARY

The size and complexity of the project required setting up four operational "sub-projects". These covered tunnels, pre-casting, the terminal at Coquelles and the provision of electrical and mechanical services. In addition, five administrative directorates were established covering, administration and finance, commercial operation, progress, human resources and quality assurance.

Each of the four operational sub-projects have responsibilities which, by normal standards would be major construction projects in themselves.

The general objectives of the total management system were -
1 Coordination of the work of all personnel and the creation of a team spirit.
2 A strict policy on safety
3 Quality control
4 Progress monitoring
5 Cost control

The ways by which these objectives are reached within the management system are explored and particular problems addressed. For example 75% of the non-technical work force has to be recruited from the North Pas de Calais region. This has been largely achieved but requires a considerable training scheme.

Progress planning is carried out at three levels:- 1 General overall 2 Sub-project and 3 Sector.

For cost control, each sub-project is an automonous profit centre. Total cost forecasts are revised every three months.

The general management objectives and the way they are to be obtained are set down in a document which is revised periodically. An organogram has been established showing the relationships between the general directorate, the sub-projects and the administrative directorates. The interfaces and the responsibilities of each section are explained.

Establishing a communication network between the elements of the management system, has to cope with the particular difficulties of such a large project, especially the intense work programme (7 days a week), and the different locations of the work force. Communication is based on regular meetings, documentation and audio-visual presentations. A network of 25 TV terminals is used to keep the work site informed of progress, safety measures, training and health. A 20 minute TV news programme is prepared every month, showing life on the project.

13A.   UK tunnels construction

A. R. BIGGART, BSc, FICE, Assistant Construction Director,
J. R. J. KING, BSc, FICE, Construction Manager — Production,
R. D. MacKENZIE, BSc, Construction Manager — Technical,
G. A. MOORE, Construction Manager — Tunnel Support Services
and J. A. MILES, Construction Manager, Precast Yard

SYNOPSIS. The paper describes the construction work involved
in tunnel driving from the UK. The scope of the works and
the design of temporary works is included. Major sections of
the work are illustrated in some detail. This includes the
forming of marshalling tunnels by the NATM method, the
driving of the main Tunnel Boring Machine (TBM) tunnels and
the manufacture of precast concrete segments. The support
services are included in the description. Progress to the
end of the first quarter of 1989 is reported. Geology and
surveying are briefly mentioned.

SCOPE OF THE WORKS

1.   The main task is to drive 90Km of TBM tunnels from
an underground development complex, called the marshalling
area, situated below the main construction site at
Shakespeare Cliff. Machine driven tunnels include twin 7.6m
i.d running tunnels and a 4.8m i.d service tunnel 22Kms out
to sea and 8Kms inland to the terminal. In addition, there
is a half kilometre cut and cover section at Holywell at the
inland end of the machine driven tunnels together with a half
kilometre length of tunnels using NATM primary lining through
Castle Hill terminating at the permanent UK portal. Included
in the works are the equipment room cross passages connecting
the service tunnel to the running tunnels and also the piston
relief ducts, connecting the running tunnels. There are
additional major tasks which include the crossover
construction 7Km offshore and two large capacity sumps formed
at tunnel low points. The crossover will be the largest sub-
aqueous chamber ever formed so far from the shore. The basic
statistics are given in Table 1.

GEOLOGY

2.   All the data available shows the tunnelling to be in
the Chalk Marl. This is a clayey carbonate mudstone in a
series of layers which vary between a moderately strong
carbonate bed to a weak clay rich layer. The material was
considered to be essentially impermeable with any water
movement restricted to discontinuities such as faults or

fissures. The material was also considered to be an ideal
tunnelling medium, easy to excavate with a stand up time
suitable for the use of an unbolted tunnel lining.

Table 1. Main statistics for UK Tunnels

Description	Quantity	Description	Quantity
TBM driven tunnels	90Km	Grout volume	60,000m^3
Total tunnel spoil	4.31mil.m^3	PCC rings - no.	55,750
TBM tunnels spoil	4.03mil.m^3	PCC segments - no.	474,810
Crossover spoil	341,000m^3	- weight	1.66 mil.t.
Cross passages-no.	306	SGI rings - no.	17,000
- length	2.69km	(750wide) - weight	49,400t
Piston Relief Ducts - no.	119	Shotcrete volume	16,000m^3
- length	2.56km	Concrete volume	150,000m^3

DESIGN OF TEMPORARY WORKS
Introduction
3. The development philosophy of the marshalling area
underground complex consists of three parallel NATM lined
tunnels approximately half a kilometre long with a TBM
erection chamber at either end of each tunnel. The three
tunnels at -41m o.d are intersected by two sloping adits
driven on a downward slope of 1:7 from the lower Shakespeare
Cliff site which is at 16 m.o.d. The upper site is connected
to the lower site by a vehicular access tunnel. In addition,
there is a 110m deep shaft connecting the marshalling area to
the upper Shakespeare Cliff site. There are also various
conveyor tunnels formed between and below the marshalling
tunnels that connect to one of the adits. The philosophy has
been to simplify the logistics by separation. Hence,
personnel enter and leave the underground works by a
personnel hoist in the shaft, muck removal is by a series of
conveyors feeding a main conveyor in the smaller of the two
adits (A1) and all construction materials and plant, except
for major TBM parts, enter the works down the large adit (A2)
which contains five 900 gauge rail tracks. Major TBM parts
are lowered down the shaft.

Surface
4. The Site is split between the upper site (Fig.1) on the
cliff top housing management offices, labour facilities, main
stores and workshops and the lower site (Fig.2) devoted to
handling segments, construction materials, tunnel spoil,
tunnelling plant, with workshops for construction, rolling

stock and maintenance installations. The site has also to function as the working area for the sea wall behind which reclamation is carried out.

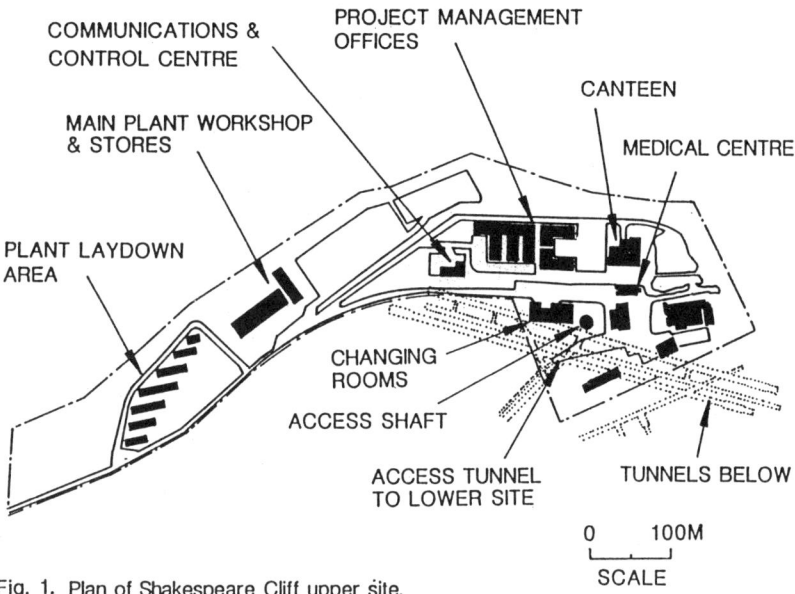

Fig. 1. Plan of Shakespeare Cliff upper site.

Sea Wall

5. Lower Shakespeare started as a 5 hectare site at commencement of the work but this is insufficient for peak operations. Tunnel arisings are used to reclaim land to its final 23 hectares.

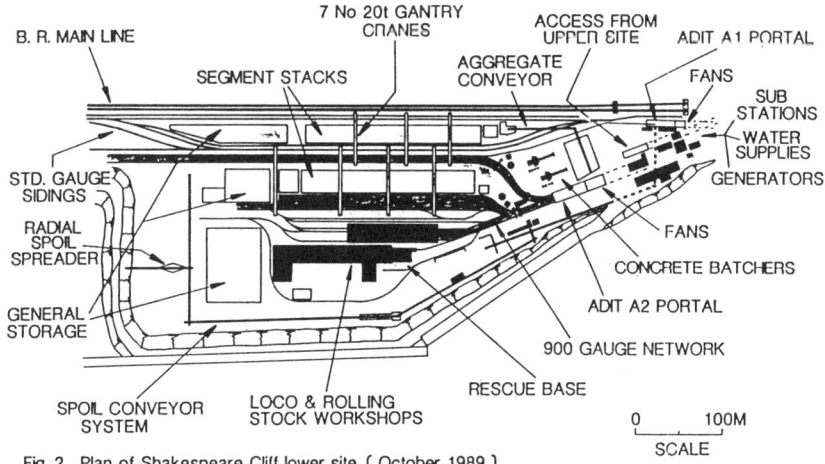

Fig. 2. Plan of Shakespeare Cliff lower site. (October 1989)

6. The timescale of land reclamation spans 4 years (Fig.3). To prevent escape of chalk fines, fill is placed in still water lagoons formed by the temporary cross walls. The design is based on a mass concrete wall (Fig.4) with the sheet piling limited to a temporary role.

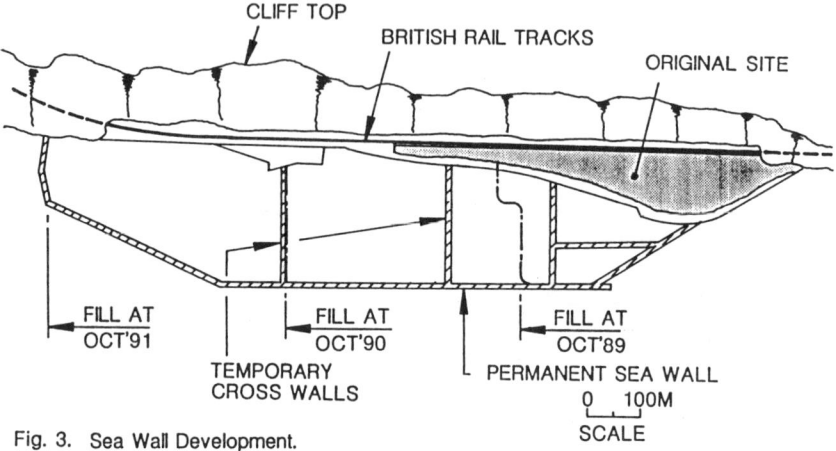

Fig. 3. Sea Wall Development.

Below Ground

7. Figure 5 shows the layout of the underground development works at tunnel level below Shakespeare Cliff. For each of the six tunnels a dedicated train marshalling area, spoil bunker, sidings for inbound material trains, and TBM erection chamber, are provided.

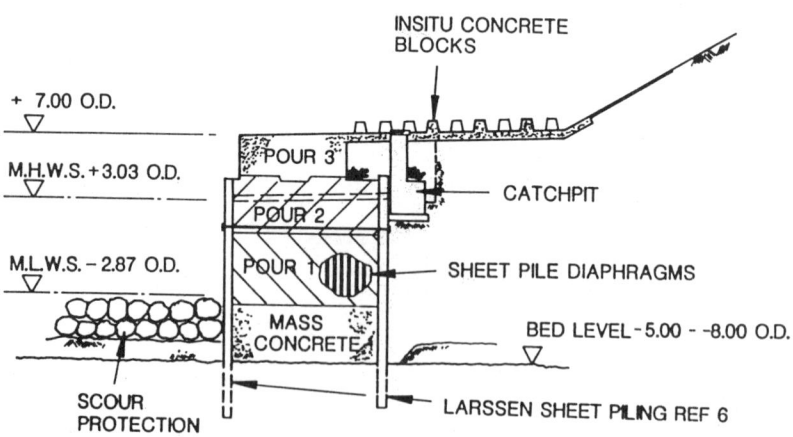

Fig. 4. Section through Sea Wall.

8. The service tunnel between points 'P' and the TBM chamber at 'Q' together with Adit A1 remain from the 1973 scheme and had been maintained in good condition by the D.O.T. This initial access was invaluable in achieving an early start to the marine service tunnel.

9. The access provided by Adit A1 on its own was clearly insufficient to service the five drives which will be driven at any one time.

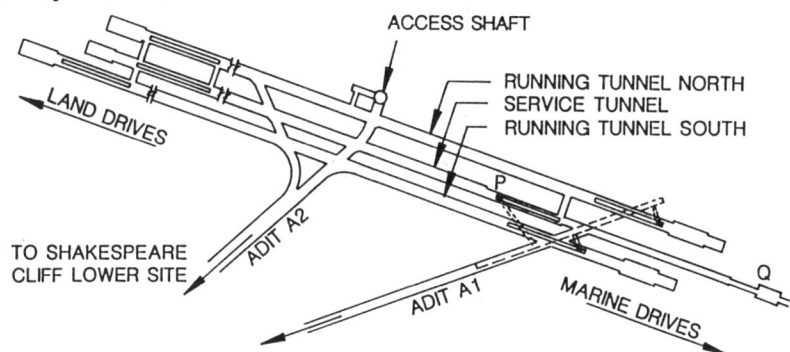

Fig. 5. Shakespeare Cliff Underground Development.

10. Adit A1 was extended to provide conveyor tunnels below the main tunnel horizon and devoted entirely to spoil removal. A new larger adit, A2 was introduced of sufficient size to accommodate five rack and pinion 900mm gauge rail tracks, together with other services. This is used to supply tunnel linings and other materials. From upper Shakespeare a 10m diametre shaft, 110m deep connects to tunnel level to provide man access. In the early stages this served as access for TBM erection.

11. An early decision was to provide temporary support by NATM (Fig.6). This decision has been fully justified in terms of the progress rates and versatility achieved in tunnelling through ground ranging from the chalk fill at lower Shakespeare to chalk marl at tunnel level. In particular the ease of forming junctions is of prime importance. Deflection and loading measurements were carried out during construction and suitable adjustments to optimise the basic design were made.

Spoil Conveyors
12. Underground spoil gathering arrangements for each drive are provided by a moving bed tipping bunker adequately sized to accommodate a complete spoil train (Fig.7). Spoil is discharged from these bunkers through metering conveyors on to the main belt conveyor to the surface. Discharge rates

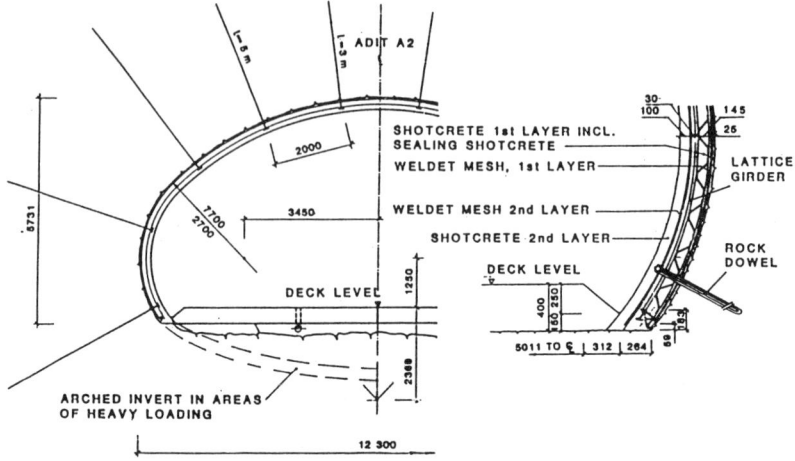

Fig. 6 ADIT A2 – Typical Section and Detail.

from the tipping bunkers are sized to the maximum advance capability of the TBM's. Transfer conveyors located below deck level carry spoil from the land side tipping bunkers through the marshalling areas to the main belt.

13. The conveyor system will handle a total of 4.1 million m^3. The majority of this material will be placed behind the sea wall. Bunkers have a capacity of 200m^3 and a discharge rate varying between 600t/hr and 1200t/hr. The metering conveyors are sized in relation to the bunker discharge rates and the main conveyor to the surface is 533m long with a maximum capacity of 2400t/hr. A system of surface conveyors feeds spoil to a 54 metre radial spreader which advances with the reclamation face (Fig.2).

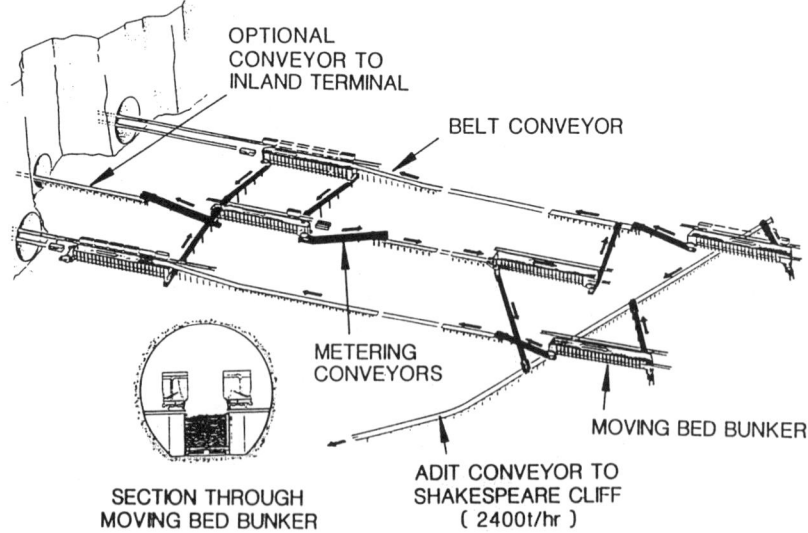

Fig. 7. Underground Conveyor System.

Crossover

14.   At the UK crossover the running tunnels come together in a common chamber 20 metres wide, 15.5 metres high and 156 metres long.   To allow access the service tunnel dips below running tunnel grade, diverts in alignment and is enlarged locally (Fig.8). Cover to the seabed at this point is 36 metres.

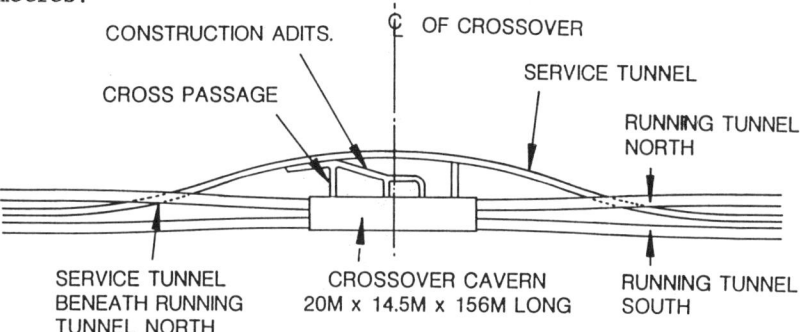

Fig. 8. Plan of U.K. Tunnel Crossover.

15.   The excavation sequence is designed to limit the amount of exposed ground.   Sidewall drifts are excavated full length followed by a full width top heading.   This arrangement is preferred to conventional heading and bench as the bench sidewalls are secured in advance of the excavation.   A reduced span top heading is envisaged if conditions are more onerous than expected (Fig.9).

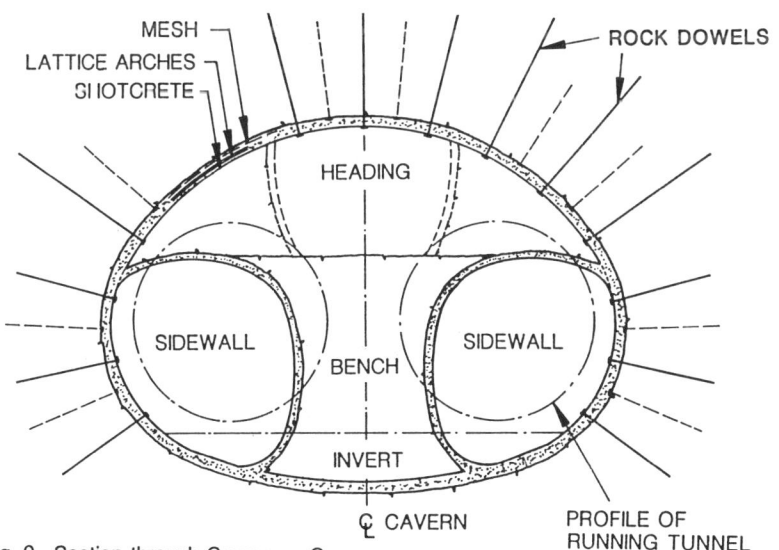

Fig. 9. Section through Crossover Cavern.

CONSTRUCTION METHODS
Hand Tunnels

16.   The speed and flexibility of the NATM has been well demonstrated in the marshalling tunnels at Shakespeare and the half kilometre lengths of tunnels in glauconitic marls and gault clay through Castle Hill. Generally major tunnels in these complexes have been constructed with top headings supported by lattice girder arches, welded mesh and shotcrete with 3m or 4m long rock dowels at 1.5 to 2.5m centres depending on ground conditions (Fig.6). In the unstable marls and clays at Castle Hill arches are raked forward to stabilise the face.

17.   Shotcrete 150mm thick in the top arches and 200mm thick in the invert sections is mixed with an accelerator and applied as a dry mix. Shotcrete materials are batched at the surface, transported in specially converted dump trucks and fed to the shotcrete machines by metering conveyor. Rock dowels using high yield re-bar are set into a cementitious grout. Access for fixing arches, mesh and dowels is provided by hydraulically operated baskets.

18.   Construction of the crossover under careful monitoring controls will provide a safe and rapid form of construction. An independent risk analysis has been carried out to analyse categories of risk to personnel and the works. The crossover will be constructed as described in paragraph 14 and 15 in an eight month time window before the arrival of the running tunnel TBMs. Maintaining logistical support to both the crossover and the TBM beyond will provide a major challenge.

Cross Passages

19.   Cross passages between tunnels (Fig.10) are provided at 375m centres for passenger escape and equipment rooms. A majority will be constructed from the service tunnel to enable installation of temporary services. These will be used as temporary electric sub-stations and staged pumping stations. The junction with the running tunnels will be constructed from the running tunnels.

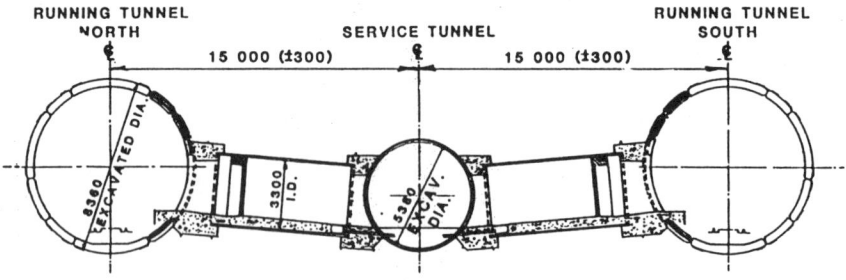

RUNNING TUNNEL NORTH   **SERVICE TUNNEL**   **RUNNING TUNNEL SOUTH**

15 000 (±300)   15 000 (±300)

Fig. 10   Cross Section showing Cross Passage

20. The cross passages are generally formed in machined cast iron bolted segmental linings excavated by traditional hand mining methods. Junction lengths are taken out in a single excavation and concrete is placed in a single pour using a collapsible shutter.

21. Construction of cross passages is serviced by materials and plant trains with track possessions closely integrated into the freight and personnel rolling stock movements for the TBM drives ahead. Spoil is carried to muck cars by conveyor and tunnel materials man handled into the cross passages.

22. Piston relief ducts (Fig.11) will be formed in a similar manner to the cross passages at 250m centres.

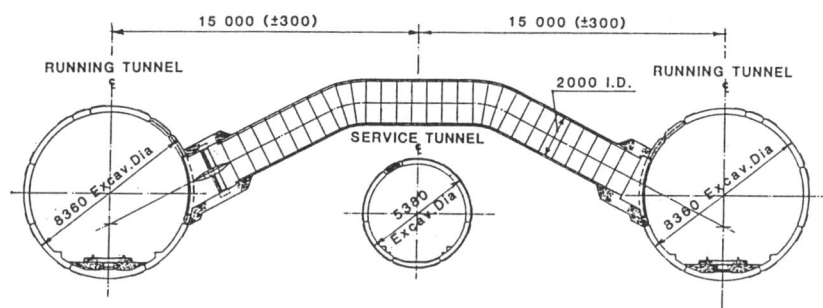

Fig. 11 Cross Section showing Piston Relief Ducts.

Machine Driven Tunnels

23. Six TBM's are employed. Two service tunnel machines (Fig.12) for marine and land drives at 5.38m o.d and 5.76m o.d respectively. Two land running tunnel machines at 8.72m o.d (Fig.13) and two marine running tunnel machines at 8.36m o.d (Fig.14). The combined weight of TBMs and back-up is 6500 Tonnes with a total value of approximately £40 million.

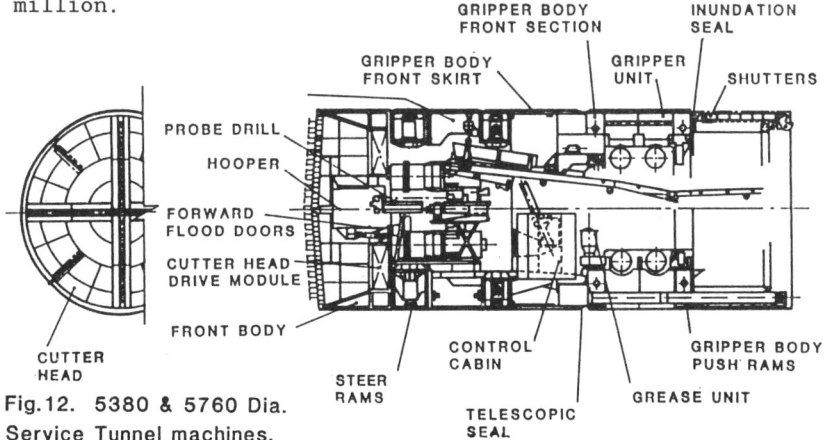

Fig.12. 5380 & 5760 Dia. Service Tunnel machines.

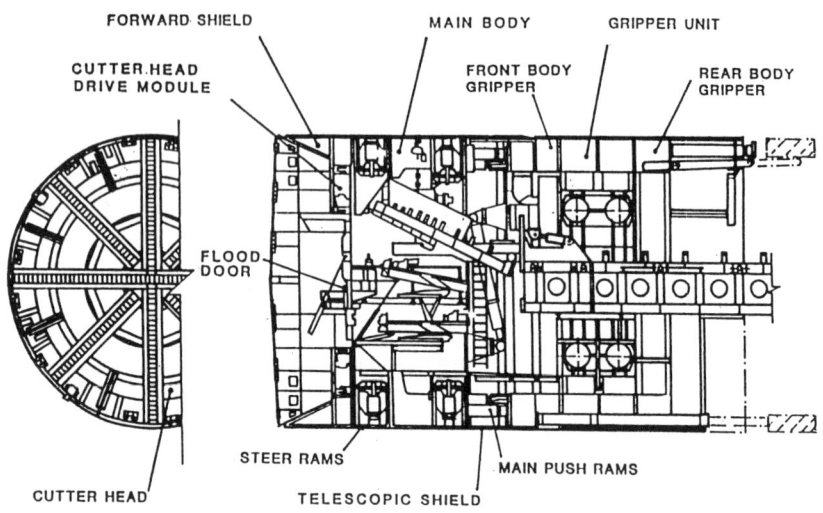

Fig.13. 8720 Dia. Land Running Tunnel

24. The open face TBMs are all of the same generic type.
There is a front body section with a full face cutting head
mounted on a large diameter multi directional main bearing.
Behind this a gripper section forms an anchor from which the
cutter section is thrust forward by the main thrust rams.
Following the gripper unit is the back-up equipment which is
divided into two basic modules. Firstly the segment handling
and building area immediately behind the gripper and secondly
a number of sledge units holding the muck disposal conveyors
and other ancillary equipment such as electro hydraulic power
packs, water and fire suppression systems, ventilation
equipment, drainage pumps, electrical distribution panels and
pipe and cable handling plant. Overall length of the TBMs
plus equipment sledges is typically 200 metres. All the TBMs
were launched from erection chambers in the marshalling area.

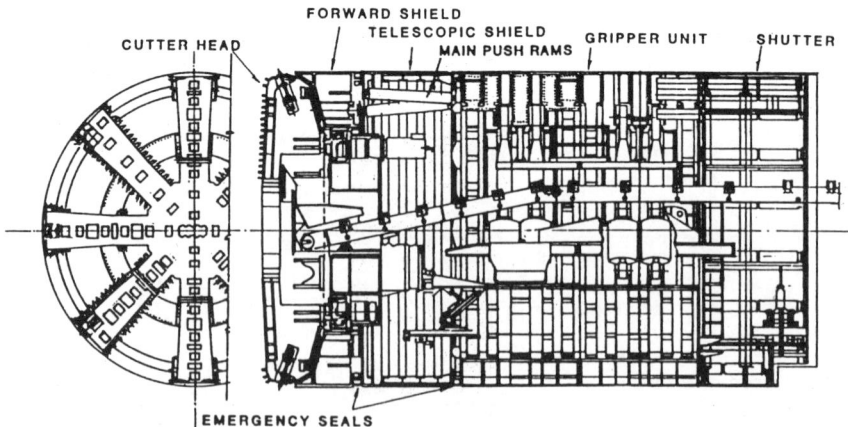

Fig.14. 8360 Dia. Marine Running Tunnel.

25. The design of the TBMs allows concurrent excavation and lining erection. Lining is the critical operation which dictates tunnel progress. Service tunnel machines are designed for a cycle time of 18 minutes whilst running tunnel machines have a cycle of 22.5 minutes. The lining is a precast concrete unbolted wedge block with pads on the extrados to allow for a 20mm grout layer. There are 7 segments per ring in the service tunnel and 9 segments per ring in the running tunnels, internal diameter being 4.8m and 7.6m respectively.

26. After the initial learning curve is over, the rate of advance is programmed to peak at 1Km per month. These rates have already been achieved in the service tunnel. In order to maintain the high average rates of progress all machine tunnels are driven 24 hours per day, 7 days per week. This effort is sustained using 4 gangs of 25 men each working 8 hour shifts. Maintenance is given top priority with fitters and electricians as part of the team and a back-up team ready at all times for major repair tasks. On all tunnel drives there is a maintenance shift once a week and on the marine service tunnel a probe shift every 200m. An unusual feature of the TBMs is that the majority of mechanical and electrical equipment is duplicated in order to minimise downtime. The overall aim is to keep TBM downtime below 10%. Table 2 gives the main statistics for the TBM's.

Table 2. TBM Data

Description		Marine Service Tunnel	Marine Running Tunnel	Land Running Tunnel
Cut diameter	m	5.38	8.36	8.72
Length overall	m	200	200	200
Weight	t	700	1250	1300
Cutting Head				
Direction of rotation		Reversible	Reversible	Reversible
No. of picks	No.	79	197	231
Rotation speed high	rpm	4.5	3.33	2.86
low	rpm	3.0	1.66	1.9
Main bearing type (All internal gear)		3row roller	3row roller	3 row roll.
Bearing life. In excess of		20,000hr.	20,000hr.	20,000hr.
Drive motors		4No/190Kv	12No/110Kw	6No/190Kv
Torque - high speed	KNM	26,880	63,088	63,483
- low speed	KNM	40,456	126,177	95,550
Total installed power	Kw	1702	2487	2120
Instantaneous penetration rate		8m/hr	6m/hr	6m/hr
Guaranteed advance rate		5m/hr	4m/hr	4m/hr
Main thrust rams. 175 bar		16No/110t	20No/126t	24No/110t
Total thrust. 175 bar		1755t	2520t	2640t
Thrust for cutting with picks 360t			Not quoted	765t

Roll correction rams	2No.	None	2No.
Auxilliary rams 175 bar	8No/110t	16No/105t	8No/110t
Gripper rams.250bar/300bar	4No/387t	8No/288t	4No/700t
Gripper locking force	3096t	Not quoted	5600t
Gripper ground bearing pressure	21 bar	21 bar	21 bar
Segment erector	Semi rotary ring+crane	Twin rotary +twin crane	Twin rotary ring
Steer Shoes	8	None	9
Conveyor - capacity	1No/380m^3/hr	3No/720m^3/hr	1No/850m^3/hr

Note: Land Service Tunnel similar to Marine Service Tunnel

Logistics and Railway

27. Tunnel materials are unloaded, stacked and reloaded from
the pit head area at the lower Shakespeare site. From here
dedicated railway tracks connect to the five drives via Adit
A2. Supplies coming in by road or by rail to the BR siding
at the lower Shakespeare site are a major operation.
Segments are unloaded by 20 tonne gantry cranes capable of
stacking 7.5m high. At peak production, segments must be
supplied at a rate of 1000 per 24 hours together with all
other back-up materials. The development of the space
required to store materials depends on the rate of land
reclamation using tunnel spoil.

28. Underground trains tip spoil into the bunkers and pick
up return loads of tunnel segments brought down by rack
locos. Turn around times including battery changes are
geared to production needs using four trains per tunnel, each
presently hauled by twin 25 ton electric trolley locos
augmented by a diesel loco. The electric locos operate from
overhead DC power lines whilst travelling to and from the
face and by battey power for shunting operations at the TBM
and at the spoil bunker. Batteries are recharged during
travel. Design and development, of the locos and overhead DC
power line to a sufficient standard has stretched the present
technology.

29. Movements of trains for each TBM drive and the ancillary
tunnel and service works are controlled by radio contact with
drivers and a simple BR freight type signalling system.

Safety

30. Safety underground has called for an enormous effort in
written procedures, inspections, training and development of
safety awareness. All personnel are subject to a 4-day
induction and orientation course before going underground.
Field safety talks are a major feature as are specialised
skill training courses heavily geared to safe practices and
safety awareness. Forward planning is discussed with the HSE
and with the MdO followed by close monitoring of performance.

SEGMENT MANUFACTURE

31. The 29ha site on the Isle of Grain (Fig.15) was chosen
for environmental reasons and because it was served by an
existing rail system and deep water jetty facilities. It was
also one of the few available flat areas in Kent.

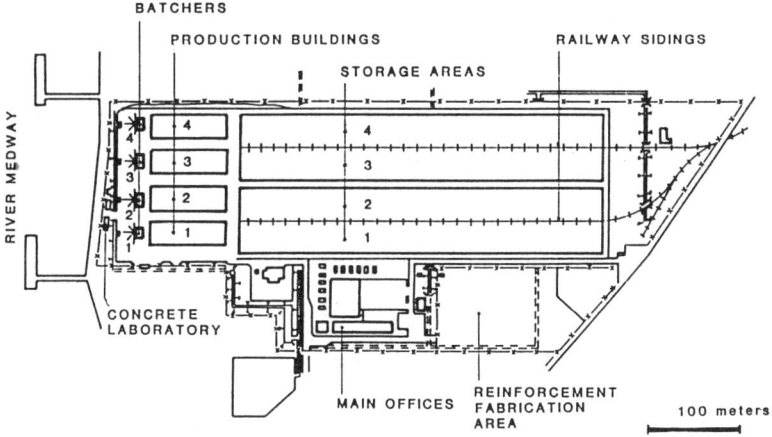

Fig. 15 THE ISLE OF GRAIN – Segment Manufacturing Site.

32. The process chosen to manufacture the half million
precast concrete segments in three years, at a peak rate in
excess of 1000 per day, required a more sophisticated system
than the conventional static mould, hence the adoption of a
continuous powered loop or carousel.

33. Four production buildings, each 106m long and 37m wide,
are used to house twin carousels located in each building
(Fig.16) A total of 45 moulds form a carousel, 9 at work
stations and 36 in the curing tunnel. They travel through
the work stations at 10 minute intervals for stripping,
cleaning, placing reinforcement, fitting inserts and
concreting before moving to the curing tunnel, where they are
steam cured at 50°C for 6 hours prior to being lifted at
10N/mm^2 compressive strength onto trailers for
transportation to the storage area. Before leaving the
building each segment is covered by a special jacket against
thermal shock. The potential output of a carousel is 144
segments in 24 hours, therefore eight carousels working for 5
days are sufficient to meet programme requirements. Weekends
are used for maintenance of plant.

34. A batching plant supplies concrete by conveyor to the
work station on the production line. Cement is brought in by
rail, whilst stone (20mm and 10mm granite) and sand (crushed

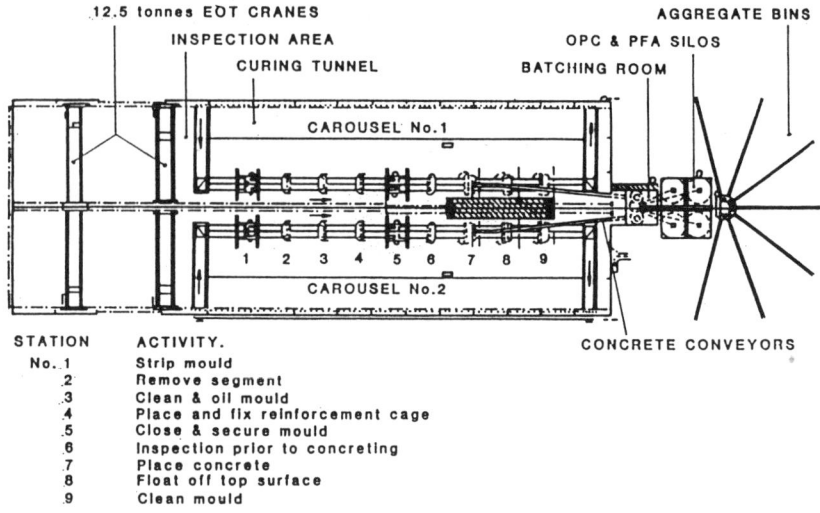

STATION | ACTIVITY.
No. 1 | Strip mould
2 | Remove segment
3 | Clean & oil mould
4 | Place and fix reinforcement cage
5 | Close & secure mould
6 | Inspection prior to concreting
7 | Place concrete
8 | Float off top surface
9 | Clean mould

Fig. 16 THE ISLE OF GRAIN – Typical production building
with two carousels.

granite) is brought in by sea from the west coast of
Scotland. Fully welded reinforcement cages are manufactured
on site to very high tolerances.

35. The storage area consists of four bays 50m wide x 500m
long with four 12½T portal cranes to each bay.

TUNNEL SUPPORT SERVICES

36. A separate construction team provides the tunnel infra
structure of water supply, drainage, electrical distribution,
ventilation, communications and monitoring.

Pumping and Drainage

37. The pumping system installed in the marine tunnels
(Fig.17) is designed to carry 645 1/sec. Allowances are
included to contain unexpected flows from faults or unsealed
boreholes. TBM's are equipped with operational and standby
pumps capable of discharging to temporary pumping stations
built in tunnel cross passages up to 3.5km to the rear of
each face. Pumping stations are constructed alternately at
375m and 750m centres.

Power

38. Six separate 11kV feeder cables provide electrical power
to the site. This is distributed from a primary substation
to meet a maximum demand of 23 MVA. Diesel generators (4No
at 1.25MVA) provide an emergency back-up for essential
services. Separate 11kV ring mains feed the TBM's and
services in each marine tunnel (Fig.18). The land tunnels

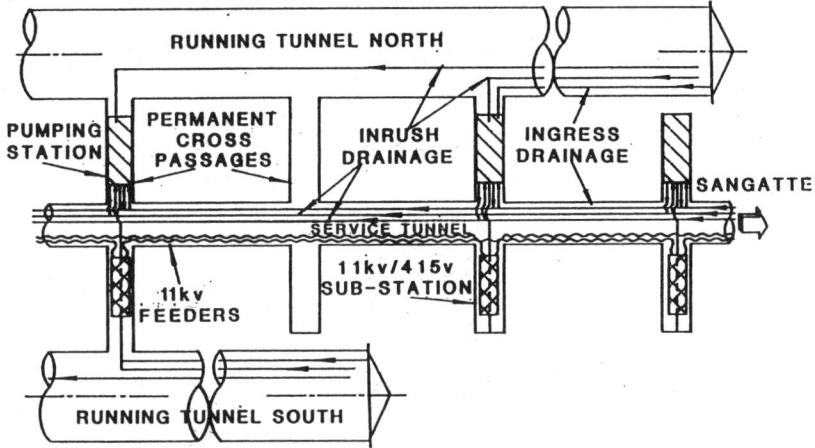

Fig. 17 MARINE TUNNELS - Pumping Arrangements.

are supplied by radial feeders. Secondary substations in
tunnel cross passages provide 415v services supplies every
1.125km along each tunnel. These provide power for lighting,
pumping, ventilation, DC traction, signalling and
communications.

Ventilation
39. Ducted fresh air systems will provide quantities in
excess of $9m^3$/min. per square metre of face area. Due to the
length of each tunnel, special consideration is being given
to air losses resulting from leakage and duct friction.

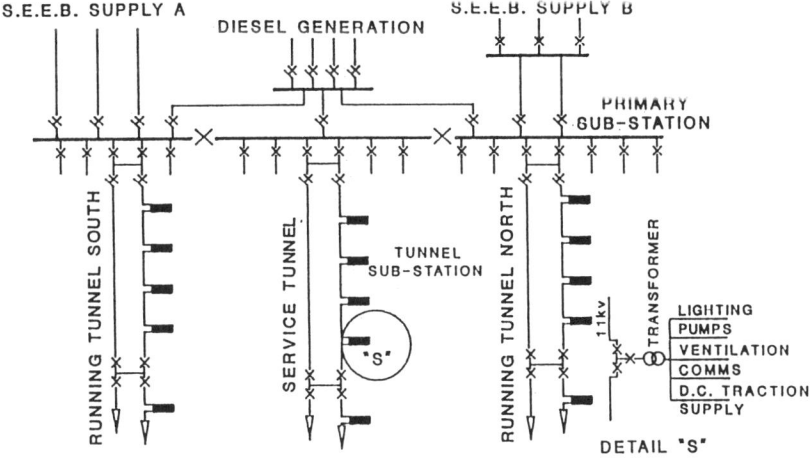

Fig. 18 MARINE TUNNELS - 11kv Supply & Distribution

CONSTRUCTION MANAGEMENT

Communications and Monitoring
40. A seven channel radio system forms the basis of the site communication system above and below ground. Back-up is provided by telephone, paging and public address systems. An auxilliary fire safe telephone system is provided. All systems are self monitoring and are connected to a communications and control centre on the surface. The environment and all essential underground equipment is monitored on VDU screens.

SURVEY

41. Considerable work has been carried out in correlating the survey of both line and level between the UK and France. In 1987 global position by satellite was carried out and errors found in the original survey were corrected.

There is a tunnel survey team responsible for initial survey and setting out. There is an in-house back-up team who come in on a regular basis to check the main stations below ground. In addition, an outside team comes in two or three times a year to carry out a further independant check.

PROGRESS

42. To date 90% of the underground development works associated with the Shakespeare marshalling tunnels have been completed. Installation of bunker and feeder conveyors has kept pace with the launching of TBMs for each drive.
43. At present the marine service tunnel drive is averaging over 800 metres per month having completed 7km of the 23km total. Inland the service tunnel has completed 2.7Km out of its 8km total to Folkestone and is achieving similar rates of progress. Delays due to earlier TBM breakdowns and difficult ground conditions in the marine drive are rapidly being recovered.
44. Recently the first of the two 8.36m marine running tunnel TBMs was launched. Commissioning trials have demonstrated the equipment's ability to maintain the rates of production necessary to hold programme.
45. Planning and procurement for the crossover is well advanced enabling work to commence in June 1989 when the marine service tunnel TBM is sufficiently clear.
46. Construction of cross passages from the service tunnels is keeping pace with the TBMs. Construction of piston relief ducts is due to commence in June 1989.
47. With three TBMs working, two being assembled and the remaining one due on site shortly mechanised tunnel driving is well underway.

RESUME

Ce document décrit la construction des tunnels à partir des
rives de La Manche du côté Britannique. Le site opérationnel
est à Shakespeare Cliff près de Douvres, à partir duquel sont
effectués le creusement sous terre vers le terminal de
Cheriton et le creusement sous mer vers la France. Le
chantier de Shakespeare Cliff est aménagé sur deux niveaux.
En partie supérieure en haut des falaises sont établis les
bureaux, les ateliers etc. En partie inférieure, au niveau
de la mer s'effectue la mise en dépôt des déblais, la
réception, le stockage et la distribution des revêtements des
tunnels, la fabrication du béton etc. Un puits d'accès relie
le site supérieur aux travaux des tunnels en bas et le site
inférieur est connecté aux travaux par deux decenderies. Les
déblais des tunnels sont utilisés pour allonger le site
inférieur par la mise en oeuvre d'une plate-forme en
direction de l'ouest. Les déblais sont déposés dans l'eau
calme retenue par la digue. Plus de 4 million m3 de déblais
devront être mis en dépôt.

Le creusement est effectué dans la craie, considérée matériau
idéal. Les tunnels ferroviaires et les tunnels de service
sous mer et sous terre sont réalisés par des Tunneliers. Les
chambres des gares à Shakespeare Cliff sont construites en
utilisant des machines excavatrices spécialisées. Les 306
rameaux de communication et 119 rameaux de pistonnement
seront effectués par les méthodes de creusement
traditionnelles manuelles. Les détails des Tunneliers sont
décrits. Les déblais sont transportés de l'arrière des
Tunneliers par des locomotives électriques jusqu'à
l'extrémité de galerie où ils sont basculés dans des trémies
à partir desquelles des convoyeurs les remontent au site
inférieur de Shakespeare Cliff jusqu'aux installations de
dépôt des déblais.

La chambre de 'Cross Over' (communication ferroviaire) est un
ouvrage particulier et très important situé à 7 km sous mer.
Ses dimensions sont 156 m de longueur, 20 m de largeur et
14.5 m de hauteur. La chambre demande à ce que le tunnel de
service, normalement entre les deux tunnels principaux soit
détourné sous et au nord du tunnel ferroviaire nord.

Les voussoirs préfabriqués du revêtement des tunnels sont
fabriqués dans une usine spécialisée située sur l'Ile de
Grain dans le Kent. Elle a une capacité de production
journalière de plus de 1 000 voussoirs. La préfabrication
est effectuée en système continu à carrousel avec des moules
passant par différents postes de travail qui, après avoir été
remplis de béton à consistance humide, traversent le tunnel
d'étuvage pendant 6 heures.

Le document décrit également les différents moyens

logistiques tels que installations électriques et de pompage, ventilation et systèmes de communication, moyens topographiques. L'avancement de la construction à ce jour est résumé.

13B.   Le creusement des tunnels côté français

J. FERMIN, Ingénieur des Mines, Directeur Adjoint Construction France

RAPPEL DU PROJET

1.   Le lien fixe TRANSMANCHE est constitué de 3 tunnels
parallèles à l'entraxe de 15 m en section courante : 2
tunnels ferroviaires à voie unique de 7,60 m de diamètre
intérieur encadrant 1 tunnel de service de 4,80 m.

2.   Chaque tunnel, d'une longueur de 50 km dont 37 km
sous la mer, est relié aux autres par des galeries
transversales permettant l'évacuation du tunnel,
l'installation des équipements électriques et des stations
de pompage et la recirculation de l'air sous l'effet de
pistonnement des navettes.

3.   De plus, 2 ouvrages de jonction implantés au tiers du
parcours - soit à 15 km des portails d'entrée - permettront
aux trains de changer de tunnel si nécessaire.

LES CONDITIONS GEOLOGIQUES

4.   La partie française des tunnels débute au portail de
Beussingue à 3 200 m à l'intérieur des terres, franchit la
côte au droit du village de Sangatte. La jonction avec les
travaux britanniques est prévue au point "M" situé au large
à 15 800 m du Puits de Sangatte.

5.   Les tunnels sont forés dans la craie qui constitue
entre Douvres et Sangatte une couche continue affleurante
de grande épaisseur ; elle est séparée des terrains crétacés
(sables) sous-jacents par une couche d'argile étanche
(argile du Gault).

6.   Le tracé du tunnel est localisé pour la partie sous
marine dans la couche de craie bleue qui, du fait de la
présence d'environ 25 % d'argile est un matériau plastique,
faiblement perméable, favorable à la réalisation du
creusement par des tunneliers.

7.   A proximité des côtes françaises, la couche de craie
bleue s'enfonce sous la côte et le tracé remonte vers le
portail au travers des couches supérieures de craie grise et
blanche perméables.

8.   D'une manière générale, la géologie du sous-sol de la
Manche côté français est plus difficile que du côté anglais:

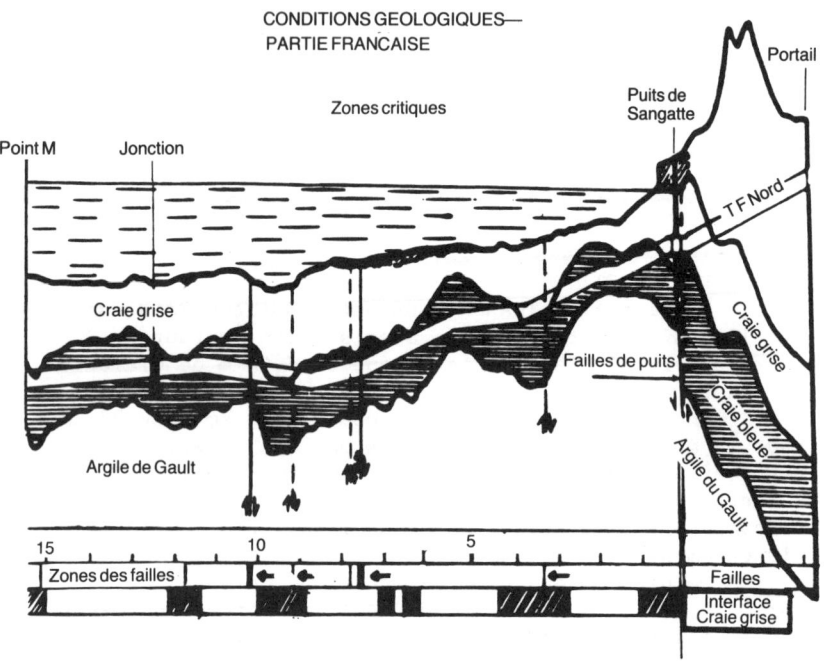

CONDITIONS GEOLOGIQUES—
PARTIE FRANCAISE

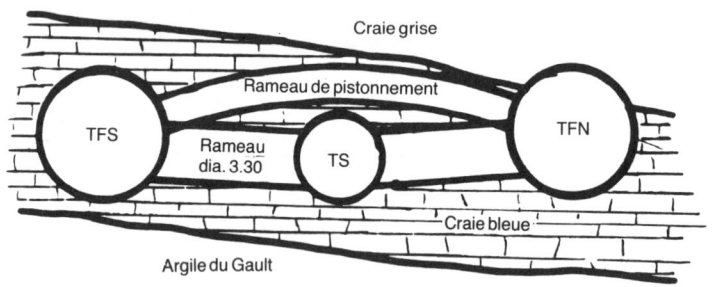

TABLEAU 1

CHOIX DES TUNNELIERS

	Sous mer	Sous terre
<u>Contexte géologique</u>	- craie bleue imperméa- ble. - nombreuses failles aquifères. - passage local dans la craie grise aquifère.	- craie grise et blan- che très aquifère. - bancs de silex.
<u>Impératif de planning</u>	- sur le chemin criti- que. - cadence élevée : 500 m/mois.	- non critique. - cadence moyenne : 330 m/mois.
<u>Choix techniques</u>	- tunneliers étanches évitant les traite- ments de terrain. - tunneliers mixtes pouvant progresser à cadence élevée en terrain sec.	- tunneliers étanches évitant les traite- ments de terrain. - tunnelier adapté au terrain fracturé et à la présence de silex.
<u>Tunneliers retenus</u>	- tête télescopique permettant la fora- tion en mode ouvert par grippage sur le terrain. - foration en mode fermé avec poussée arrière sur le revê- tement. - bouclier articulé.	- progression par pous- sée arrière sur le revêtement. - bouclier monolithique
<u>Constructeur choisi</u>	- ROBBINS/KAWASAKI.	- MITSUBISHI.
<u>Vitesse de foration</u>	- ouvert 12 cm/mn - fermé 9 cm/mn.	- 8 cm/mn.
<u>Cadence horaire</u>	- ouvert 4,4 m/h. - fermé 2 à 3 m/h.	- tunnel de service : 3,5 m/h. - tunnel ferroviaire : 3 m/h.
<u>Cycle de pose des voussoirs</u>	- 15 mn = 2 érecteurs.	- 24 mn = 1 érecteur avec 2 bras.

- forte tectonisation avec présence de nombreuses failles, d'où traversée de zones aquifères dans la partie sous-marine du tracé localisé dans la craie bleue.
- accroissement du pendage descendant général des couches vers le Nord et réduction de l'épaisseur de craie bleue en s'approchant des cotes françaises.
- traversée de couches perméables à proximité de la cote et dans la partie sous-terrestre du tracé.

LES CHOIX TECHNIQUES

9. Compte-tenu des aléas géologiques et des cadences élevées requises par le programme général - creuser 16 km en 30 mois, soit 500 mètres par mois en moyenne - les options techniques suivantes ont été prises côté français :

- Creuser les tunnels à partir d'une attaque intermédiaire unique située à 300 mètres de la mer à Sangatte, ce qui permet de concentrer le chantier en un seul site et de bien maitriser sa logistique.
- Assurer la fiabilité et la sécurité de cette attaque en réalisant un puits de 55 mètres de diamètre et de 65 mètres de profondeur, creusé à l'intérieur d'un écran étanche en bentonite-ciment pour assurer la mise hors d'eau des travaux de démarrage ; ce puits permet l'installation de moyens de manutention industriels automatisés.
- Choisir des tunneliers étanches à confinement capables de progresser dans des terrains de mauvaise qualité ou aquifères sans nécessité de traitement de terrain à l'avancement.
- Avoir, côté mer, des tunneliers "mixtes", capables de forer aussi en mode ouvert à grande cadence dans les terrains secs et de bonne qualité que constitue la craie bleue en dehors des zones de failles.

LES TUNNELS

10. Le creusement des 6 attaques depuis le puits de Sangatte est réalisé par 5 tunneliers :

- Côté mer : 3 tunneliers devant creuser chacun 15 800 m jusqu'à la jonction sous la Manche avec les tunneliers britanniques.
- Côté terre : 2 tunneliers, le premier devant creuser les 3 200 m du tunnel de service jusqu'au portail, le second creusant successivement les 2 tunnels ferroviaires soit 6 500 m.

11. Le tunnel de service sert de galerie pilote ; une reconnaissance du terrain à l'avancement par forage destructif avec enregistrement de paramètres est faite depuis le tunnelier, lorsque les conditions géologiques

l'exigent ; des reconnaissances sont réalisées en arrière du tunnelier pour reconnaître le niveau de l'argile du Gault sous-jacent ainsi que celui du toit de la craie bleue.

12. Si nécessaire, en cas de problèmes géologiques ou mécaniques graves, il est possible d'injecter tous les tunnels à l'avancement ; de plus, il est aussi possible d'injecter les tunnels ferroviaires depuis le tunnel de service en utilisant les équipements de forage et d'injection prévus pour l'exécution des galeries transversales.

13. Le revêtement du tunnel est constitué par des voussoirs préfabriqués en béton armé assemblés par anneaux successifs de 1,4 m de long (tunnel de service) et de 0,32 m d'épaisseur ou 1,6 m (tunnels ferroviaires) et de 0,4 m d'épaisseur ; ces éléments sont équipés de joints périphériques néoprènes à proximité de l'extrados qui assurent l'étanchéité ; l'étanchéité entre le tunnelier et le revêtement est réalisé par plusieurs rangées de brosses métalliques appuyées sur l'extrados des anneaux entre lesquelles est injectée de la graisse sous pression.

14. Après mise en place dans la jupe du tunnelier, les anneaux sont maintenus par la poussée des vérins arrières qui assurent aussi la compression des joints d'étanchéité. Le boulonnage des anneaux entre eux assure le maintien du serrage de ces joints en cas de suppression momentanée de la poussée.

15. Le vide annulaire entre le revêtement et le terrain est rempli à l'avancement par un mortier de bourrage assurant le calage de l'anneau dans son trou ; ce mortier est injecté directement à l'arrière du tunnelier depuis des pipes installées dans la jupe.

LA FORATION

16. Les tunneliers côté français sont des tunneliers étanches à bouclier pleine face dimensionnés pour assurer des cadences d'avancement instantanées élevées :

- 4,4 m/h en mode ouvert côté mer avec une cadence minimum de 2 m/h en mode fermé sous pression hydrostatique maximum (11 bars).
- 3,5 m/h (tunnel de service) et 3 m/h (tunnels ferroviaires) côté terre pour une pression hydrostatique maximum de 3 bars au droit du puits d'attaque de Sangatte.

17. Le creusement est réalisé par une roue de coupe pleine section mise en rotation par des moteurs électriques; les 8 bras de la roue de coupe sont équipés d'une part de molettes centrales à simple disque qui appuient et découpent le front, et d'autre part de pics sur les bords des bras qui raclent le terrain et ramènent les déblais dans la chambre de coupe. Les tunneliers côté terre ont été équipés d'un

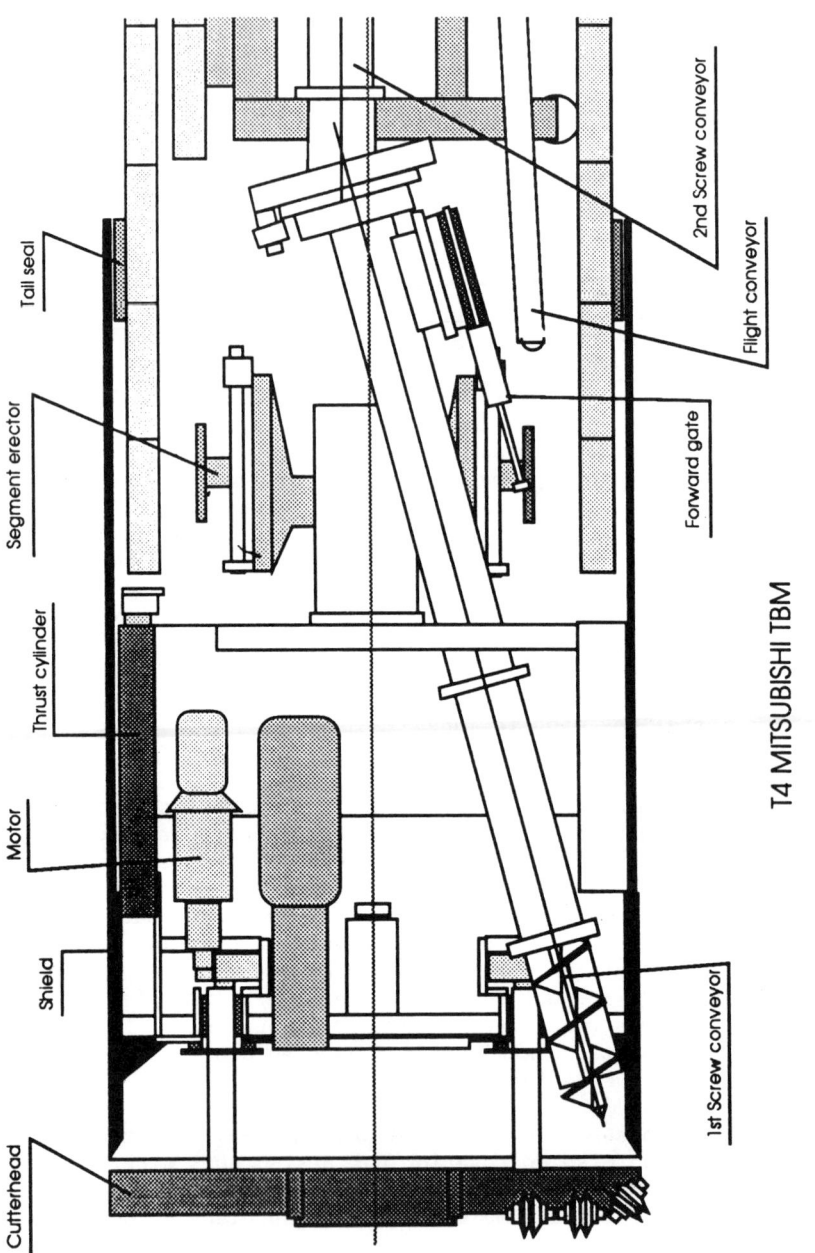

T4 MITSUBISHI TBM
Service tunnel land

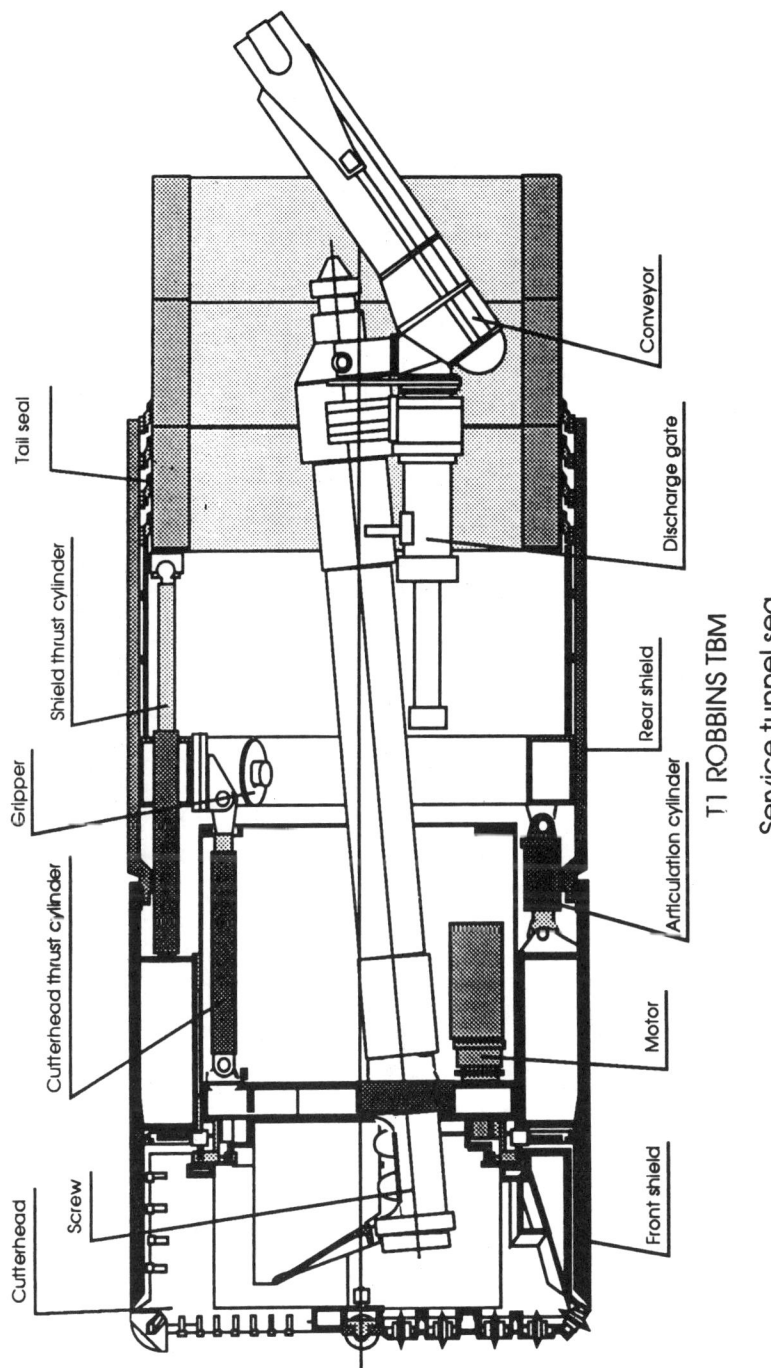

Tail seal

Shield thrust cylinder

Gripper

Cutterhead thrust cylinder

Cutterhead

Screw

Conveyor

Discharge gate

Rear shield

Articulation cylinder

Motor

Front shield

T1 ROBBINS TBM

Service tunnel sea

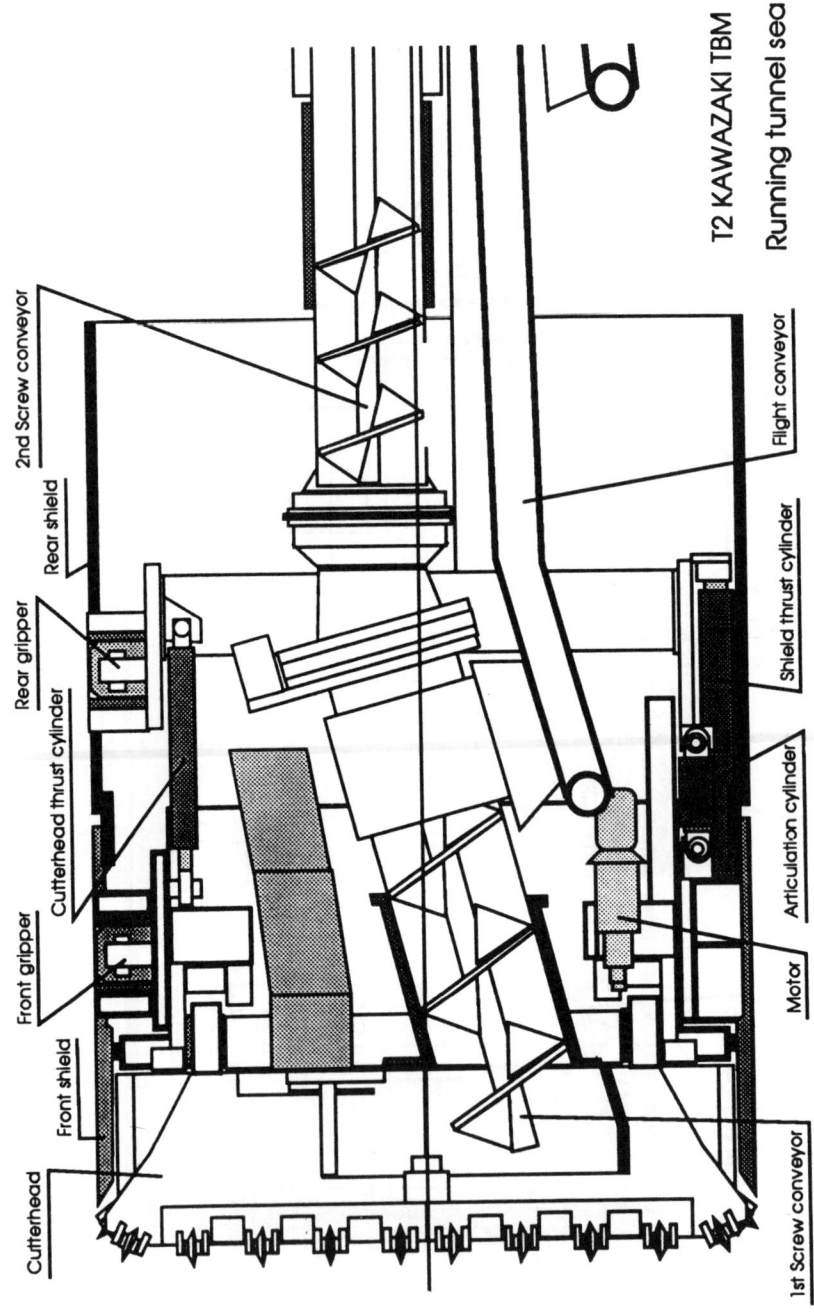

T2 KAWAZAKI TBM
Running tunnel sea

cône central facilitant la dislocation des bouchons de craie dans les zones humides.

18. Côté terre, la roue de coupe est portée par un bouclier monolithique assurant l'étanchéité ; ce bouclier est prolongé par une jupe arrière cylindrique à l'intérieur de laquelle sont montés les anneaux préfabriqués du revêtement. L'avancement de l'ensemble : roue de coupe - bouclier - jupe pendant le creusement est assurée par la poussée de vérins hydrauliques sur le revêtement déjà posé à l'intérieur de la jupe.

19. Côté mer, la roue de coupe est portée par une tête télescopique guidée par la partie avant du bouclier et prenant appui par des vérins hydrauliques sur la partie arrière du bouclier ; les 2 parties du bouclier sont articulées entre elles ; le bouclier arrière est équipé de vérins de grippage latéral sur le terrain et est prolongé par une jupe cylindrique à l'intérieur de laquelle sont montés les anneaux du revêtement. Le creusement se fait selon 2 modes:

- En mode fermé, en présence de pression hydrostatique ou de mauvais terrain, le tunnelier avance d'un seul bloc en prenant appui par des vérins hydrauliques sur le revêtement déjà posé dans la jupe. Ces vérins de poussée arrière sont fixés sur le bouclier avant, sur lequel la tête télescopique vient en butée mécanique.
- En mode ouvert, (bon terrain et pression hydrostatique faible ou nulle), le tunnelier prend appui sur le terrain par ses grippeurs arrières et le creusement se fait par avancement de la tête télescopique seule ; la moitié des vérins de poussée arrière reste appuyée sur le revêtement pour assurer la stabilité des anneaux posés. A la fin de chaque cycle de poussée, les grippeurs sont rétractés et le bouclier est avancé en prenant appui sur le revêtement. La pose du revêtement se fait simultanément au creusement.

L'EXTRACTION DES DEBLAIS

20. Les déblais sont évacués de la tête de coupe par une vis d'extraction équipée d'un dispositif permettant d'abaisser la pression pour permettre leur transfert de la tête sous une pression pouvant atteindre 11 bars jusqu'au convoyeur, à la pression atmosphérique.

21. Côté terre où les tunneliers sont toujours dans des terrains aquifères, le déblai pâteux ou liquide est récupéré dans le fond de la chambre de coupe par une vis à extrémité libre ; l'abaissement de pression est assuré par 2 dispositifs installés sur les 2 vis : le "casing rotator", partie de l'enveloppe de la vis pouvant tourner pour **accroître la perte de charge** ; et le contrôle de l'ouverture

TABLEAU 2

CARACTERISTIQUES DES TUNNELIERS COTE FRANCE

	T1	T2/T3	T4	T5
BOUCLIER				
Diam. (m)	5,74	8,72	5,59	8,62
Longueur (m)	11,00	13,75	10,54	11,86
Poids (t)	470	1250	350	660
MODE OUVERT				
Poussée sur la tête (t)	1 800	2 000	NON	NON
Nb et poussée vérins	7 x 258 t	12 x 166 t	/	/
MODE FERME				
Poussée arrière (t)	4 000	11 500	4 000	9 000
Nb et poussée vérins	20 x 200 t	26 x 442 t	20 x 200 t	30 x 300 t
ROUE DE COUPE				
Couple maximum (tm)	360	1 300	400	1 300
Vitesse rotation (t/mn)	2,4/4,8	1,5/3	0,9/1,8	1/2
Nb et puissance moteur	7 x 126 kw	12 x 180 kw	10 x 75 kw	16 x 90 kw
GRIPPEURS				
Nb	4	6 AV + 4 AR	NON	NON
Poussée en t	4 x 112	6x300+4x500	/	/
PUISSANCE INSTALLEE				
(Transfo. MT/BT)	2 700 kva	6 400 kva	2 500 kva	5 000 kva
CARACTERISTIQUES CREUSEMENT				
Diam. foration (m)	5,77	8,78	5,61	8,64
Diam. bouclier (m)	5,72	8,72	5,59	8,62
Vide annulaire (cm)	16,00	19,00	8,5	12,00
Ep. voussoir (cm)	32,00	40,00	32,00	40,00
Long. voussoir (m)	1,40	1,60	1,40	1,60
Diam. int. tunnel (m)	4,80	7,60	4,80	7,60
CONSTRUCTEUR	ROBBINS	ROBBINS KAWASAKI	MITSUBISHI	MITSUBISHI
TUNNEL	TS MER	TF MER	TS TERRE	TF TERRE

de la trappe de sortie arrière de la vis sur le tapis
convoyeur. De plus, pour le tunnelier du tunnel de service,
une deuxième vis a été rajoutée en série avec la première.
    22. Côté mer, la majeure partie du tracé est située dans
des terrains secs. Les déblais sont remontés par des godets
situés à la périphérie de la roue de coupe pour retomber
dans la trémie d'entrée de la vis située juste sous l'axe du
tunnelier. Deux dispositifs distincts ont été installés pour
abaisser la pression :

-   Pour le tunnel de service (ROBBINS - KOMATSU), deux
    pistons de décharge installés à la sortie de la vis
    constituent chacun un sas dont la marche alternative
    permet d'assurer un débit semi-continu ; le recul du
    cylindre du piston avec la porte de sortie fermée
    permet l'admission des matériaux depuis la vis par
    gravité. Après l'avancement du cylindre isolant le
    piston de la vis, la guillotine de la porte s'ouvre
    et le piston pousse les déblais à l'extérieur du
    cylindre sur le tapis ; après fermeture de la
    guillotine, le cycle recommence.
-   Pour les tunnels ferroviaires, KAWASAKI a installé un
    système de deux vis en série ; la diminution de la
    vitesse de rotation de la deuxième vis et son recul
    par rapport à la première permettent la formation
    d'un bouchon dans l'espace ménagé entre les deux vis
    ; lorsque le bouchon est créé, la deuxième vis est
    réavancée en accroissant sa vitesse, ce qui permet
    d'extraire le matériau au fur et à mesure de son
    arrivée dans la zone du bouchon ; les déblais sont
    recueillis sur un tapis à la sortie de la 2ème vis ;
    lorsqu'il n'y a pas ou peu de pression dans la
    chambre, les déblais sont extraits directement à
    l'extrémité de la 1ère vis.

    23. Les déblais sont évacués par tapis convoyeur
jusqu'aux berlines des trains de travaux stationnés sous les
remorques du train suiveur du tunnelier.

LA POSE DU REVETEMENT
    24. Chaque anneau est composé de 5 voussoirs et d'une clé
mise en place par l'avant ; les voussoirs sont amenés au
tunnelier par palettes d'un anneau, déchargés sous une
remorque du train suiveur et transportés à front l'un
derrière l'autre sur un convoyeur en voûte (au-dessus du
convoyeur à déblais).
    25. L'ensemble des opérations de levage est réalisé par
un système de ventouse à vide :

-   Prise sur le wagon porte-palettes et dépose sur le
    convoyeur.

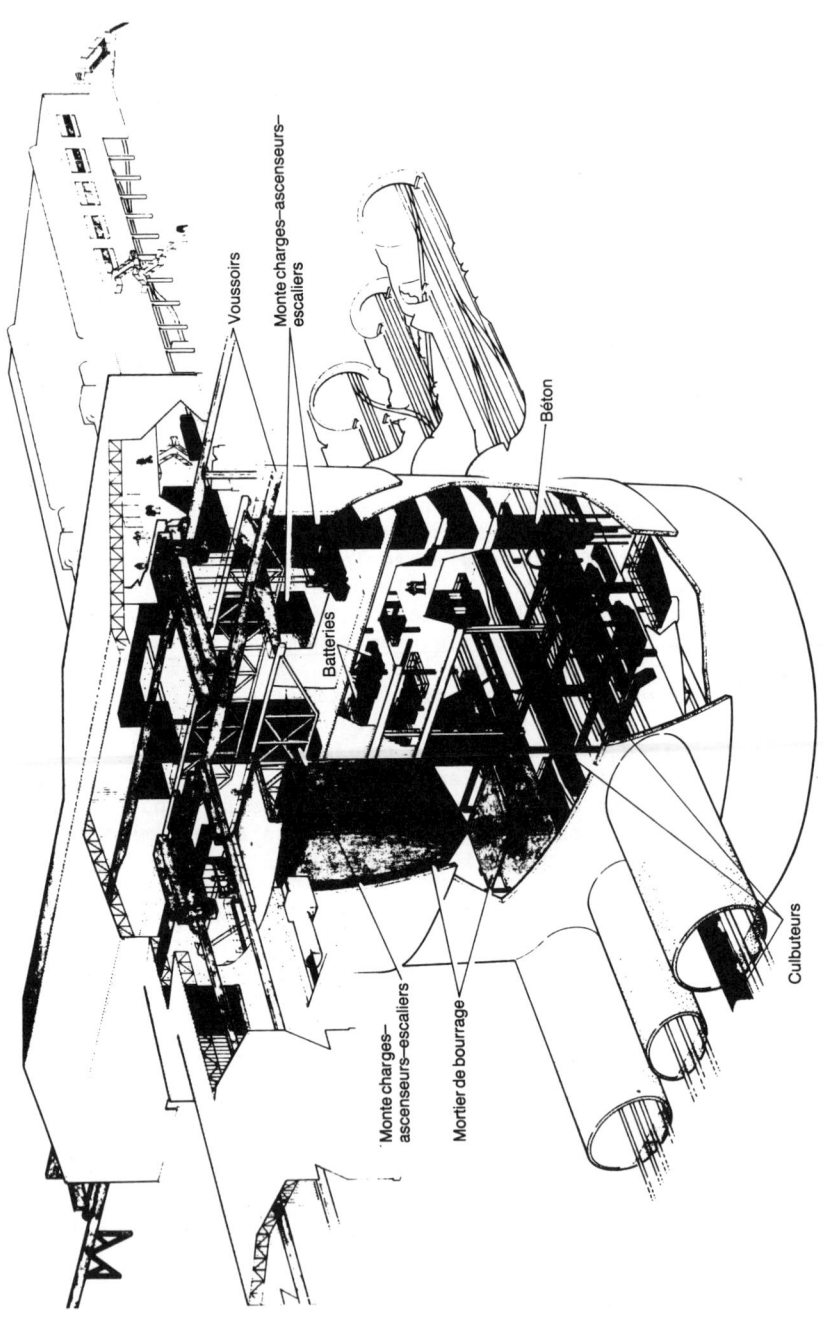

Voussoirs

Monte charges—ascenseurs—
escaliers

Béton

Batteries

Monte charges—
ascenseurs—escaliers

Mortier de bourrage

Culbuteurs

- Prise sur le convoyeur par l'extrados et dépose de l'intrados sur l'érecteur.
- Mise en position et assemblage aux voussoirs précédents par l'érecteur.

26. Après serrage des boulons de sécurité, les vérins de poussée arrière sont remis en appui sur le voussoir pour assurer la stabilité de l'anneau.

27. Côté terre, le nombre et l'extension des vérins permettent éventuellement de poser les voussoirs pendant la foration ; les tunneliers sont équipés d'un érecteur double; le temps de pose théorique d'un anneau est de 24 mn.

28. Côté mer, il est possible en mode ouvert de poser les voussoirs pendant la foration, le tunnelier s'appuyant sur le terrain par l'intermédiaire des grippeurs latéraux ; en mode fermé, la foration et la pose ne peuvent être simultanés ; les tunneliers sont équipés de deux érecteurs indépendants ; le temps de pose théorique d'un anneau est de 15 mn.

TOPOGRAPHIE ET GUIDAGE

29. Les problèmes posés par le creusement du Tunnel sous la Manche sont de deux sortes :

- Le problème topographique de la jonction des tunnels français et britanniques sous le détroit du Pas-de-Calais du fait de la très grande longeur de creusement sans attaque intermédiaire depuis la cote - 16 km côté français et 21 km côté anglais -.
- Le problème du guidage des tunneliers est de suivre le tracé imposé en respectant le gabarit de circulation et les critères de rayon et de pente ; ce problème est particulièrement critique dans le cas des tunnels ferroviaires où circuleront les trains de British Rail et de la SNCF, le futur train à grande vitesse, et les navettes EUROTUNNEL de transfert des voitures et camions ; compte-tenu des tolérances de fabrication et de pose des voussoirs, des déformations du revêtement, la tolérance de $\pm$ 15 cm sur le gabarit conduit à avoir une précision de pilotage du tunnelier de $\pm$ 5 cm.

30. La précision requise de quelques décimètres au percement du tunnel de service a nécéssité dans un premier temps le raccordement précis des réseaux géodésiques de base français et anglais ; un système de coordonnées spécifique a été créé et son imprécision a été réduite aux environs du décimètre par l'utilisation du réseau NAVSTAR de positionnement par satellite. De plus, afin de réduire la dispersion liée à la longueur du cheminement sans recoupement en tunnel, l'utilisation de théodolite gyroscopique permet de ramener l'imprécision au percement à environ 3 dm

par la référence de chaque mesure au nord géographique.

31. Avant percement du tunnel de service, des forages de contrôle permettront de rectifier la trajectoire du tunnelier sur la dernière centaine de mètres en respectant les critères de tracé du véhicule de service (moins sévère que pour les tunnels ferroviaires). Le percement des tunnels ferroviaires ne posera pas de problèmes spéciaux, la chaîne de mesure topographique ayant été refermée par le tunnel de service.

32. Le guidage du tunnelier est lié à la conception du revêtement préfabriqué :

- Celui-ci est constitué d'anneaux "pincés" plus courts à gauche (anneaux "rouges") ou à droite (anneaux "bleus") ; le pincement permet au revêtement de suivre les courbes du tracé, la position du pincement pouvant être déplacée vers le haut ou vers le bas par rotation de la position de la clé de l'anneau (4 positions possibles pour le TS et 6 positions pour les TF).
- Le guidage des tunneliers côté mer est réalisé à l'aide d'un système type ZED qui, à l'aide d'une cible laser et d'un programme de calcul, donne en temps réèl la position du tunnelier par rapport à sa trajectoire théorique, et la projection de cette position sur l'avancement futur.
- A partir de la position réèlle du tunnelier par rapport à la trajectoire théorique, un programme spécifique détermine la trajectoire de correction à suivre en prenant en compte les caractéristiques géométriques du revêtement (anneaux rouges, bleus, position imposée de la clé) et les diverses contraintes (anneaux déjà approvisionnés, anneaux spéciaux) ; ce programme a pour objectif de centrer au mieux l'anneau du revêtement dans la jupe du tunnelier pour éviter la détérioration du joint d'étanchéité arrière de la jupe ainsi que tout effort parasite entre la jupe et le revêtement.

L'ETAT DU CHANTIER FIN MAI 1989

33. Les installations de surface et les équipements du puits seront opérationnels à 100 % en juillet ; il reste actuellement à terminer l'installation du Tunnel Ferroviaire Sud et les équipements définitifs d'alimentation en produits d'injection.

34. La mise en place des équipements de construction en tunnel a débuté dans le tunnel de service : installation de pompage, signalisation ferroviaire, portiques de reconnaissance et d'injection de terrain au droit des rameaux de liaison entre tunnels.

35. Les tunneliers ont tous été mis en service et le premier tunnel a été percé côté terre en avril :

- Le tunnelier T4 (tunnel de service terre) a réalisé le dernier mois avant le percement un avancement record de 887 m après avoir démontré sa capacité à réaliser en toute sécurité des cadences de creusement satisfaisantes en zone aquifère sous une pression de 3 bars.
- Le tunnelier T1 (tunnel de service mer) après une mise au point délicate dans des conditions géologiques difficiles est monté en cadence depuis septembre 88 pour atteindre un avancement moyen supérieur à 500 m/mois sur les 3 derniers mois.
- Le démarrage des tunneliers T2 et T3 (tunnels ferroviaires sous mer) a confirmé les difficultés constatées sur T1 et T4 à réaliser un confinement satisfaisant de la craie grise et une extraction pâteuse des déblais sans bloquer la roue de coupe. Cependant, du fait de l'expérience acquise dans le tunnel de service des avancements satisfaisants ont pu être obtenus : 350 m/mois côté Nord (T2) le 6ème mois de creusement et 200 m/mois dès le 2ème mois côté Sud (T3).

36. En conclusion, l'expérience d'un peu plus d'une année et de 8 km de creusement a montré la difficulté à maîtriser le comportement de la craie aussi bien à l'abattage qu'à l'extraction :

- ce comportement est très différent selon la teneur en argile, la teneur en eau, et l'activité hydraulique de la fracturation.
- de nombreuses adaptations ont du être faites aussi bien dans la disposition des outils de la tête de coupe que dans le mode opératoire du creusement.
- d'autres adaptations seront encore nécessaires, en particulier si les terrains rencontrés sont suffisamment secs pour permettre une extraction "au rocher" analogue à ce qui est actuellement possible dans la craie bleue du côté anglais.

RESUME

1.   On the French side the works include the boring of 3 tunnels each 19 km long, of which approximately 16 kms are under the sea to meet the the UK Works.

2.   The tunnels are driven in chalk by 5 tunnel boring machines from an intermediate outwards attack point at Sangatte near the coast :

- on the land side, through layers of permeable grey, then white chalk;
- on the sea side, mainly through blue chalk which is relatively watertight, but containing watery ground in areas of faults or grey chalk.

3.   The initial choice of full face watertight TBM's with shield led to the selection of two designs :

- On the land side, two MITSUBISHI TBM's with a single shield, Archimedes screw containment and "casing rotator".
- On the sea side, three TBM's with articulated shield which can function in closed mode in watery zones or in open mode at high speed in zones of dry blue chalk; containment is carried out at the end of the spoil extraction screw either through a dual discharge pump system for the ROBBINS TBM (service tunnel) or through a second screw which creates a plug in the case of the two ROBBINS-KAWASAKI TBM's (running tunnels).

4.   On the 1st June 1989, surface installations and shaft equipment were in operation, the five TBM's were commissioned and the underground service tunnel has been underway since April; more than 6 km of the service tunnel and 2 km of the running tunnels have been bored, the objective being to bore 15 km of tunnel within a year.

5.   The first two TBM's have proved their ability to achieve the prescribed rates in complete safety, under pressure of up to 3 or 4 bars :

- TBM T4 (underland service tunnel) made record progress of 887 metres during the month preceding breakthrough.
- For the past three months the TBM T1 (undersea service tunnel) has actually been progressing in open mode at an average speed of over 500 metres per month.

6.   The two ROBBINS-KAWASAKI TBM's are currently progressing at a satisfactory speed despite difficult geological conditions at the outset: 350 metres/month in the North running tunnel and 200 metres/month towards the south one month after start up.

14A.   UK terminal construction

J. S. RUSHTON, BEng, FIHT, Constructipon Manager Terminals UK

SYNOPSIS. This section of the paper describes the basic
location, geography and geology of the UK Terminal Area, and
gives brief details of the work in progress and methods of
construction.

THE UK TERMINAL CONSTRUCTION SITE

General
1. The site for the UK Terminal at Folkestone was
identified during the last Channel Tunnel attempt which was
halted in 1975. The land has been reserved in planning terms
since that date, and therefore has remained mainly
agricultural. Despite the proximity to Folkestone only 19
properties have had to be demolished for the Works. Most of
these properties were in a ribbon development along Danton
Lane, a small country lane that crossed the site.

The Environment
2. The site is in a prominent position alongside the
motorway and is visible from large areas of Folkestone, parts
of the M20 motorway, and the road and public footpath along
the top of the North Downs. Great efforts are being made
throughout the design and construction phases to reduce the
visual impact of the works, and blend in where possible with
the dramatic backdrop of the escarpment to the North Downs.

During the Parliamentary Select Committee stages of the
Channel Tunnel Bill a lot of time and debate was devoted to
full consideration of the environmental problems associated
with the scheme. Eurotunnel gave assurances on many of the
points raised by objectors which required changes to the
design. Three of the major changes deserve comment.

(a) The line of the access and egress roads was changed
from the proposed North West approach across open farmland,
to the Southern Access entering the Terminal on a line
parallel to the railway. This entails constructing the works
in a very restricted area between the A20 and M20, additional
bridges, and a change to the earthworks balance which

requires a much larger amount of extra imported bulk fill.

(b) The surface water from the Terminal was originally to be discharged via a pipeline from the west end of the stabling siding, under the M20 motorway and railway to a balancing lagoon to be constructed on the Seabrook Stream.

Detailed ecological surveys in the area during the early stages of the Select Committee hearings established that the proposed lagoon area has a uniquely high number of species of crane fly. After representations by nature conservation groups, Eurotunnel agreed to finance a new tunnel outfall beneath Folkestone to the sea, to be designed and constructed by the Southern Water Authority.

The head manhole for the new outfall is at the east end of the terminal. The direction of the main drain in the terminal has been reversed to connect into it. This requires a tunnel 1.8m diameter, 1.2km long at depths of up to 18m below finished ground levels for the main run.

(c) When the original scheme was presented to Parliament some 700,000 cubic metres of imported bulk fill were required, which was anticipated to be minestone imported from the East Kent Coalfield. Several pressure groups, formed by residents living near the collieries, put up a vigorous fight during the Select Committee proceedings against the transport of minestone by road. In response, Eurotunnel gave an assurance that minestone would only be moved by rail to the site. After further discussion on the overall problem of transport of all bulk materials for the construction of the terminal, the Committee agreed to accept a railhead constructed off-site, on the proviso that the route for transport from the railhead would be along the M20. Eventually a site already owned by Eurotunnel at Sevington just east of Ashford was identified, and a railhead was constructed which came into operation in June 1988. This facility now handles all the bulk fill and aggregate requirements for the terminal.

Geology
3. The terminal is on the edge of the Weald of Kent formed by the erosion of the South East England anticline. The chalk strata which form the North Downs escarpment, and the underlying chalk marl have been eroded away leaving a tapering layer of gault clay up to 20m thick over the eastern end of the site. The underlying Folkestone sand beds outcrop over the west end of the terminal and some of the area of the access roads. Over these base layers there are areas covered with a variable depth of superficial deposits, which are either recent hillwash deposits or older soliflucted material. There are also four ancient landslips slumped against the escarpment which extend into the site.

The landslips have had a major influence on the design of the earthworks for the Terminal. A comprehensive geotechnical survey of the area has confirmed that the landslips, though stable under existing conditions have low factors of stability. The landslips are large and therefore the economic solution is to avoid excavation wherever possible, and build up the required flat plateau for the shuttle operations by importing fill, rather than carry out a sidelong cut to fill operation, as this would require the construction of massive retaining walls along the north edge of the site.

BASIC DESCRIPTION OF THE WORKS

4. The area of the whole site including access roads, egress roads, and the Dollands Moor siding for British Rail, is some 140 hectares. The main terminal area is 2½km long, 800m wide at the west end tapering to 150m wide at the east end, and lies between the M20 motorway and the North Downs escarpment. The villages of Newington and Peene form the western boundary of the Terminal complex, and the tunnel portal at Castle Hill marks the eastern limit.

The Folkestone terminal Sub-Project splits conveniently into three main areas for the purpose of this discussion.

(a) The Southern access
(b) The Western Third of the Terminal
(c) The Eastern Two-Thirds of the Terminal

Southern Access
5. The access and egress roads and rail tracks outside the main body of the Terminal consist of: 8 structures, 4km of mainly three lane carriageway plus a hardshoulder, and 13km of railtrack, with all the necessary drainage, fencing, earthworks, landscaping etc together with the formation drainage and ballast for the new Dollands Moor siding for British Rail freight use. British Rail will complete the construction of the sidings from top of ballast using their own resources.

Western Third of the Terminal
6. The western third of the terminal contains the main building complex and is split by the embankment for the continental main line (CML) south of which is the large cutting for the shuttle stabling area and the arrival loop lines. These loop lines run in the cut and cover tunnel under the CML then re-emerge in the approach fan to the platforms. On the large area contained within the Loop will be built the administration, amenity and control buildings surrounded by the border control, toll and parking areas forming the main complex.

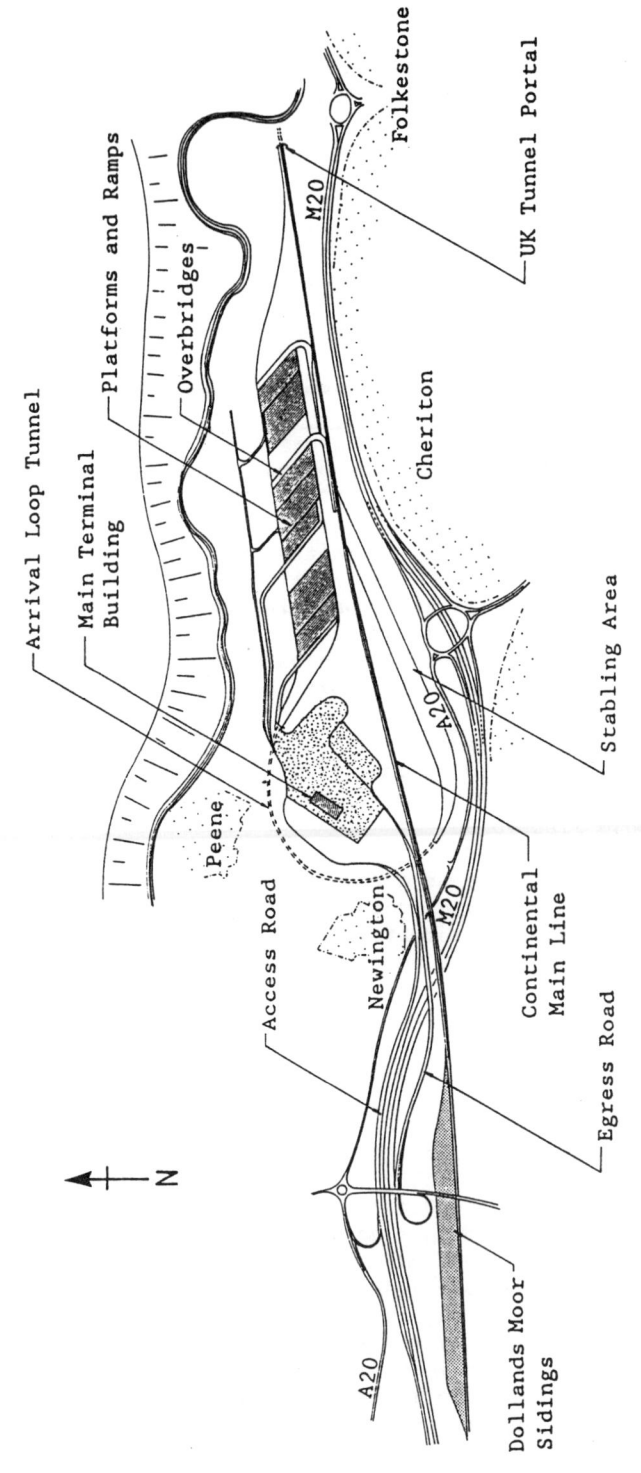

Fig 1. Folkestone Terminal - Outline Plan

Eastern Two-Thirds of the Terminal

7. The remainder of the terminal, is dominated by the embankment for the three tracks of the CML along its southern boundary, and the fill behind it to support the platforms. The existing ground level falls along the length from west to east, but the tracks rise in level to reach the Castle Hill tunnel portal. Therefore the embankment increases rapidly in height to a maximum of 14m above the level of the adjacent M20 motorway.

On the north side of the CML there will be four large overbridges feeding the shuttle loading and unloading platforms. This platform area can be described as a displaced parallelogram with sides 1000m by 300m, and will be virtually flat. The existing ground profile slopes from north to south so large fill, triangular in cross section, is required to achieve the design concept of minimum excavation along the north boundary of the site in the ancient landslips.

Fig 2. Aerial View of Southern Access Bridges over M20.
        Looking West, Feb '89

Fig 3.   Aerial view of Terminal
          Looking East, Feb '89

Fig 4. M20 Bridges Steelwork in Progress

Fig 5. Movable sofit shutters under erection for loop tunnel.

Fig 6. Loop tunnel centre section

CONSTRUCTION PROGRAMME

8. The construction programme for the terminal, which drives the design programme, connects into the critical path for the total project by three main activities.

(a) The need to provide early road and rail access for the supply of material to the fixed equipment fit-out operations in the tunnels.

(b) The construction of the control building in the West terminal area to allow early commissioning of the computer systems.

(c) The supply of power via the main substation to the test track for rolling stock, associated with the completion of the land drive north running tunnel.

TML are contracted to design and build a complete working transport system by June 1993 and to achieve this, the main thrust of the civil engineering work and a substantial amount of the building work for the control and computer systems, must be complete by late 1991.

GENERAL CONSTRUCTION PROGRESS

9. Possession of the majority of the terminal site was given in January 1988 and work began in earnest in March. The initial works were hampered by very wet conditions which caused delays, particularly on the areas of gault clay and head deposits, which became very sticky when wet. Progress improved rapidly from mid summer and continued through one of the driest winters for twenty years.

Southern Access Earthworks

10. The dry weather allowed the earthworks to the Southern Access roads and Dollands Moor to continue throughout the winter, and they are now virtually complete apart from two small cuts which support the existing B2065. Roadworks operations including drainage and sub-base, are now in hand.

Southern Access Substructures

11. Substructures for the Southern Access bridgeworks are also progressing well and steelwork erection is underway. The bridges over the motorway have all been designed with composite decks of in situ concrete, cast on precast concrete slabs used as permanent formwork supported on steel plate girders. This form of construction is particularly useful on the railway bridge and road bridge, which cross the motorway on a heavy skew, where the traffic flow has to be maintained during construction. Seven traffic management phases to allow traffic to flow smoothly are required on the motorway over the fifteen month construction period. The last phase is required to permit the demolition of the existing B2065 bridge over the motorway. This structure, which was opened in 1979, probably has the distinction of being the shortest lived permanent motorway bridge in the UK.

The three bridges carrying the access and egress roads
together with the railway tracks over the diverted A20 into
the terminal, are also constructed in steel with in situ
concrete decks. They are however, three span steel portal
frames with inclined legs, to give the required sight
distances for motorists passing below, whereas the bridges
over the M20 are supported on multiple vertical concrete
columns 1.2m diameter.

Terminal Construction

12.   Construction work inside the terminal has concentrated
on two main activities, earthworks and buried structures.
The most interesting earthworks operations have been those to
supply the imported bulk fill requirements. As mentioned
previously under paragraph 2(c), when initially conceived the
earthworks required 700,000 cubic metres of imported bulk
fill. The change from the north west access to the southern
access considerably reduced the amount of cut available to be
used as fill. Further changes to layouts to provide extra
flexibility in the control and platform zones, and the
decision to extend the service tunnel into the terminal,
increased the amount of fill required. The total imported
bulk fill requirement is now some 2,000,000 cubic metres.
The initaial fill operation began in June 1988 using
minestone from the East Kent coalfield when the Sevington
railhead was commissioned.

Sevington Railhead

13.   The railhead which was built between November 1987 and
May 1988 handles concrete aggregates and dry stone supplies
from existing commercial sources, as well as minestone. The
minestone is loaded at Snowdown colliery, East Kent, onto
railwagons standing in a siding rebuilt for the operation.
It is then hauled in a train of eighteen 88 ton gross box
wagons to Sevington and unloaded by two backhoes fitted with
1,000 litre clam shell grabs. A train carries a payload of
1,200 tonnes and makes two trips a day. The material is
then offloaded into lorries at Sevington and transported by
road to the terminal site down the M20. Its journey by train
and lorry is some 65km. Snowdown colliery is only 16km away
from the UK terminal by road using the A2 and A260.

Marine Sand Fill

14.   Minestone deliveries from the date of the application
for Planning Permission for the railhead took about 8 months
to bring into commission. An alternative proposal to suppply
marine sand as bulk fill, was pursued in parallel to the
minestone option. The process of obtaining the necessary
permits from the authorities, to dredge sand and pump it into
the site via a steel pipeline laid through twenty three
separate landowners' property, took 10 months to complete and
a further 4 months to bring into operation.

Under this scheme, sand is dredged from the North Goodwins in the English Channel by a trailing suction hopper dredger, then transported 60km in 3,500 cubic metre loads to a second dredger moored 700m off the beach at Hythe which is used as a booster pump. It is then pumped as a 1:4 mixture with sea water through a 800mm diameter steel pipeline 5.3km across country, including crossings of 4 local roads, a railway, the M 20 and the A20, against a head of 65m, to two large reception lagoons built on the gault clay area of the site. The sea water is drained off into a third lagoon and then pumped back to sea through the delivery pipeline at the end of the delivery cycle. When weather permits steady operation, a dredger load of sand is delivered every 7 hours, giving a maximum possible production rate of 70,000 cubic metres per week.

The sand is removed from the reception lagoons by a fleet of 10 N° motorscrapers and taken to the fill area where it is compacted by self propelled vibrating rollers. This fill operation operated throughout the 1988-1989 winter and will be completed in June 1989. The minestone operations were stopped in November 1988 due to weather problems, the material quickly becomes unsuitable when wet, a problem which does not affect the sand.

Loop Tunnel

14. The loop tunnel for the shuttle trains is the major buried structure under way in the west terminal area. The tunnel, which takes three tracks, is 1,060m long and is designed in three distinct sections. The central section, which is founded on the Folkestone sand beds is a reinforced concrete portal on spread footings. The south section is similar but is a double portal, the extra span carrying the site service road under the CML. The north section is being constructed as a full box because it is within the gault clay strata and at its north end is within the area of the Danton Pinch landslip, which imposes large lateral loads on the structure.

15. Good progress has been made on the early works for the terminal, but the pace of construction has to increase in 1989 and 1990 to ensure that hand over dates for the start of the Fixed Equipment installation in 1991 are met.

RESUME

UK TERMINALS

1.   Ce chapitre fournit les données de base sur la
implantation, la géographie et la géologie du terminal UK
ainsi que de quelques éléments sur les travaux en cours et
les méthodes de construction.

2.   Les travaux représentent une surface totale de 140
hectares y compris les routes d'accès et les lignes de
chemin de fer. La partie principale du Terminal s'étend de
l'autoroute M 20 au pied de l'escarpement du North Downs. Ce
site est bien en vue et tous les efforts ont été faits afin
d'essayer d'améliorer l'impact visuel des travaux.

3.   En ce qui concerne la protection de l'environnement,
Eurotunnel a effectué trois modifications majeures aux
propositions lors du passage du "Channel Tunnel Act" au
Parlement.

(a)   Le tracé des routes d'accès et de sortie du Terminal
      a été modifié de manière considérable en augmentant
      la quantité de remblais nécessaire à ces travaux et
      en augmentant le nombre de ponts.

(b)   La décharge du collecteur principal du terminal a été
      modifiée. Elle ne se fera plus dans un bassin se
      déversant dans un cours d'eau local mais dans un
      nouveau tunnel passant sous Folkestone. Ce tunnel
      sera réalisé par le "Southern Water Authority" mais
      financé par Eurotunnel.

(c)   Le transport jusqu'au Terminal des remblais provenant
      du terril d'une vieille mine de charbon du Kent ne se
      fera plus par route mais par voie ferrée, ce qui
      exige la construction d'une nouvelle antenne
      ferroviaire à Sevington près de Ashford.

4.   Des problèmes géologiques complexes sont associés à
la zone du Terminal dûs principalement à le superposition de
quatre glissements de terrain contre l'escarpement du North
Downs. Afin d'éviter de creuser dans des terrains d'un
niveau de stabilité très faible, une grande quantité de
remblais est nécessaire afin de réaliser un grand plateau
pour les quais rapportés de chargement.

5.   Les matériaux importés proviennent de deux sources:
remblais en schistes du terril du Kent et sable de mer. Le
sable est dragué de North Goodwins au bord de la Manche et
transporté ensuite sur 60 km jusqu'à un deuxième dragueur
ancré à 700m au large de Hythe. Le deuxième dragueur est
utilisé pour pomper le sable mélangé à l'eau de mer dans un
pipeline terrestre en acier de 800 mm de diamètre d'une
longueur de 5,3 km et d'une dénivellation de 65m. Ce mélange
arrive ensuite dans des lagunes construites à l'intérieur du
Terminal. L'eau de mer est alors drainée et repompée dans la
mer par l'intermédiaire du tuyau de déchargement à la fin du
cycle. Le sable est transporté dans les zones devant être

remblayées par toute une flotte de décapeuses automotrices. La quantité de sable transportée par semaine est de l'ordre de 70 000 m^3.

6. A ce jour, les principales zones d'activités sont les suivantes :

(a) Ponts des routes d'accès entrée et sortie de le liaison ferroviaire principale qui va en Terminal.
(b) Travaux de terrassement pour les routes d'accès et de sortie etc.
(c) Travaux de terrassement majeurs à l'intérieur du Terminal et en particulier les remblais en provenance de l'extérieur.
(d) Le tunnel de la boucle dans le Terminal, structure souterraine comprenant 3 voies ferrées d'une longueur de 1060 m une fois achevée.

14B. Construction du terminal de Coquelles

O. COLIN, Ingénieur Civil des Ponts et Chaussées, Directeur Travaux
Terminal France

INTRODUCTION. La construction du Terminal Français du Lien
Fixe Transmanche ne se caractèrise pas d'abord par la mise
en oeuvre de techniques originales ou inhabituelles. Les
rares aspects quelque peu particuliers du chantier, tels que
la consolidation des sols ou la gestion des eaux de
drainage, ne justifieraient pas en soi une intervention
spécifique dans le contexte de la conférence sur le Tunnel
sous la Manche. Le seul point qui mérite de retenir
l'attention du public réside dans les conditions dans
lesquelles le chantier s'est déroulé. Nous voulons parler
ici de l'étroite imbrication et de la superposition des
activités d'étude, de construction, de montage et de mise en
service du système, et de l'impact de cette imbrication sur
la conduite des travaux.

RAPPEL DESCRIPTIF DU PROJET TERMINAL
Le Terminal : de la fonction à l'objet
   1. Comme l'a bien montré l'exposé présenté par
l'Ingénierie "l'objet" Terminal doit remplir les trois
fonctions principales :

   (a) la fonction d'un entonnoir qui permettra de faire
       monter, en phase ultime, le trafic d'une autoroute à 3
       voies (3455 véhicules/heure) sur un système de
       navettes ferroviaires, après accomplissement des
       diverses formalités (douane, police, péage)
   (b) la fonction d'un injecteur qui introduira dans le
       tunnel les navettes dans des conditions telles
       qu'elles puissent effectuer le trajet quai à quai en
       moins de 35 minutes, tout en assurant le passage des
       trains et TGV venant du réseau SNCF (20 trains et
       navettes par heure dont 5 TGV). La conjugaison des
       fonctions (a) et (b) doit permettre une rotation des
       navettes en 110 minutes pour les véhicules légers et
       130 minutes pour les poids lourds
   (c) la fonction d'un centre d'entretien pour le matériel
       ferroviaire.

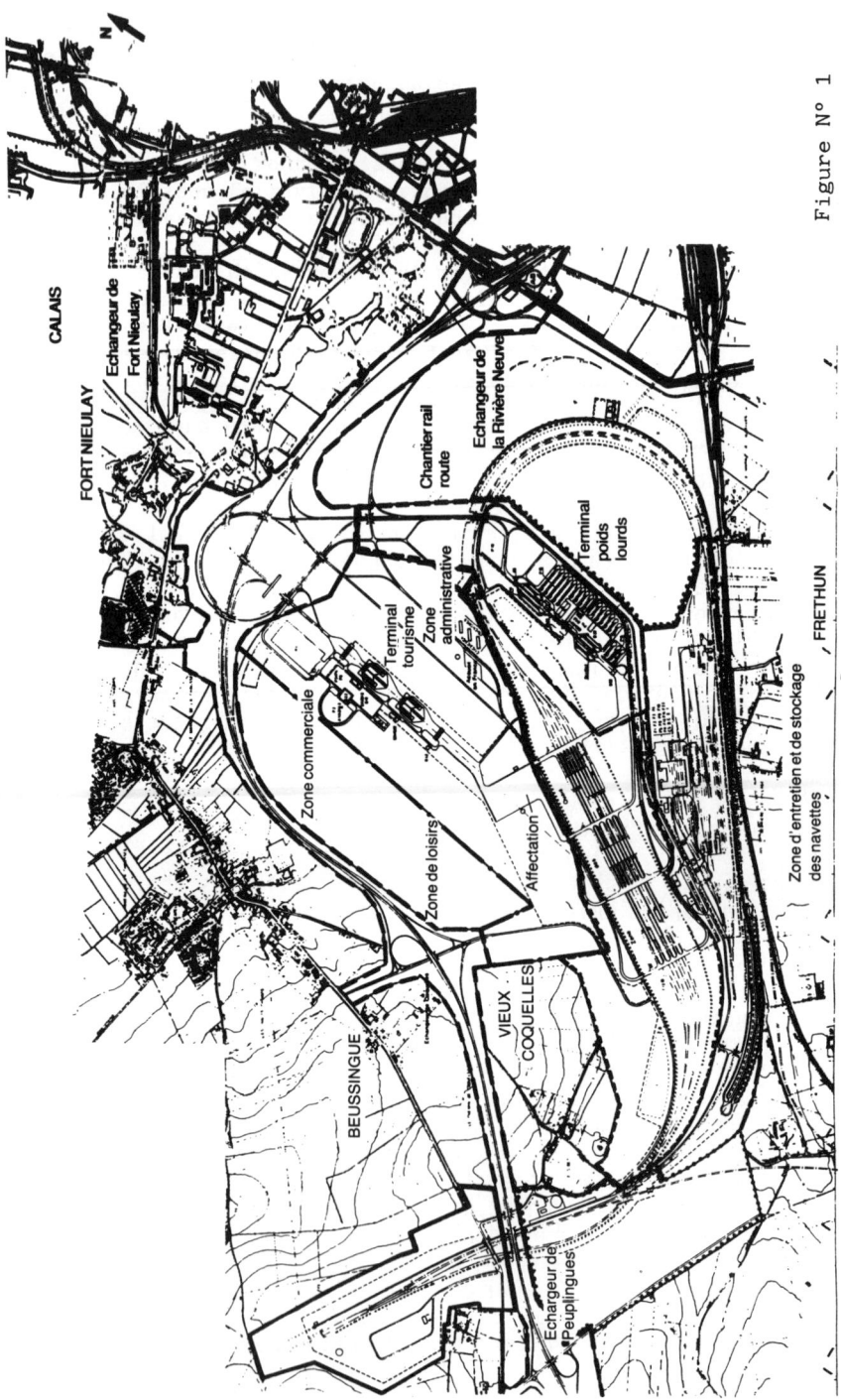

Figure N° 1

2. Ces trois fonctions ont déterminé l'organisation générale du Terminal (fig. 1).

(a) la fonction "entonnoir" a dimensionné les accès et les routes, les terminaux tourisme et poids lourds et les quais d'embarquement

(b) la fonction "injecteur" a défini la position géographique du Terminal et a dimensionné la boucle ferroviaire

(c) la fonction "centre d'entretien", à travers le nombre et la nature des opérations d'entretien et l'organisation du travail, a déterminé le plan masse de la zone d'entretien.

3. A partir de cette organisation générale, et en prenant en compte la configuration du terrain dans un souci de minimiser la quantité de matériaux d'apport, rares et chers dans cette région, le plan masse a pû être dessiné.

Le facteur temps

4. Avant de décrire les travaux à réaliser, il est nécessaire de donner quelques indications sur le programme de l'opération. Le "facteur temps" a en effet joué un rôle plus important dans les choix principaux qui ont structuré la construction que des contraintes purement techniques.

5. Le planning global du Terminal (fig. 2) ne présente rien d'extraordinaire : 5 années complètes pour réaliser les travaux constituent une performance tout à fait normale sur un chantier de 700ha où des attaques multiples sont possibles.

6. Par contre la présence d'un certain nombre de chemins critiques, induits par l'imbrication des tâches de conception, de construction et de montage des équipements, mérite d'être soulignée. On peut citer :

(a) la nécessité de démarrer le chantier de terrassement dès le début de l'année 1988, alors même que la conception était à un stade embryonnaire, pour pouvoir consolider les zones de sol compressible suffisamment tôt pour que le génie civil des ouvrages puisse débuter en temps utile.

(b) l'optimisation économique de la consolidation dont le coût diminue lorsque la durée augmente

(c) la nécessité de démarrer, sur terrain consolidé, le chantier de la zone des quais (ponts traversant et rampes d'accès) dès le 2ème trimestre 1989. L'achèvement du génie civil de la zone des quais, 26 mois plus tard, conditionne le démarrage du chantier de pose des voies ferrées et ensuite les essais du matériel roulant (ref.(d)).

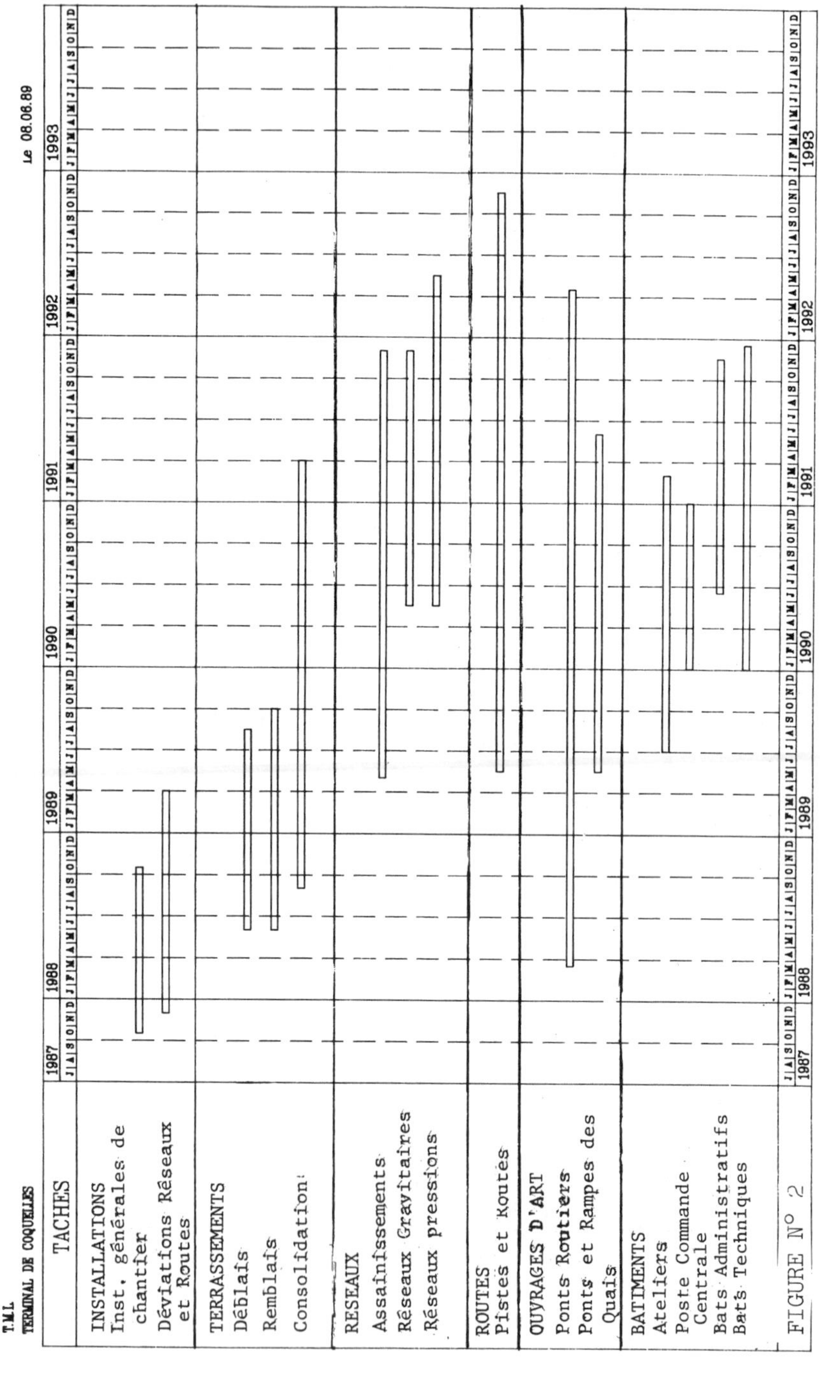

T.M.L.
TERMINAL DE COQUELLES

Le 08.06.89

TACHES	1987	1988	1989	1990	1991	1992	1993

INSTALLATIONS
Inst. générales de chantier
Déviations Réseaux et Routes

TERRASSEMENTS
Déblais
Remblais
Consolidation

RESEAUX
Assainissements
Réseaux Gravitaires
Réseaux pressions

ROUTES
Pistes et Routes

OUVRAGES D'ART
Ponts Routiers
Ponts et Rampes des Quais

BATIMENTS
Ateliers
Poste Commande Centrale
Bats Administratifs
Bats Techniques

FIGURE N° 2

(d) la nécessité de démarrer, sur terrain consolidé, la construction de l'atelier principal d'entretien du matériel roulant, dès la mi-1989, car cet atelier doit être disponible, avec ses équipements, dès Juillet 1991, pour l'assemblage des navettes.

(e) la mise à disposition de 18km de voies ferrées, à la même date, pour le stockage, puis pour les essais de matériel roulant

(f) l'achèvement, dès la fin 1990, du génie civil du poste central de contrôle, pour permettre la réalisation des équipements puis les essais et la mise en service du système de contrôle communication en temps utile.

7. Ces contraintes externes, conjuguées avec l'optimisation économique des moyens de l'entreprise, ont donc figé un planning ambitieux dans la réalisation d'objectifs partiels du type :

(a) fin 1988 : désenclavement de la zone de Beussingue et fin des terrassements de la voie de service (nécessaire à l'accueil du tunnelier T4 à Beussingue)

(b) fin 1989 : livraison au chantier de montage électromécanique de la zone comprise entre le portail de Beussingue (accès aux tunnels) et l'ouvrage de croisement des voies ferrées (ouvrage E1 dit "saut de mouton")

(c) fin 1990 : livraison au chantier de montage électromécanique de la boucle ferroviaire (une voie minimum) raccordée à l'atelier d'entretien

(d) etc....

LES TRAVAUX A REALISER

8. Nous l'avons déjà dit, les travaux à réaliser ne se caractérisent, à l'échelle du projet du Lien Fixe, ni par leur importance, ni par leur technicité. Ils comprennent sur un site de 700ha :

(a) des terrassements : environ 10 millions de m^3 de mouvements de terre ont été ou vont être réalisés, dont 4.5 millions de m^3 de déblais crayeux ou limoneux 3.0 millions de m^3 d'apports extérieurs, essentiellement en sables (tapis drainants, remblais et couches de forme), et 2.5 millions de m^3 de mouvements de surcharge

(b) des réseaux : environ 180km de canaux, caniveaux, fossés et canalisations

(c) des chaussées : environ 1 million de m^2 (1 200 000T de matériaux) dont 2/3 pour les routes et pistes et 1/3 pour les voies ferrées

(d) des ouvrages d'art : 51 000m^2 de ponts et 4300m de quais.

(e) des bâtiments   : 60 bâtiments pour 31 000m^2
(f) des voies ferrées : 60km de voies et 90 appareils.

9. Sur le plan technique, nous mentionnerons, pour mémoire, deux aspects particuliers : la consolidation des sols et la gestion des eaux de drainage.

La consolidation des sols

10. Une grande partie du Terminal est située sur des sols compressibles (tourbe et vase) : l'échangeur d'accès et une partie du Terminal tourisme, le Terminal poids lourds et la boucle, la zone d'entretien et la voie d'urgence et la quasi-totalité de la zone des quais. Il a donc été nécessaire de consolider les sols avant la réalisation des ouvrages définitifs. La méthode de consolidation, mise au point dès 1986 avec l'aide du bureau d'études TERRASOL, et confirmée par la réalisation, début 1987, de remblais d'essai, est celle de l'éponge pleine d'eau sur laquelle on vient placer une brique :

  (a) une couche de sable drainant de 50cm d'épaisseur est mise en place directement sur le terrain naturel non décapé. La capacité drainante de cette couche peut être améliorée, si la configuration l'exige, par un réseau de drains horizontaux
  (b) des drains verticaux sont alors mis en place, à travers la couche drainante, sur la hauteur du terrain compressible (de 3.40m à 11.50m environ) avec un maillage d'environ 2m X 2m
  (c) le remblai proprement dit est ensuite réalisé avec les matériaux du site (limons puis craie), complétés éventuellement par des matériaux d'apport jusqu'au niveau de la couche de forme (niveau final avec intégration des tassements prévisibles)
  (d) la surcharge nécessaire à une durée de consolidation compatible avec le tassement prévisible et avec les contraintes de planning, est alors mise en place. Cette durée est de l'ordre de 6 mois mais peut être

      inférieure si le planning l'exige. Ces contraintes, ainsi que l'optimisation des mouvements de terres ont conduit à approvisionner environ un quart du volume nécessaire total, et à le mettre en oeuvre par phases successives sur quatre zones du Terminal.

11. Pour conclure sur ce point, il faut retenir qu'aujourd'hui, les tassements constatés sont généralement très proches des tassements prévisibles (± 10%) et que, globalement, sur la totalité du chantier, 1 million de m^3 va ainsi disparaître.

La gestion des eaux de drainage

12.   Le problème à résoudre ici était celui de la collecte et de l'évacuation des eaux de drainage :

(a)   sur un terrain plat dont le sous-sol est constitué en majeure partie de tourbe ou de vase
(b)   avec un niveau de nappe très élevé.

13.   La solution classique, qui consistait à recueillir l'eau en des points bas et à la refouler, a dû être écartée: les points bas n'existaient pas et il aurait fallu aménager des bassins enterrés, très onéreux et peu fiables.   Il a donc été retenu une solution plus originale, dont le principe est le suivant :

(a)   aménagement de bassins suspendus
(b)   collecte gravitaire des eaux en pied de bassin
(c)   relevage de l'eau dans le bassin
(d)   écoulement gravitaire (canal ou collecteur) depuis le déversoir du bassin n jusqu'au pied du bassin n+1
(e)   etc...

14.   La figure n°3 montre l'application de ce principe au site :

(a)   5 bassins suspendus ont été créés sur une surface totale de près de 100.000m^2
(b)   le cas particulier de la tranchée de Beussingue est résolu par un drainage de la nappe et des eaux de surface, vers une station de pompage située au portail.   Les eaux sont relevées en crête de talus et s'écoulent gravitairement jusqu'au pied du bassin n°1
(c)   une capacité de pompage de 85.000m^3/h va être mise en place en 6 stations
(d)   l'écoulement principal entre bassins et vers l'exutoire principal (canal des Pierrettes et Rivière Neuve) est assuré par 10km de canaux de grande section ou de collecteurs.

15.   La grande capacité de stockage des bassins (250.000m^3 hors bassin n°5) et leur interconnection assurent une sécurité optimale au système et une flexibilité importante.

LE VRAI CHALLENGE

16.   Nous l'avons donc vérifié : la vraie difficulté dans la réalisation du Terminal Français ne réside pas dans la complexité des techniques mises en oeuvre.   L'étendue des travaux à réaliser est inhabituelle mais n'est pas non plus extraordinaire, elle est celle ni plus, ni moins d'un grand aéroport international.   Le délai global de construction - 5 ans environ - est lui aussi très raisonnable.

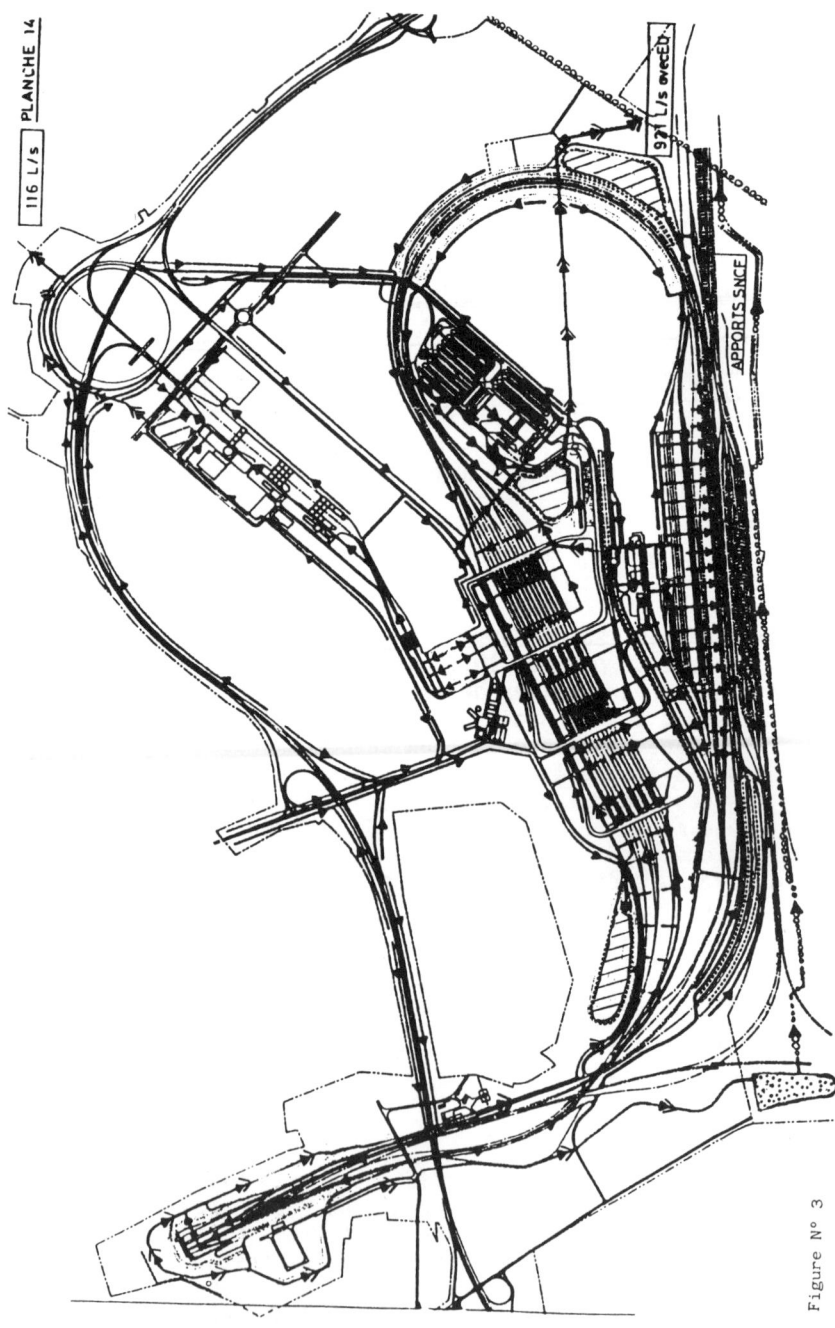

Figure N° 3

17. Le vrai challenge est ailleurs. Nous l'avons entrevu à plusieurs reprises dans l'exposé qui précède, essayons maintenant de le cerner avec plus de précision.

18. Il réside d'abord dans l'intégration du Terminal Français dans un ensemble beaucoup plus vaste, celui de la réalisation du Lien Fixe Transmanche, qui n'est pas d'abord un énorme chantier de construction, mais qui est avant tout un système de transport. Il réside également dans le fait que ce système est un prototype. Dans la plupart des cas, les grands projets d'ingénierie (métros, complexes sidérurgiques etc...) sont conçus et réalisés par rapport à un projet de référence. Les adaptations sont certes importantes, mais pour chaque choix majeur, il y a un exemple, un modèle auquel on peut se rapporter. Ce n'est pas le cas du Lien Fixe. L'aventure est donc comparable, par exemple, à celle du TGV, le Train à Grande Vitesse Français. Et il ne faut pas oublier que, pour le premier TGV, le TGV Sud-Est, entre la décision de lancer les études détaillées de conception et la mise en service, il s'est passé quelques 10 ans. Le TGV Atlantique a pris environ 7 ans et le TGV Nord, dont la réalisation débute à peine, sera en service dans 4 ans. Le Lien Fixe doit, lui, être conçu, réalisé et mis en service en 7 ans. C'est donc un pari extraordinaire, qui a des répercussions fondamentales jusque sur la construction d'un objet aussi simple que le Terminal Français.

19. Prenons un exemple, déjà évoqué, celui de l'atelier d'entretien du matériel ferroviaire. Il s'agit d'un bâtiment de type industriel de 11.000m^2, qui n'a donc en soi rien d'extraordinaire. Dans un contexte classique les études et la réalisation en demanderaient à peu près un an.

20. Dans le contexte du Lien Fixe :

(a) ce bâtiment ne peut être conçu, dans son détail, avant la définition des équipements (ponts roulants, tours d'entretien etc...) qui y seront installés

(b) ces équipements ne peuvent être choisis avant que le type de matériel ferroviaire (navettes etc...) soit retenu et que ce matériel soit suffisamment étudié

(c) ce bâtiment doit être opérationnel, avec ses équipements, dès le mois de juillet 1991, pour l'assemblage des premières navettes.

21. Quand on sait en plus que le choix du matériel ferroviaire n'a pu intervenir avant l'été 1989 et que la fabrication et le montage des équipements de l'atelier prendront 22 mois, on réalise l'ampleur du problème.

22. Pour le résoudre, il a fallu, en mai et juin 1989, avant le choix définitif du matériel roulant, réaliser un avant-projet de l'atelier et lancer le processus de consultation des entreprises et de construction du bâtiment. Un bâtiment, qui bien sûr, dans ces conditions, ne représente sûrement pas la solution optimale, techniquement

et économiquement, au problème posé. Mais un bâtiment qui, construit dans les délais, remplira une condition nécessaire à la réalisation, dans les délais du Lien Fixe, puisque les premières navettes pourront être assemblées dès la fin 1991 et le système essayé en 1992 et progressivement mis en service à partir du début 1993.

23. L'exemple cité illustre clairement que la réalisation d'un projet aussi complexe que le Lien Fixe n'est pas possible si chacun de ses constituants est conçu et réalisé dans le souci exclusif de sa perfection : l'optimisation de l'ensemble n'est pas la somme des optimisations de chacune des parties.

24. C'est bien l'application de ce principe qui a permis entre autres :

(a) le démarrage des travaux de terrassement dès le début de l'année 1988, alors que le plan masse était loin d'être définitivement arrêté. Ce démarrage, on l'a vu, était vital pour la réussite du projet dans les délais. Il n'a été possible qu'au prix d'un certain nombre de choix délibérés, qui, pris chacun séparément, étaient contraires à l'optimisation du chantier :

. multiplicité des accès pour désenclaver à coup sûr le chantier, compte-tenu de la mise à disposition tardive et progressive des terrains et des incertitudes du plan masse

. choix d'une solution flexible pour le matériel de chantier, compte tenu des aléas de planning qui pesaient sur la disponibilité des plans, et des contraintes économiques

. provenance diversifiée des matériaux d'apport

. etc...

(b) la réalisation, en amont, d'un certain nombre de travaux et d'ouvrages qui ont contribué, en outre, à figer le projet :

. ponts sur la rocade

. ouvrage d'interconnection avec le trafic SNCF

. travaux de drainage

. etc...

25. Il faut également mentionner, toujours dans le souci d'augmenter la flexibilité et la rapidité de réaction aux évènements :

(a) la mobilisation des équipes, en particulier en ce qui concerne l'encadrement. Cette mobilisation a été précoce, afin de donner aux responsables le temps de prendre la mesure des problèmes qu'ils auraient à résoudre. Elle a été aussi légèrement surabondante afin de permettre à chacun de prendre sa place et de pourvoir aux remplacements nécessaires, compte tenu de la durée du chantier

(b)   la définition des installations de chantier. Là
      encore, les installations de chantier ont été conçues
      très tôt, à une époque où, par exemple, la politique
      de sous-traitance n'était pas arrêtée. Les
      installations ont donc été dimensionnées largement
      afin de faire face à toute éventualité
(c)   la politique de sous-traitance où il a été choisi de
      limiter le nombre d'entreprises sous-traitantes et de
      les fidéliser.

CONCLUSION. S'il y a un enseignement à tirer de la
construction du Terminal Français, c'est bien celui-là :
l'optimisation d'un projet aussi complexe et imbriqué par le
Lien Fixe Transmanche ne passe pas par l'optimisation de
chacun de ses parties. L'application de ce principe a
permis au chantier de tirer en avant l'ensemble du projet en
figeant, à un stade où la conception était encore hésitante,
un nombre croissant de points fixes.

FIGURES
1. Plan de zones
2. Planning général
3. Système de drainage

SYNOPSIS

1. Apart from the substantial ground consolidation and the complex drainage system, the French Terminal construction works are nothing out of the ordinary. The infrastructure and building works all use traditional techniques and the construction period of 5 years would seem ample. Indeed, the construction works from a technical point of view could be compared with a large international airport project.

2. The construction works are however distinguished by the fact that they are executed in the context of an ambitious fast-track and multi-disciplinary international transport system project the likes of which has no equivalent.

3. Confronted with this unique challenge, it was necessary to organise the execution of the work giving priority to flexibility with other considerations taking a secondary role.

4. If a conclusion can be made from the French Terminal construction works, it would be that the development and optimisation of a project as complicated and multi-disciplinary as the Cross-Channel Link does not pass by the optimisation of each individual discipline. The application of this principle has enabled the construction work to advance the whole project at a point in time when the project conception was still uncertain.

15. Réalisation du système de transport

C. AGARD, Ecole Polytechnique, Ecole Nationale Supérieure du Génie Maritime, Directeur du système de transport, Transmanche — Link joint venture

REALISATION DU SYSTEME DE TRANSPORT

Synopsis

La réalisation du Système de Transport a nécessité la mise en place d'une structure et d'une organisation spécifique.
Les différents aspects de la mission de réalisation , ses points originaux et les moyens mis en oeuvre sont passés en revue.

1 - Avec l'aide du Groupe Ingénierie et en utilisant les études faites par celui-ci, le Groupe système de Transport de Transmanche-Link a la mission de réaliser les Equipements Fixes et le Matériel Roulant du Système de Transport du Tunnel sous la Manche. Cette réalisation doit se faire dans le respect du Programme, des coûts et des performances retenues par Eurotunnel et Transmanche-Link.

Il sera successivement examiné :

- Points originaux de la mission,

- L'Organigramme Technique - Organisation,

- La Politique de réalisation particulière aux Equipements Fixes d'une part, au Matériel Roulant d'autre part,

- La mise en Service,

- Les moyens de Gestion, les Hommes.

2 - Points originaux de la Mission

Le Système de Transport est "unique", en ce sens qu'il n'a pas de références et que tout en mettant en oeuvre des techniques classiques, il le fait avec des exigences extrêmes de performance, de sécurité et avec une interaction avec l'infrastructure (tunnels) rarement connue jusqu'à présent.

C'est un projet Franco-Anglais animé par des Français et des Anglais. L'Histoire est là pour nous rappeler les caractères de ces 2 nations ! Ce projet est financé par des capitaux privés internationaux.

C'est un projet appartenant à un client neuf, sans historique, mais le Système de Transport doit constituer une liaison entre 2 réseaux de transport bien établis, à culture historique très différente, et qui, jusque-là, n'avaient pas d'échanges pratiques.

Le Projet a un programme très tendu. De plus, l'installation des Equipements fixes du Système de Transport doit se faire en coexistant avec les activités de creusement de Tunnel et à partir de deux accès, l'un en France, l'autre au Royaume-Uni, distants de 50 km environ.

Le Matériel Roulant est totalement original, son gabarit nécessite la recherche de zones d'essais très spéciales.

3 - Organigramme Technique - Organisation

a - L'organigramme technique a repris, avec quelques légères modifications, le découpage des Etudes du Groupe Ingénierie. Cette approche permet de limiter le réaménagement des documents d'Ingénierie. On retrouve donc, regroupés sous forme de disciplines les différents "Primary System Dossiers" déjà rencontrés.

Les disciplines retenues ont été :

Pour les Equipements Fixes :
- Alimentation Electrique et Caténaires,
- Voies Ferrées et Système de Transport du Tunnel de Service,
- Systèmes de Contrôle et de Communication,
- Mécanique.

Pour le Matériel Roulant :
- Les locomotives (Electriques et Diesel)
- Les wagons (Touristes, Autocars, Poids Lourds).

b - A cet organigramme technique va correspondre une organisation basée sur la notion de Directeur de Projet chargé de diriger chacune des disciplines citées ci-dessus dans les domaines suivants :

- Négociation : attribution des sous-contrats
- Direction des sous-contrats, gestion des interfaces
- Relation avec Eurotunnel
- Responsabilité d'ensemble dans les domaines coûts, programme, performance.

Les "Directeurs de Projet" s'appuient sur des services
fonctionnels communs participant fortement à la cohésion et
l'homogénéité du Système. Ces services sont :

- Le Service Contrat spécialisé dans la gestion contractuelle
  des sous-contrats et les relations avec Eurotunnel.

- Le Service Programme/Contrôle des Coûts chargé d'assumer la
  tenue à jour du programme de toutes les disciplines, de
  leurs interactions, des interfaces avec le Génie Civil ou
  des événements extérieurs. Le Service assure aussi le
  contrôle des coûts et élabore avec les "Directeurs de
  Projets" les prévisions et tendances.

- Le Service Administration et Finance.

- Le Service Assurance Qualité.

  c - L'installation des Equipements Fixes dans les Tunnels et
  Terminaux, l'insertion du Matériel Roulant dans les
  Equipements Fixes nécessitent d'autres actions et d'autres
  coordinations. Le Groupe Système de Transport s'appuie donc
  dans les deux sites France et UK sur des Moyens chargés de
  fournir les moyens logistiques aux sous-traitants retenus,
  de surveiller l'exécution des travaux et de traiter
  directement certaines activités. Une fonction Installation
  est donc présente dans l'organisation pour assurer la
  liaison avec les 2 sites.

  d - L'annexe 1 montre l'organisation existante le 1er avril
  1989.

4 - Le Politique de réalisation des Equipements Fixes

C'est essentiellement une politique de sous-traitance.

4a. Les études du Groupe Ingénierie, associées à des clauses
contractuelles reproduisant tout ou partie des clauses du
Contrat existant entre Eurotunnel et TML constituent les
bases d'appels d'offres internationaux faisant l'objet de
publicité dans le Journal Officiel de la Communauté
Européenne (J.O.C.E.) et dans certaines revues. Après
préqualification et mise en compétition, les contrats sont
attribués. Ces contrats sont des contrats dits de performance
dans lesquels les performances à atteindre par chacun des
sous-systèmes sont définies le plus clairement possible, les
moyens pour les obtenir étant, dans leurs détails, laissés à
l'imagination des sous-traitants. Ce type de contrat est
relativement peu usuel et s'apparente beaucoup aux contrats
"Design and Build".

Ces contrats comprennent les phases suivantes :

Ingénierie détaillée, fourniture, formation du futur exploitant, installation ou sa supervision, mise en service, période de garantie.

Quelques mots sur l'Installation :
Les Equipements Fixes sont installés dans les zones terminales et dans les tunnels. L'installation dans les zones terminales relève des méthodes classiques mais l'installation dans les tunnels pose un challenge particulier. En effet, environ 3x50 km de tunnels sont à équiper à partir de leurs 2 extrémités. Ces tunnels comportent environ 500 salles souterraines. De plus les travaux de Génie Civil vont, étant donné le Programme Général du Projet, coexister avec la mise en place des Equipements Electro Mecaniques.

En fait, un fantastique système de transport provisoire est prévu pour pouvoir tout à la fois traiter les besoins du Génie Civil, assurer le transport des équipements électromécaniques et leur installation. La méthodologie d'installation doit aussi pouvoir s'adapter à des imprévus sur le programme des tunneliers.

L'accès aux différents points du tunnel est donc soumis à des aléas et il a paru bon de limiter les sous-traitants intervenant dans les tunnels. C'est pourquoi beaucoup de sous-contrats contiennent une alternative pour la phase Installation :
- Supervision du montage réalisé par un tiers,
- ou montage réalisé directement.

4b. Quels sont les sous-contrats prévus ?
Ils sont au nombre de 50 environ. On y distingue des contrats de plusieurs types :

- des contrats dits "Système" dont le domaine d'application s'étend d'un terminal à l'autre après avoir transité par les Tunnels. On peut citer dans ce cas : la signalisation ferroviaire, le système de transmission de données, etc.

Ces contrats posent certains problèmes juridiques en raison de leur caractère binational.

- des contrats pour des installations clairement localisées et qui sont alors des contrats plus simples dans leur approche juridique.

L'annexe 2a/b jointe donne, en même temps que leur planning, la liste des principaux Contrats.

5 - **La politique de réalisation du Matériel Roulant**
Le cheminement a été légèrement différent de celui suivi pour les Equipements Fixes. A partir d'une définition des performances à atteindre, dans un cadre très serré de

contraintes (dimensions, charge à l'essieu, environnement...)
les Groupement de constructeurs préqualifiés ont eu à
présenter leur solution technique dans le cadre d'une
"Request for Proposal", un concours en quelque sorte. Après
examen et adpatation de ces propositions, TML a demandé aux
Groupements ayant répondu leur proposition financière. C'est
ce processus qui a conduit à la sélection des fournisseurs
que l'on connaît aujourd'hui.

Cette démarche a été retenue en raison de l'absence de
références pour les matériels en question.

6 - La Mise en Service

La mise en service du Système de Transport constitue une
phase très importante et très délicate dans la vie du Projet.
C'est pendant cette période que vont se matérialiser les
interfaces techniques et fonctionnelles des différents
systèmes en même temps que la prise en main graduelle par le
Client, Eurotunnel. C'est dire que les problèmes techniques
et humains vont faire de cette période limitée dans le temps
(6 mois environ) une période dense et tendue. Ce sera aussi
le parachèvement de la formation reçue par le Personnel
d'Eurotunnel dans le cadre des sous-contrats. Une logique et
le programme correspondant ont été préparés et sont tenus à
jour en fonction de l'évolution des programmes, des systèmes
individuels et de l'infrastructure. Un projet d'organisation
spécifique est également en cours d'étude.

Pour être plus concret, précisons que cette mise en service
va, étant donné l'aspect "fast track" du projet et sa
complexité, démarrer à des périodes différentes de la vie du
projet : le matériel roulant va commencer à être testé dès la
fin 1991 pour les têtes de série livrées, des tests plus
particuliers d'endurance, utilisables aussi pour la formation
du personnel d'Eurotunel, sont prévus mi 1992 ; les
locomotives seront testées en endurance sur les réseaux
nationaux à partir des mêmes dates.

De la même manière les sous stations électriques principales
verront leur mise en service partielle être faite vers la mi
1991. Ce ne sera que 6 mois environ avant l'ouverture que les
différents éléments du puzzle seront en place afin que
l'ensemble des interfaces puisse être validé. Pendant cette
période seront également testées toutes les interfaces prises
en compte au niveau Ingénierie avec les deux réseaux
nationaux, British Rail et SNCF, à la fois dans le domaine
des trains classiques et du train à grande vitesse (TGV).
Bien entendu également, cette période fera l'objet d'une
attention particulière de la Commission Intergouvernementale.

7 - Les Moyens de Gestion

Etant donné le critère "temps", le programme et sa gestion

permanente constitue la pièce maîtresse de l'outil de Direction du Projet. Le Programme de l'ensemble du Projet est sur ordinateur avec les tâches et leurs interfaces entre les différentes parties du Projet : Tunnel, Terminaux, Equipement Fixe Electromécanique, Matériel Roulant. La partie Equipement Fixe et Matériel Roulant comprend à l'heure actuelle un nombre de 900 tâches dont 155 interfaces et 1700 contraintes. La phase "Mise en Service" y est évidemment intégrée avec les interfaces nécessaires. A partir de cet outil, auquel s'ajoutent, petit à petit, les événements principaux des plannings des sous traitants, le projet est également à même d'assurer certaines prévisions des flux financiers associés. Des tableaux de bord correspondant à chacun des sous contrats permettent de les gérer vis-à-vis des sous traitants et d'Eurotunnel.

8 - Les hommes

Je ne terminerai pas sans rappeler l'importance des hommes. Ceci est vrai dans tous les projets, c'est encore plus vrai dans un projet de cette nature : exceptionnel par sa taille, exceptionnel par son caractère binational, par ses difficultés. L'équipe du Groupe du Système de Transport est binationale et est basée à Sutton, près de Londres. Après un certain apprentissage de la vie en commun, cette équipe Franco-Britannique (environ 40 Français et 50 Britanniques) a appris à travailler ensemble au rythme du Programme, et dans une langue qui n'est plus ni tout à fait l'anglais, ni tout à fait le français. Je tenais donc à lui rendre un hommage mérité.

ANNEXE 1

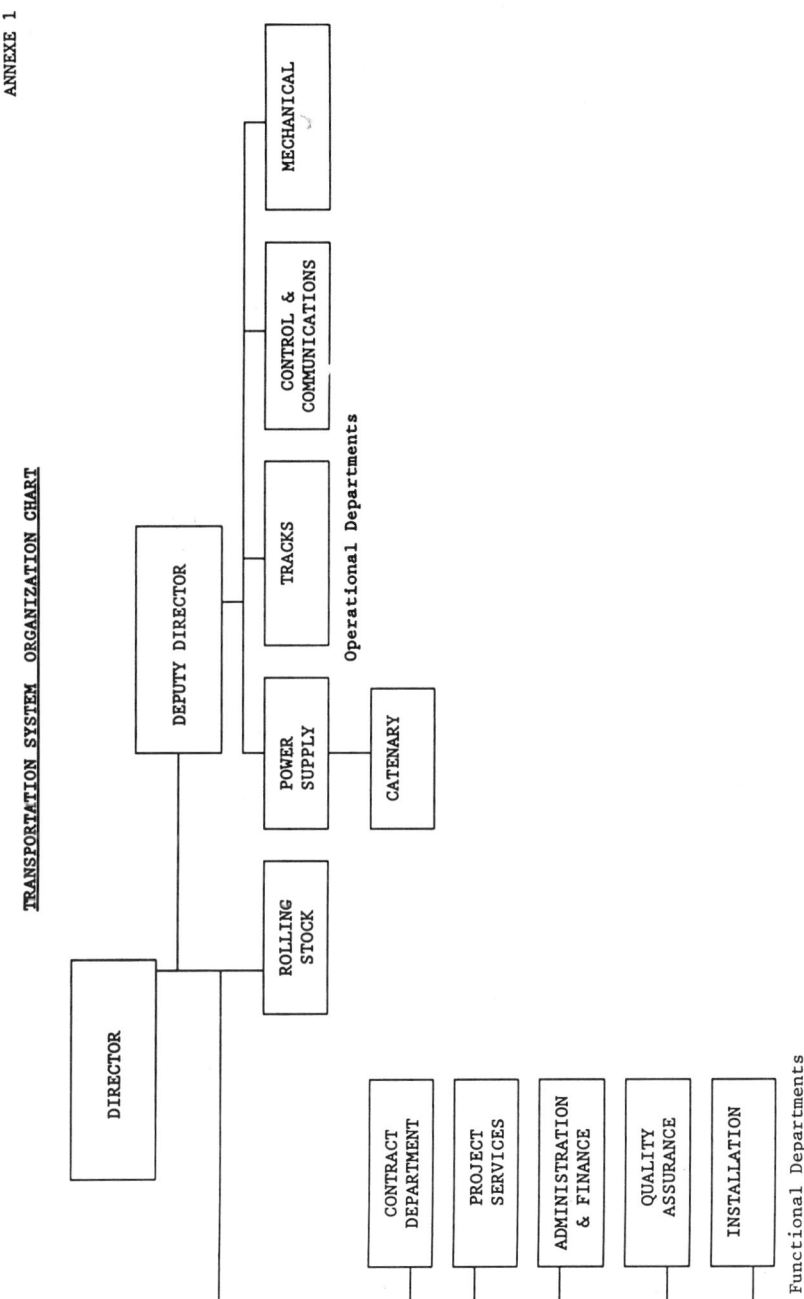

TRANSPORTATION SYSTEM ORGANIZATION CHART

DIRECTOR

DEPUTY DIRECTOR

ROLLING STOCK

POWER SUPPLY

CATENARY

TRACKS

CONTROL & COMMUNICATIONS

MECHANICAL

Operational Departments

CONTRACT DEPARTMENT

PROJECT SERVICES

ADMINISTRATION & FINANCE

QUALITY ASSURANCE

INSTALLATION

Functional Departments

TRANSPORTATION SYSTEM
SUBCONTRACT LISTING

ROLLING STOCK

1	LOCOMOTIVES	1
2	DIESEL LOCOMOTIVES	2
3	WAGONS	3

POWER SUPPLY

4	GRID CONNECTION EDF	4
5	MAIN SUBSTATION (FRANCE)	5
6	TERMINAL SUBSTATIONS (FRANCE)	6
7	SANGATTE SUBSTATION	7
8	EXTERNAL LIGHTING (FRANCE)	8
9	GRID CONNECTION SEEB	9
10	MAIN SUBSTATION (U.K)	10
11	TERMINAL SUBSTATIONS (U.K)	11
12	EXTERNAL LIGHTING (U.K)	12
13	SHAKESPEARE CLIFF SUBSTATION	13
14	TUNNEL SUBSTATIONS HV	14
15	TUNNEL LV & LIGHTING	15
16	POWER SYSTEM EARTHING	16

CATENARY

17	CATENARY (FRANCE & U.K)	17
18	TUNNEL EARTHING & BONDING	18

CONTROL & COMMS

19	CONTROL CENTRES (FRANCE & U.K)	19
20	ROAD TRAFFIC MANAGEMENT (FRANCE & U.K)	20
21	RAILWAY SIGNALLING (FRANCE & U.K)	21
22	RADIO SYSTEMS (FRANCE & U.K)	22
23	DATA TRANSMISSION, TELEPHONE & PUBLIC ADDRESS	23
24	ACCESS CONTROL (FRANCE & U.K)	24
25	FIRE DETECTION (FRANCE & U.K)	25

MECHANICAL

TRACKWORKS

Page 1 of 1

Annexe 2b

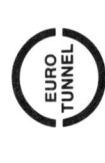

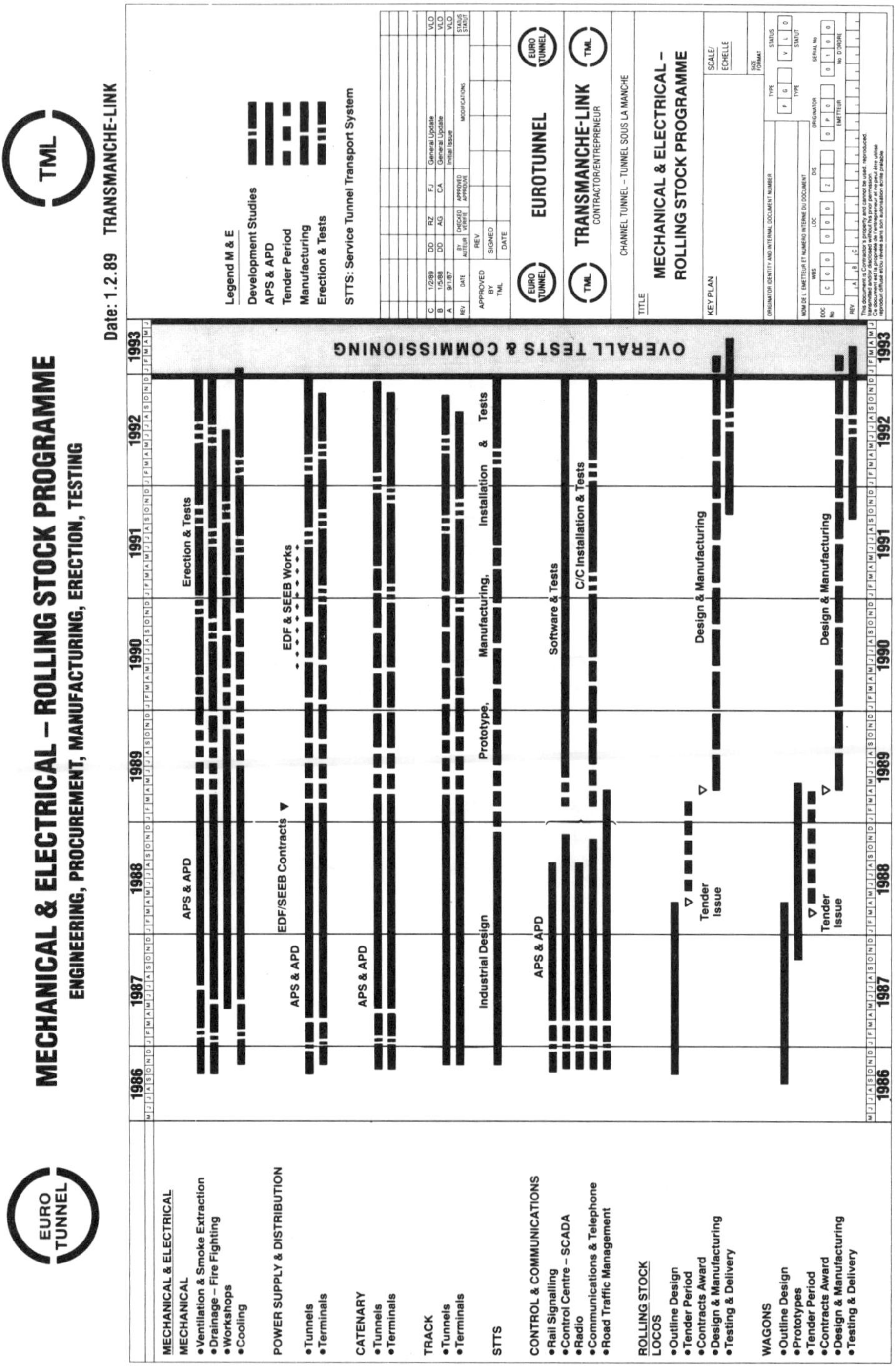
MECHANICAL & ELECTRICAL – ROLLING STOCK PROGRAMME
ENGINEERING, PROCUREMENT, MANUFACTURING, ERECTION, TESTING
EURO TUNNEL
TML
Date: 1.2.89
TRANSMANCHE-LINK
MECHANICAL & ELECTRICAL
MECHANICAL
•Ventilation & Smoke Extraction
•Drainage – Fire Fighting
•Workshops
•Cooling
POWER SUPPLY & DISTRIBUTION
•Tunnels
•Terminals
CATENARY
•Tunnels
•Terminals
TRACK
•Tunnels
•Terminals
STTS
CONTROL & COMMUNICATIONS
•Rail Signalling
•Control Centre – SCADA
•Radio
•Communications & Telephone
•Road Traffic Management
ROLLING STOCK
LOCOS
•Outline Design
•Tender Period
•Contracts Award
•Design & Manufacturing
•Testing & Delivery
WAGONS
•Outline Design
•Prototypes
•Tender Period
•Contracts Award
•Design & Manufacturing
•Testing & Delivery
1986
1987
1988
1989
1990
1991
1992
1993
APS & APD
Erection & Tests
EDF/SEEB Contracts
EDF & SEEB Works
Industrial Design
Prototype, Manufacturing, Installation & Tests
Software & Tests
C/C Installation & Tests
Tender Issue
Design & Manufacturing
OVERALL TESTS & COMMISSIONING
Legend M & E
Development Studies
APS & APD
Tender Period
Manufacturing
Erection & Tests
STTS: Service Tunnel Transport System
EUROTUNNEL
TRANSMANCHE-LINK
CONTRACTOR/ENTREPRENEUR
CHANNEL TUNNEL – TUNNEL SOUS LA MANCHE
MECHANICAL & ELECTRICAL – ROLLING STOCK PROGRAMME

SYNOPSIS. The implementation of the Transportation system of the Channel Tunnel has requested a specific organisation in the General Project. This organisation is quite original in the project in that way that its approach is on a System basis and so links French and British sides. The Author reviews The Technical Organigramme based on Discipline as Power Supply and Catenaries, Tracks, Mechanical, Control and Communications, Rolling Stock. The role of the Project Manager in charge of each discipline is emphasized and the role of functionnal services (Contract, Project, Administration and Finances, Quality Assurance) is described.

The implementation policy is based on sub-contracting according to competitive process which includes advertisements in EEC publications and specific periodics. Fixed Equipment sub contracts are on a design and build basis and so include detail design, supply, installation or supervision, commissioning, training, warranty. Rolling Stock sub contracts are based on a request for Proposal formula followed by a financial offer. The list and general planning of this sub contracts are given in annexe. Commissioning is also expressed in general terms with an emphasis on the "fast track" aspect of the project which compels to have a complex logic. Time being the essence of the project, the programme is managed on an integrated dedicated software system on computer. Finally emphasis is drawn on people who are the most important part of this specific organisation which integrates French and British.

16. The Channel Tunnel — the effect in the UK

J. HENES, Co-Chairman, Channel Tunnel Intergovernmental
Commision

SYNOPSIS

1.   For through-rail services, British Rail are developing
their plans for investment in infrastructure and rolling
stock to meet the forecast demand for passenger and freight
services. Proposals for over £600m of expenditure have
already been approved in principle by the Government. BR are
currently consulting local authorities and other interested
parties to identify commercially viable opportunities for
services throughout the UK. For roads, the Tunnel will have
a considerable impact on the network in East Kent. The
M20/A20 will be improved to become the main route for
cross-Channel traffic, and will provide a direct link to the
UK's motorway network.

INTRODUCTION

2.   The UK Government's policy for international transport
is to increase consumer choice and promote efficiency by
encouraging competition and innovation. That approach lay at
the heart of the decision that the Channel Tunnel project
should go ahead. The Tunnel will compete on fair terms with
air and sea for the growing traffic in the vital
cross-Channel market. Travel and trade will reap the benefit
of lower costs and improved services. Better quality
transport links will draw the whole of the UK economy to the
European Market, and will offer new opportunities for the
generation of industrial and commercial development. In
addition to diverting traffic from other modes, the Tunnel is
expected to create new traffic, increasing the size of the
total market, which is already growing steadily as a
reflection of the UK's closer trade links with Europe. This
paper outlines the provision being made for new rail and road
infrastructure to cater for Channel Tunnel traffic.

RAIL

3.   The Channel Tunnel offers British Rail the biggest
opportunity for expansion in many years. For passengers, BR
- jointly with SNCF and SNCB - will offer new high-quality,
high-speed trains linking London to Paris and Brussels with

Table 1. Traffic forecasts - summary of international
traffic forecasts

(Millions passenger trips and freight tonnes a year)

Year	Forecast	Passenger	Freight
1993	BR (MVA)	13.4	6.1
	Eurotunnel	15.4	7.4
	SNCF	16.5	7.2
2003	BR (MVA)	17.4	7.0
	Eurotunnel	19.8	11.4
	SNCF	21.4	10.6
2013	BR (MVA)	21.2	7.7
	Eurotunnel	22.4	16.4
	SNCF	26.2	13.4
2023	BR (MVA)	25.9	8.5
	SNCF	31.9	16.4

Notes: 1   These figures do not allow for extra traffic being
            generated by a high speed link in Britain, but
            they do allow for TGV operation in France, Belgium
            and beyond at appropriate dates.

        2   BR's forecasts assume that about 12% of trips
            will be for business purposes, with the remainder
            for leisure.

        3   The regional spread of the BR (MVA) passenger
            forecasts is shown in Table 2. It is not possible
            to provide such a breakdown of the freight
            forecasts, but BR expect that about 75% of freight
            traffic will have origins/destinations beyond
            London.

Table 2. Potential passenger demand - regional spread (BR forecasts - not available for Eurotunnel and SNCF forecasts)

Single Visitor Journeys (in thousands)	VISITORS (1 & 2)			TOTAL (3)		
	Actual 1985	Demand Forecast 1993	Demand Forecast 2003	Actual 1985	Demand Forecast 1993	Demand Forecast 2003
To/From						
Greater London	1045	4245	5520	1360	5775	7510
Home Counties - South East	205	530	685	285	905	1175
- South West	145	580	750	260	1240	1605
- North West	40	240	310	170	665	865
East Anglia	70	150	200	230	625	815
South West/South Wales	170	625	815	260	1065	1385
Midlands	55	270	350	110	785	1020
North West/North Wales	55	365	475	115	895	1165
North East/Yorkshire	55	300	390	190	930	1210
Scotland	65	315	410	120	515	675
TOTAL	1905	7620	9905	3100	13400	17425

Notes: 1 Visitors - mainland European residents plus non-Europeans visiting Britain

2 Single visitor journeys do not equal total number of visitors, since some will make single journeys while others will travel in both directions by rail

3 Total demand forecast - visitors plus Great Britain residents

Table 3. Freight - rail share of total market

Forecast by	1993	2003
BR	6.7%	6.3%
Eurotunnel	8.1%	8.4%
SNCF	7.9%	12.1%

Table 4. Rail freight - modal split
(BR forecasts, expressed in millions of tonnes)

Mode	Now	1993	2003	2013	2023
Trainload	-	1.8	2.0	2.3	2.6
Wagonload	1.2	1.4	1.5	1.6	1.7
Container	0.8	2.9	3.4	3.8	4.1
Total*	2.0	6.1	6.9	7.7	8.4

* differences from Table 1 due to rounding

frequent services and journey times competitive with air travel. Through passenger services from points beyond London are under consideration. For freight, there will be international services between all parts of the country and mainland European destinations, offering delivery times typically 24 to 48 hours shorter than at present.

Traffic forecasts

4. Various forecasts have been produced of passenger and freight traffic likely to be attracted, on a commercial basis, to through rail services. Table 1 sets out the ranges within which BR's plans are being developed. For passenger services, Table 2 shows the forecast demand broken down as between London and the regions of the UK. For freight, Table 3 shows the estimated rail share of the overall market and Table 4 shows the forecast modal split.

Infrastructure investment

5. BR's plans for investment in the infrastructure and rolling stock needed to meet the forecast demand for passengers and freight services are being developed in three phases.

Phase 1: The infrastructure needed for the "core system", which will cover the London/Paris and London/Brussels passenger services planned for 1993 when the Tunnel opens and the new international freight services between the United Kingdom and the Continent.

Phase 2: The evaluation of plans for an international station at Ashford, and of the market for through passenger services beyond London.

Phase 3: Consideration of additional line capacity across Kent and through London, and a second international passenger terminal in central London.

Phase 1

6. Investment proposals for infrastructure and rolling stock worth over £600m at current prices have been approved in principle by the Government. They include the new high speed passenger rolling stock and a number of freight locomotives. Track and signalling between London and the Tunnel will be improved so that international trains can run on the existing lines through Kent without disrupting existing services. An international passenger terminal is to be built at Waterloo, and a rolling stock maintenance depot is planned at North Pole, near Old Oak Common in West London.

7. The track between Clapham Junction and Willesden will be upgraded and electrified, and the signalling improved. It will be connected to improved freight marshalling and customs clearance facilities at Willesden. Electrification of the line between Redhill and Tonbridge, so that international freight wagon and container trains can be hauled by electric

Table 5.   Infrastructure Investment - Phase 1
BR Phase 1 investment proposals

Infrastructure (£298 million [all costs at Q3 1988 prices])

Waterloo international passenger terminal

North Pole maintenance depot

Stewarts Lane chord line (Battersea) and associated works

West London Line electrification and upgrading

Tonbridge-Redhill electrification

Other track and signalling

Dollands Moor freight sidings

Willesden freight facilities

Other (eg reservation & ticketing system, telecoms
strengthening, advance maintenance)

Rolling Stock (£308 million)

High speed trainsets

Night services coaching stock

Class 90 AC electric locomotives

Class 92 dual voltage electric locomotives

InterCity rolling stock

TOTAL COST - £606 million

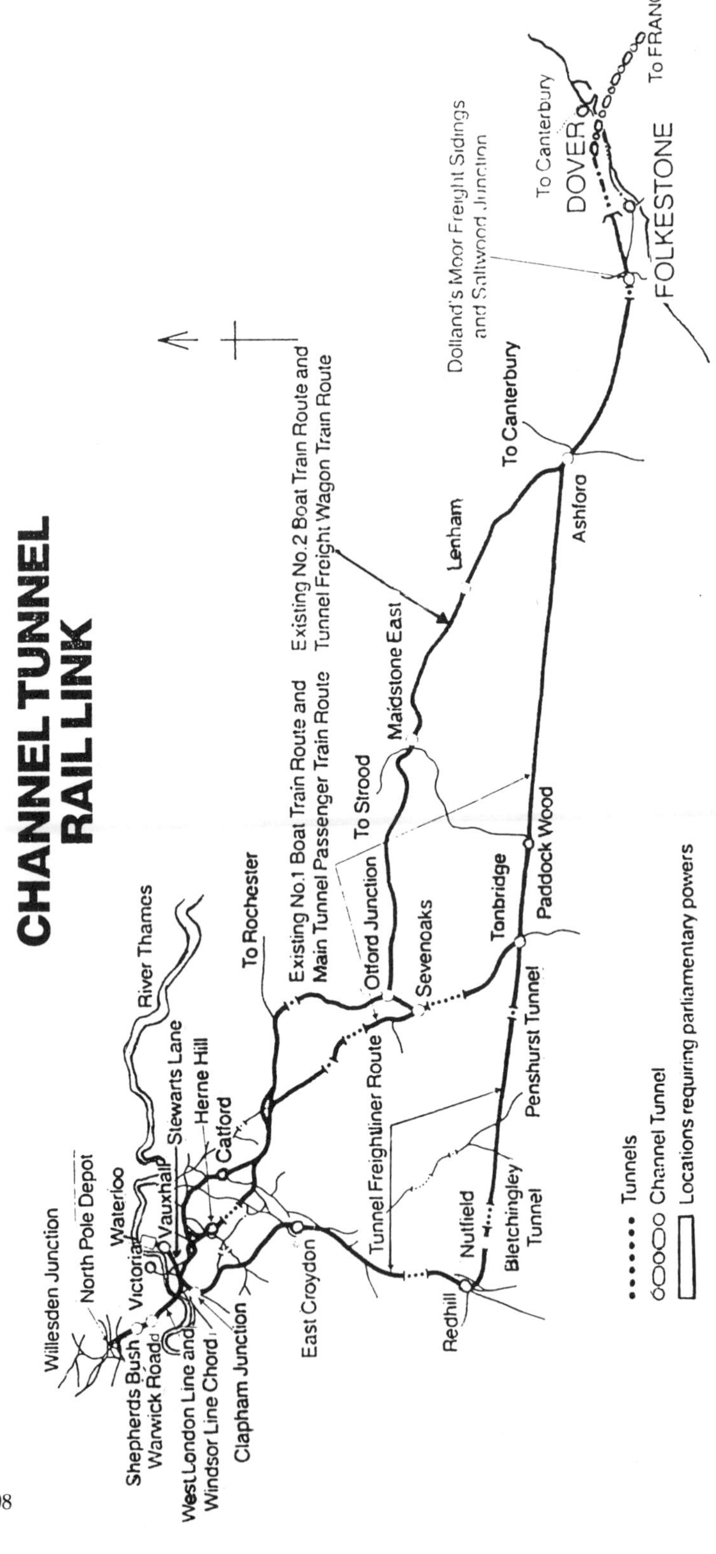

CHANNEL TUNNEL RAIL LINK

Willesden Junction
North Pole Depot
Shepherds Bush
Warwick Road
Waterloo
Vauxhall
Victoria
Stewarts Lane
West London Line and
Windsor Line Chord
Herne Hill
Catford
Clapham Junction
River Thames
East Croydon
To Rochester
Existing No.1 Boat Train Route and
Main Tunnel Passenger Train Route
To Strood
Otford Junction
Sevenoaks
Existing No.2 Boat Train Route and
Tunnel Freight Wagon Train Route
Maidstone East
Lenham
To Canterbury
Tunnel Freightliner Route
Redhill
Nutfield
Bletchingley
Tunnel
Penshurst Tunnel
Tonbridge
Paddock Wood
Ashford
Dolland's Moor Freight Sidings
and Saltwood Junction
To Canterbury
DOVER
FOLKESTONE
To France

•••••• Tunnels
ooooo Channel Tunnel
☐ Locations requiring parliamentary powers

FIGURE 1: PHASE 1 FIXED WORKS

308

locomotives between Willesden and the Tunnel, is being
examined.

8.   At Dollands Moor, near the Tunnel entrance, a new
group of sidings will be built to permit the exchange of
locomotives. Wagon safety and other checks will be carried
out there before freight trains enter the Tunnel.

9.   The full list of the Phase 1 investment proposals is
shown at Table 5. The location of the various works is shown
at Figure 1.

10.   For passengers, the new inter-capital services will
provide a London/Paris journey time of 3 hours and
London/Brussels of 2 hours 40 minutes. The trains will be
able to take electric power from 25kV ac overhead supplies in
Britain and France and within the Tunnel, 750 dc third rail
in south-east England, and 1.5 and 3kV dc overhead supplies
in France and Belgium. Each train will be around 380 metres
long and carry up to 770 passengers. It is planned initially
to run 45 trains a day in each direction to and from Paris
and Brussels.

11.   The freight services being developed for the "core
system" under Phase 1 will extend throughout the whole
country. BR are required by Section 40 of the Channel Tunnel
Act to prepare by the end of 1989 a plan for the provision of
international freight services (and passenger services - see
Phase 2 below) to all parts of the UK.

12.   At the time of writing (May 1989), BR are in the
process of consulting local authorities and other interested
parties in the regions with a view to identifying
commercially viable opportunities, and it is too early to
draw any conclusions about the pattern of services which may
emerge. But for freight it is expected that the system will
offer wagon, container and swap body services to and from
points throughout the UK and mainland Europe. Examples of
expected railfreight transit times are shown in Table 6. New
wagon designs to facilitate carriage of higher and wider
loads within UK loading gauges are being developed. In
total, the number of freight services is expected quickly to
reach 27 trains a day in each direction. Each train will be
up to 750 metres long with a maximum payload of 1,600 tonnes.
In addition to the improved marshalling facilities at
Willesden, existing Freightliner depots - many of which
already have well established customs clearance facilities -
will be improved to increase capacity where necessary.
Generally, the existing BR network of depots, private
sidings, routes and internal services is expected to provide
the basis on which traffic volume can be expanded, but
proposals have already been made for additional depot
facilities constructed by partnerships between the private
sector, British Rail and local authorities.

Table 6. Railfreight transit times - examples

Holyhead	6	London	12	Paris
Birmingham	3	London	36	Vienna
Coatbridge	8	London	24	Munich
Glasgow	8	London	24	Frankfurt
Stockton	8 1/2	London	30	Milan
Leeds	5	London	24	Basel
Liverpool	4	London	48	Barcelona
Manchester	4	London		
Cardiff	4	London		

All times shown in hours to and from London eg Holyhead to
Paris transit time 18 hours.

Phase 2

13. BR is co-operating closely with Kent County Council
and Ashford District Council while it appraises plans for a
new international station at Ashford, which may include other
commercial developments.

14. This phase also includes the appraisal of passenger
services beyond London. As in the case of freight, the
commercial opportunities are being studied in the context of
the plan required under Section 40 of the Channel Tunnel Act.
Conclusions have not yet been reached, but one possibility
under consideration is the provision of daytime services on
the two electrified main lines to the North, with some
additional overnight services not necessarily confined to the
electrified network. New trains both for daytime and
overnight services would need to be provided but little
expenditure on physical facilities is envisaged as being
necessary. Indeed were such expenditure to be required to
any great extent, it is doubtful whether the provision of a
range of through services could be financially justified.

Phase 3

15. The third phase of BR's planning covers consideration
of additional line capacity across Kent and through London
and a second international passenger terminal in London.

16. The existing railway network - with the infrastructure improvements outlined above - will have adequate capacity to handle the volumes of traffic forecast by BR's consultants in the early years after the Tunnel opens. In later years, or with higher traffic levels than forecast by BR's consultants, and bearing in mind expected increases in domestic traffic over these routes, it is generally agreed that the existing network will not be able to cope fully, although forecasts vary as to the exact time when capacity constraints will become a problem.

17. Because the constraints on capacity are not confined to local bottlenecks but extend over significant lengths of the existing routes, BR have concluded that the best way of providing the additional capacity required is to build a completely new line. The proposed route, which will link the Tunnel with a new terminal at King's Cross, is shown in Figures 2 and 3. Waterloo will also be served by this new line.

18. The new line will be 68 miles long; 35% of it in tunnel, 32% in cutting, 6% on the level, and 27% on embankment. Two thirds of the surface route will use existing transport corridors - 16 miles alongside existing railway and 14 miles along the alignment of the M20. The estimated cost is £1.7bn, of which 30% is attributable to environmental protection measures.

19. It is too early to be firm about the timing of the new line's construction. BR will first need to get Parliament's approval to their proposals in the Bill which they hope to present in November 1989. This could take perhaps two years. Once work on the new line starts, it could take a further six years before it becomes operational. BR cannot however begin work until they have satisfied themselves and the Government that the investment will earn a proper commercial rate of return. There are still uncertainties about the traffic forecasts, the costs of the project, and the revenues. BR will want to consider proposals from the private sector for building and financing the line and to hear their views on timing. There can be no Government subsidy for the new line; this is ruled out by Section 42 of the Channel Tunnel Act.

ROADS

20. Improvements to the road links in East Kent would have been necessary to meet growing traffic needs whether or not the Tunnel project had gone ahead. Nevertheless, the Tunnel will have a considerable effect on the road network in East Kent, as a result of the redistribution of traffic reflecting the change in the centre of cross-Channel movements and because of the growth in traffic which the Tunnel itself is expected to generate.

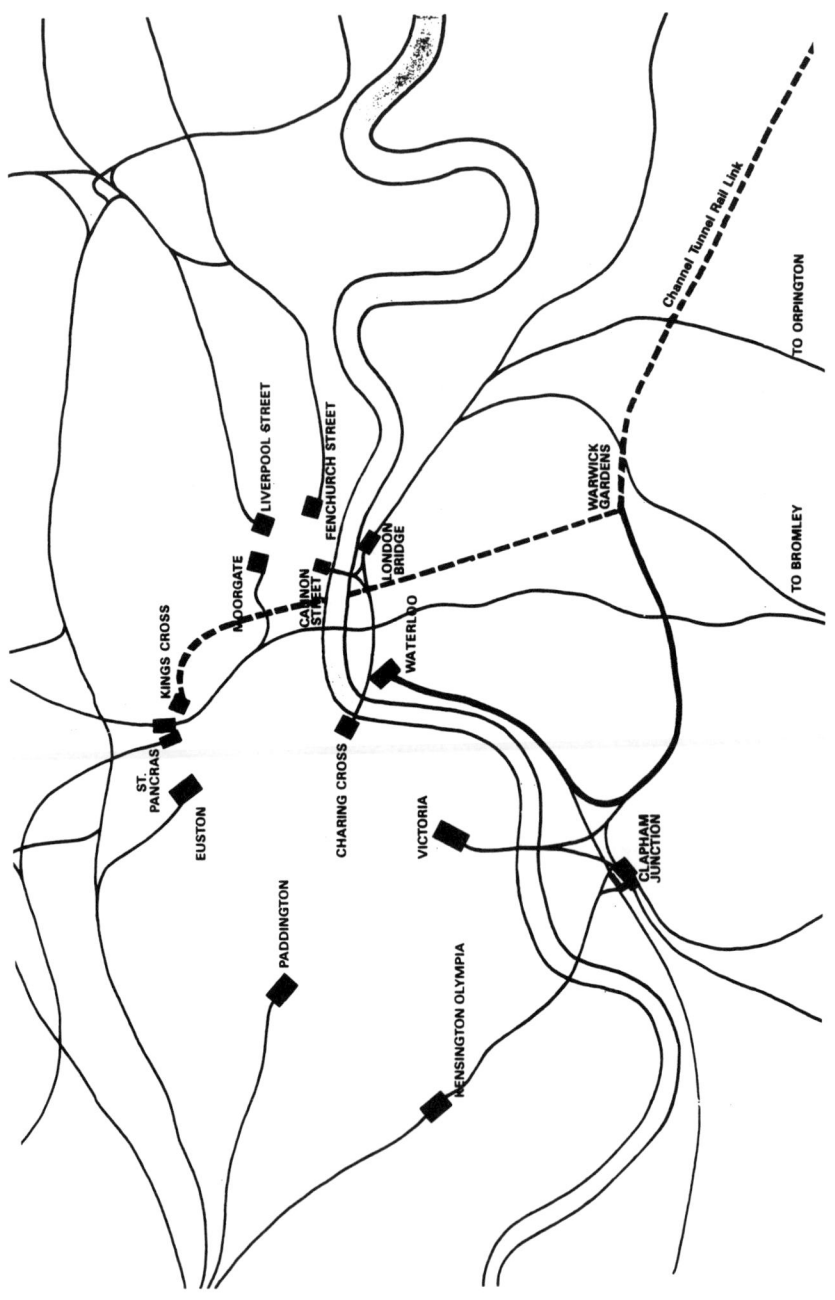

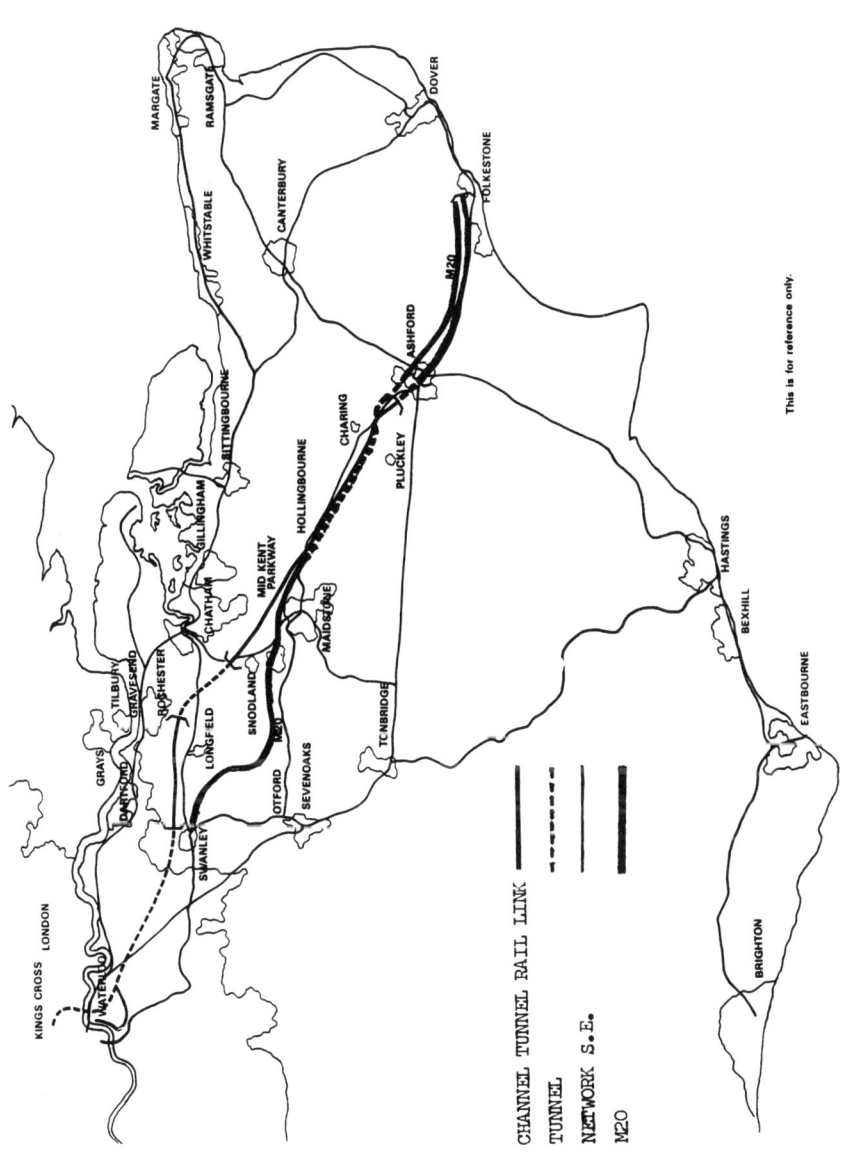

This is for reference only.

CHANNEL TUNNEL RAIL LINK

TUNNEL

NETWORK S.E.

M20

21. A joint working group, representing the Department of Transport, Kent County Council and Eurotunnel, examined in June 1986 the effect of the Tunnel on the expected traffic conditions in East Kent and the south eastern trunk road network. The Working Group agreed traffic data, taking into account Eurotunnel's forecasts, and combining Tunnel flows with those for the ferry operations to give total cross-Channel traffic with and without the Tunnel. Eurotunnel's forecasts have subsequently been revised in both 1987 and 1988. Table 7 sets out the results of a preliminary analysis of the cross-Channel traffic based on Eurotunnel's 1988 forecasts and on the group's earlier work on other flows. Even without the Tunnel, cross-Channel traffic would have been expected to rise between 1986 and the year 2008 by 75%, from 8,000 to 14,000 vehicles per day (vpd). The Tunnel is estimated to increase this by a further 6,000 vpd and to attract 8,000 vpd from the ferries. Total traffic is thus put at 20,000 vpd, of which 14,000 vpd will use the Tunnel.

Table 7. Existing and predicted cross-Channel traffic (December 1988)

Year	Facility	With Tunnel Two Way AADT flow vpd	Without Tunnel Two Way AADT flow vpd
1986	Dover and Folkestone Ferries	-	8,000*
1993	Dover Ferries	4,000)	11,000
	Tunnel	10,000) 14,000	-
2008	Dover Ferries	6,000)	14,000
	Tunnel	14,000) 20,000(21%HGVs)	-

Notes: 1 AADT flows represent total yearly traffic divided by 365.

2 Peak flows could be double these figures.

3 Assumes 20 train paths/hour in each direction - could increase to max of 30. About 50% through services, 50% shuttle trains.

4 *Includes 250 at Folkestone which is assumed to close in 1993 at Tunnel opening.

These are average flows; flows are expected to be about 50% higher during the peak summer months. Commercial traffic represents about 25% of the average flows.

22. It is estimated that 75% of the total Tunnel traffic will be destined for London and the south-east, although about half of the total heavy goods vehicle traffic will be destined for the north and west of the country.

23. To put the total cross-Channel road traffic flow into context, the figure of 20,000 vpd is considerably below traffic flows on the motorway system and less than one sixth of flows experienced on parts of the M25.

24. The main routes to the Tunnel are shown in Figure 4. The M20/A20 will carry the majority of main cross-Channel traffic rather than the M2/A2 as at present, and will provide a direct link to the M25 and the rest of the UK's motorway network. The "missing link" of the M20 between Maidstone and Ashford is currently under construction and will be completed by mid 1991. The Maidstone Bypass section of the M20 (junctions 5-8) will be widened to dual three lane, and junctions 3-5 widened to dual four lane standard. It is proposed to improve the A20 between Folkestone and Dover to dual two-lane standard on a new alignment to provide Dover with a high quality access from the M20. For both the Maidstone Bypass and A20 improvements, the Department of Transport is committed to try to complete the schemes by the time the Tunnel opens.

25. Other trunk road improvements relevant to cross-Channel traffic are being made to the A259 south coast road between Hythe and Hastings; the A261 between Hythe and Junction 11 of the M20; and the A249 between the A2 and M2.

26. Local roads in Kent require improvement to cope with the growth of traffic, including that to be generated by the Tunnel. Kent have a big programme of local road improvements, to the financing of which the Department of Transport is contributing through Transport Supplementary Grant. Only one of the 108 highway authorities is receiving more of that grant in 1989/90 than Kent. Among the improvements for which the grant is being provided are the A2070 linking Ashford to the A259, the A260 linking Folkestone to the A2, and the A256 between Dover and Sandwich.

27. Figures 5 and 6 show the forecast traffic flows on the road network for the year 2008. The extra 6,000 vpd generated by the Tunnel will diffuse widely in the network, resulting in about 2,000 vpd on the M20 between Junctions 1 and 2 and 2,000 vpd on the M26 (in each case around 5% of the total forecast traffic flow).

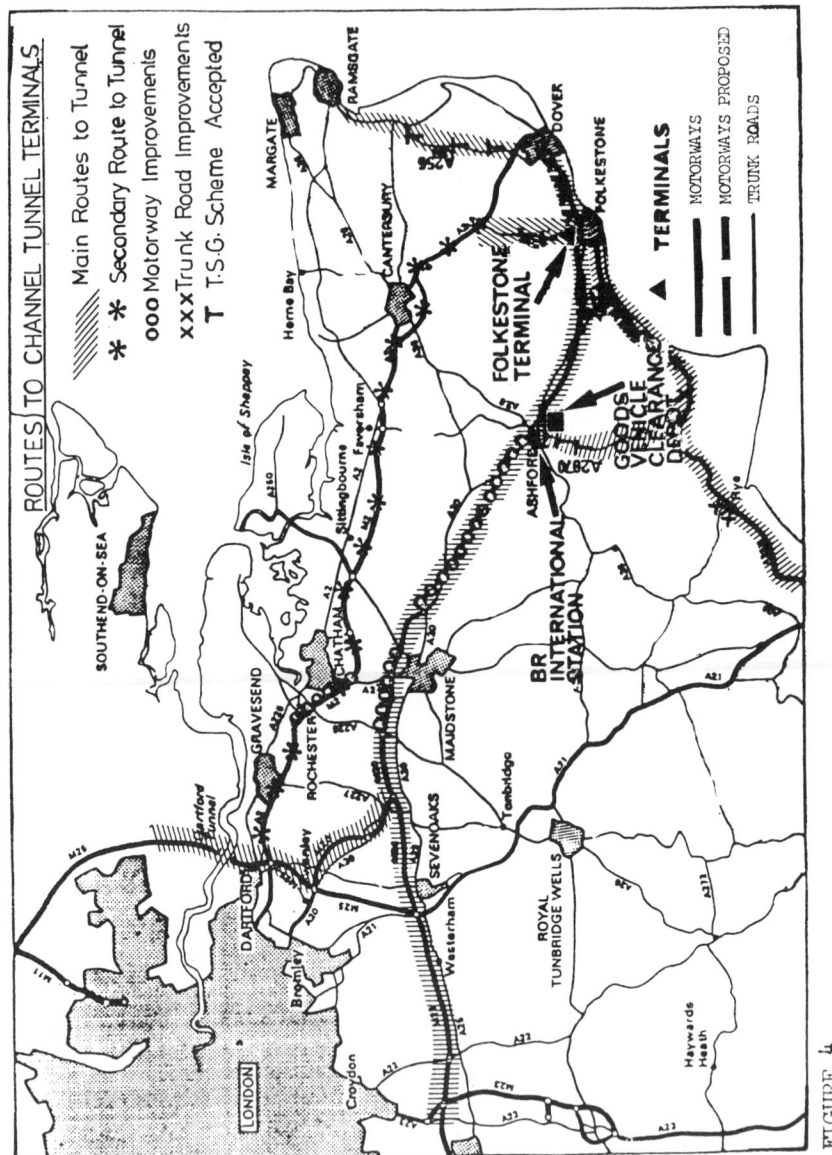

FIGURE 4

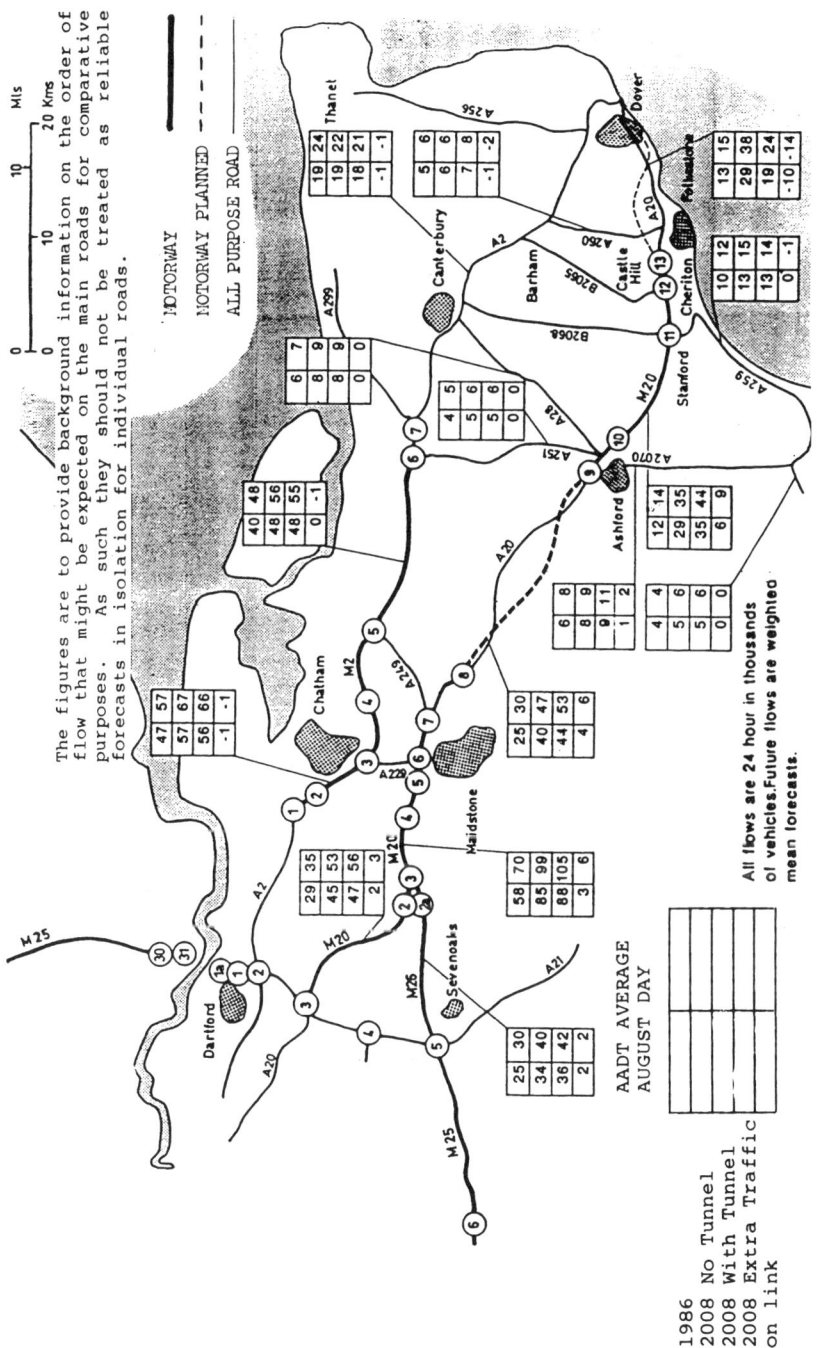

The figures are to provide background information on the order of flow that might be expected on the main roads for comparative purposes. As such they should not be treated as reliable forecasts in isolation for individual roads.

MOTORWAY

MOTORWAY PLANNED

ALL PURPOSE ROAD

All flows are 24 hour in thousands of vehicles. Future flows are weighted mean forecasts.

AADT AVERAGE
AUGUST DAY

1986 No Tunnel
2008 With Tunnel
2008 Extra Traffic
on link

317

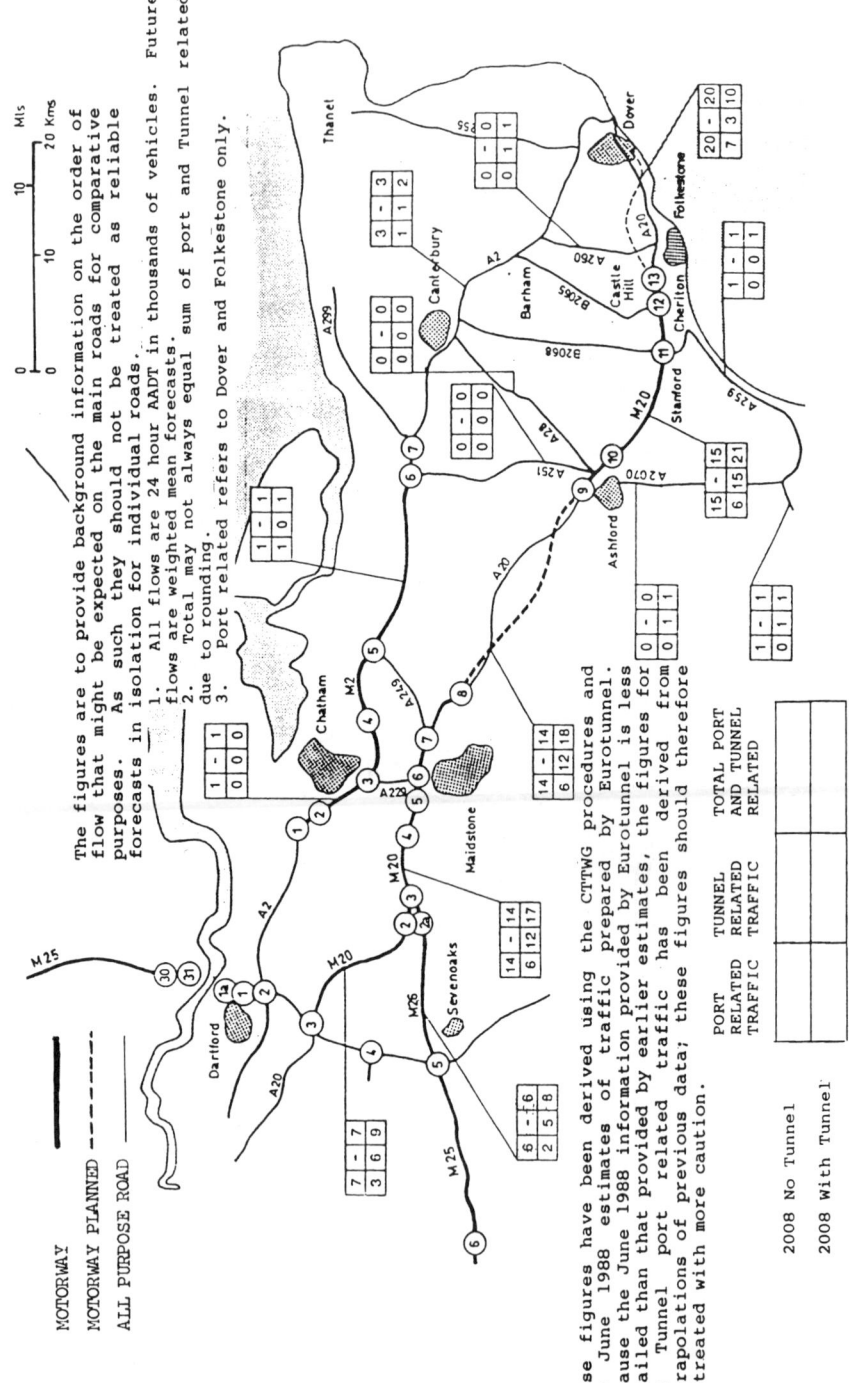

MOTORWAY
MOTORWAY PLANNED ‐ ‐ ‐ ‐ ‐
ALL PURPOSE ROAD

The figures are to provide background information on the order of flow that might be expected on the main roads for comparative purposes. As such they should not be treated as reliable forecasts in isolation for individual roads. Future

1. All flows are 24 hour AADT in thousands of vehicles.
   flows are weighted mean forecasts.
2. Total may not always equal sum of port and Tunnel related due to rounding.
3. Port related refers to Dover and Folkestone only.

These figures have been derived using the CTTWG procedures and the June 1988 estimates of traffic prepared by Eurotunnel. Because the June 1988 information provided by Eurotunnel is less detailed than that provided by earlier estimates, the figures for non Tunnel port related traffic has been derived from extrapolations of previous data; these figures should therefore be treated with more caution.

	PORT RELATED TRAFFIC	TUNNEL RELATED TRAFFIC	TOTAL PORT AND TUNNEL RELATED
2008 No Tunnel			
2008 With Tunnel			

318

28. The effects of the Tunnel were taken fully into account in the White Paper 'Roads for Prosperity' published earlier this year. Channel Tunnel related growth was one of the factors considered in drawing up the new National Road Traffic Forecasts announced in the White Paper; the most important such factor was traffic growth associated with general economic growth. Even if the increase in cross-Channel movements in 1993 is substantially greater than forecast by Eurotunnel, general traffic growth is likely to be much more significant.

29. Other programmed improvements will benefit Tunnel traffic and improve access to other parts of the country. These include the privately funded Dartford-Thurrock Crossing, widening of the three lane sections of the M25, the completion of the M40 to Birmingham and the M3 to Southampton, and widening and junction improvements on the M3, M4 and M23. Major improvements to the trunk road network in other parts of the country, especially to the M1, M6 and A1, will also benefit Tunnel traffic travelling north.

RESUME. Le Tunnel sous la Manche apportera à l'usager une possibilité de choix supplémentaire et il intensifiera la concurrence sur le marché transmanche. Il détournera une partie du trafic des autres modes de transport, et il créera un trafic supplémentaire.

1.   Pour British Rail, le Tunnel est la plus belle chance d'expansion depuis des années. Les projets de BR se développent en 3 phases. La Phase 1 concerne l'infrastructure nécessaire pour le système de base; les liaisons passagers Londres-Paris et Londres-Bruxelles et les nouveaux services de fret internationaux. Un investissement de plus de 600 millions de livres a été approuvé pour l'infrastructure et le matériel roulant. BR est en train de consulter les autorités locales et les autres groupes concernés en vue d'identifier les possibilités commerciales viables pour les services de fret. La Phase 2 concerne l'éventualité d'une nouvelle gare internationale de voyageurs à Ashford, et l'étude d'un éventual développement du trafic passager au delà de Londres. La Phase 3 concerne l'étude d'un accroissement du réseau ferroviaire entre le Tunnel et Londres, et d'un second terminus passager international à Londres.

2.   Le réseau existant aura la capacité voulue pour absorber le trafic prévu à l'ouverture du Tunnel mais on reconnaît que le réseau deviendra insuffisant à un certain moment - quant à prévoir à quel moment il se rélèvera insuffisant, les opinions diffèrent. BR a décidé que le meilleur moyen de fournir la capacité supplémentaire voulue est de construire une nouvelle ligne. Il est trop tôt pour donner avec certitude la date de sa construction.

3.   Quant aux routes, des améliorations du réseau dans l'est du Kent auraient été nécessaires pour faire face aux besoins croissants du trafic, que le Tunnel soit réalisé ou non. Le Tunnel aura un effet sur le réseau routier du fait de la redistribution du trafic entre les ferries et le Tunnel et du fait de l'augmentation de trafic que le Tunnel engendrera. La M20/A20 recevra la majeure partie du trafic transmanche que la M2/A2 reçoit à présent. On achève et élargit la M20 qui fournira une liaison directe avec la M25 et avec le reste du réseau autoroutier du RU. Est aussi en projet l'amélioration d'un certain nombre de routes nationales et départementales.

4.   En dehors du Kent, le Tunnel n'aura pas d'effet direct sur le réseau routier mais les prévisions nationales prennent en compte les besoins nouveaux créés par le Tunnel, et il y a de nombreux projets routiers qui faciliteront l'accès au Tunnel.

17. Les infrastructures d'accompagnement

M. FRYBOURG, Ingénieur des Ponts et Chaussées, Professeur Associé au Conservatoire National des Arts et Métiers

SYNOPSIS. Les infrastructures d'accompagnement du tunnel sous la Manche figurent dans des contrats Etat/Région intéressant le fer, la route et les infrastructures portuaires. Elles se justifient par la nécessité de raisonner au niveau du système de transport, par l'effet de concentration du trafic et le souci de valorisation au profit des régions concernées. Il ressort de cette présentation que l'image du "maillon manquant" est trompeuse car le tunnel n'a de sens que s'il fait parti d'un tout à l'intérieur duquel il n'est même plus majoritaire.

AVANT PROPOS

1. Le terme même d'infrastructure d'accompagnement est ambigu. Il est certes possible de lui donner une signification administrative puisqu'il existe un plan routier et ferroviaire Transmanche et que l'on peut en déduire que les infrastructures inclues dans ces plans sont les infrastructures d'accompagnement mais une telle definition part d'une vision politique dont la légitimité tient certes au processus de décision mais non aux effets socio-économiques des dits investissements. Il est donc utile de préciser les critères d'évaluation et de s'efforcer de relier les choix aux effets attendus.

LES CRITERES D'EVALUATION

2. Parler des infrastructures d'accompagnement, c'est faire référence à trois réalités:
     -la première tient au concept même de système de transport qui rattache à un objectif d'amélioration de qualité de service les composants en interaction du système à savoir: les infrastructures, les véhicules et les techniques d'exploitation;
     -la deuxième résulte de l'effet de concentration du trafic que toute infrastructure majeure ne manquera pas d'avoir et
     -la troisième se rattache aux effets externes positifs de tout investissement structurant qu'il faut d'abord amplifier ensuite valoriser au profit des régions concernées.

Le système de transport

3.   Le grand public s'est avant tout attaché au grand
ouvrage que constituerait inévitablement le lien fixe et le
débat sur le tunnel ou le pont reste très présent à la
mémoire de tous. Il n'est d'ailleurs pas clos car une fois
l'amortissement du tunnel garanti il restera à faire face aux
perspectives d'accroissement du trafic. Faire un trou dans
la craie devient une aventure lorsque ce trou a 7,20 m de
diamètre, 50 km de long et se situe sous la mer. Mêmes
soigneusement évalués et circonscrits les aléas ne peuvent
disparaître totalement et les risques, notamment de surcoût
sont loin d'être nuls avec les enjeux financiers et risques
humains qui en résultent.

4.   Sans vouloir ramener le forage du tunnel à un chantier
ordinaire de travaux publics, il serait cependant très
imprudent de focaliser sur la seule difficulté du chantier
les problèmes à résoudre. Le même tunnel plus ou moins bien
inséré dans le système de transport peut conduire à des
résultats significativement différents pour les usagers.
Paris-Londres en 2 h 1/2, 3 h 05 ou 4 h 1/2, ce n'est pas la
même chose pour le choix modal entre l'avion et le train et
pour le trafic total de passagers. Ce seul critère de temps
de parcours doit, de plus, être enrichi par les
caractéristiques de confort et de ponctualité de la liaison.

5.   Les automobilistes vivront avec plus ou moins de
satisfaction un ensemble complexe de prestations allant de
l'accès autoroutier aux formalités et délais d'accès à la
navette pour ensuite subir ou déguster une traversée qui
restera un temps fort du voyage. Quant aux chargeurs des
marchandises, ils ne connaîtront que le résultat en terme de
niveau de service et n'attacheront aucun plus à l'utilisation
d'un ouvrage exceptionnel de génie civil.

6.   Sans doute le type de financement impliquant la prise
de risque par le seul propriétaire Eurotunnel a-t-il eu pour
conséquence d'occulter cette co-responsabilité de fait sur le
système de transport qu'implique le recours à des
infrastructures d'accompagnement. Il faut cependant prendre
conscience que le coût de ces infrastructures est du même
ordre que celui du tunnel à cette réserve près que sa
justification économique ne repose pas sur le seul trafic du
tunnel.

La concentration du trafic

7.   L'effet d'entonnoir du tunnel est le plus évident. Il
reste à le chiffrer et à le valoriser en apportant aux
usagers tous les services annexes qu'ils peuvent attendre.
        -Pour les voyageurs, les services relèvent de la
nécessité, du confort ou de l'agrément, du transit ou du
tourisme. Il revient aux régions traversées d'encourager le
passage de l'utile à l'agréable. L'incidence sur la
fréquentation du tunnel de l'enrichissement des prestations

est certes impossible à chiffrer mais probablement non négligeable.

    -Pour les marchandises, il faut parler des prestations logistiques, incluant les opérations de groupage, empotage, stockage et le traitement de l'information nécessaire à la consolidation des envois et au suivi des marchandises.

L'aménagement régional

8.   Les infrastructures d'accompagnement conçues dans une optique d'aménagement sont les plus difficiles à planifier car il faut répondre à plusieurs questions:

    -la première porte sur la zone d'influence du tunnel et donc les limites de l'entonnoir;

    -la deuxième sur le réel effet d'entraînement et le bénéfice que peuvent tirer les régions traversées d'un trafic qui ne leur est pas, à priori, destiné;

    -la troisième porte sur l'effet de seuil à partir duquel un équipement souhaité avant le tunnel devient économiquement justifié grâce au tunnel et non le tunnel un prétexte pour justifier l'équipement et

    -la quatrième sur l'effet positif ou non d'une plus grande perméabilité de la région qui peut se traduire par une perte d'autonomie et une plus grande dépendance.

LE PLAN FERROVIAIRE: LE T.G.V. NORD ET LE CONTRAT ETAT/REGION
Le T.G.V. Nord

9.   Le 9 octobre 1987, le gouvernement a décidé la réalisation du TGV Nord et de la ligne d'interconnexion nord-sud en Ile-de-France via l'aéroport de Roissy - Charles de Gaulle, au même horizon que le tunnel sous la Manche (printemps 1993), ainsi que celle du prolongement de la ligne TGV Sud-Est jusqu'à Valence "le plus tôt possible".

10.   Le TGV Nord, synthèse de la partie française du projet Paris - Bruxelles - Cologne/Amsterdam et de la liaison avec le tunnel sous la Manche, est un réseau de 322 km de lignes nouvelles exploitées à 300 km/h de vitesse maximale qui reliera Paris au nord de la France (Paris - Lille en 1 h 00), à la Belgique (Paris - Bruxelles en 1 h 20) et, au delà, aux Pays-Bas, à l'Allemagne et à la Grande-Bretagne (Paris - Londres en 3 h 00), mais aussi le Tunnel à la Belgique (Londres - Bruxelles en 2 h 35). A l'horizon 1993, le trafic prévu s'élèverait à 31 millions de voyageurs et assurerait une rentabilité interne de 13% en francs constants à cette infrastructure dont le coût est estimé à 12 milliards de francs 1985.

11.   Le 26 octobre 1987, les ministres des transports de la Belgique, de la France, de la RFA et du Royaume-Uni, en présence de leur collègue luxembourgeois, ont confirmé leur volonté de réaliser le projet de liaisons ferroviaires rapides entre Amsterdam, Bruxelles, Cologne, Paris et Londres, via le tunnel sous la Manche, ce project est désigné par le sigle PBKAL. Les entrées et les traversées des villes

seront effectuées sur les lignes existantes. Les gares de
Paris-Nord, Bruxelles-Midi et/ou Bruxelles-Nord et de Londres
(Waterloo) seront aménagées et développées de façon à offrir
les capacités ferroviaires nécessaires ainsi que les
meilleures conditions d'exploitation possibles.

12. La mise en exploitation d'un réseau à grande vitesse a
deux effets sur le trafic:
      -un premier effet "statique", appréhendé par le modèle
économétrique utilisé et
      -un deuxième effet "dynamique" qui caractérise la
phase de pénétration du produit "grande vitesse", telle
qu'elle ressort de l'exemple du TGV Sud-Est. Cet effet se
traduit durant les premières années d'exploitation par une
croissance plus rapide que la croissance "naturelle"
constatée par ailleurs. Comme dans la vie de tout produit
marchand, cet effet s'atténue progressivement au fil des
années.

Le contrat Etat/Région
13. Le passage du TGV à l'intérieur de Lille sera
l'occasion d'une importante opération d'urbanisme autour de
la future gare. Ainsi la ville de Lille a-t-elle décidé de
créer un centre international d'affaires qui se développera
entre la gare actuelle et celle qui, à 400 m de là, doit être
construite pour acceuillir le TGV. Le surcoût du passage du
TGV dans Lille est estimé à 800 MF. Il sera pris en charge
par l'Etat et la SNCF à hauteur de 50% et par le conseil
régional à hauteur de 33%. Le solde, soit 17%, est à
répartir entre les autres collectivités.

14. Le conseil régional a, de plus, demandé et obtenu un
programme de raccordement et d'électrification régional
incluant l'électrification des lignes Calais-Boulogne,
Cambrai-Douai et Lille-Baisieux (Tournai) et les branchements
de la voie classique sur le TGV à hauteur de Cassel et au
nord d'Arras. Le montant d'investissement de ce programme
est de 519 MF, à répartir par tiers entre l'Etat, la Région
et la SNCF. Il deviendra ainsi possible de brancher la ligne
Amiens-Lille sur le TGV et de mettre Amiens à 1 h 40 de
Bruxelles et Arras à 1 h 05 de Bruxelles et 2 h 35 de
Londres. Quant à Dunkerque, il sera à 2 h 00 ou 1 h 30 de
Paris, selon que l'on utilisera le débranchement à Arras ou à
Cassel.

15. Le projet PBKA a clairement montré la voie pour la
mise en place d'un réseau à grande vitesse intégré. Les
cahiers des charges établis pour le matériel à grande vitesse
à gabarit britannique appelé à emprunter le tunnel sous la
Manche, d'une part, et pour le matériel continental PBKA,
d'autre part, illustrent bien le type de démarches à
entreprendre pour réaliser une véritable Communauté
européenne des chemins de fer. Il faudra poursuivre cet
effort par la définition technique et commerciale de

matériels à grande vitesse de nuit et l'harmonisation des systèmes de signalisation et de télécommunications pour les besoins de l'exploitation technique et commerciale (l'EDI des chemins de fer).

LE PLAN ROUTIER TRANSMANCHE

16. Le plan routier Transmanche a été lancé en janvier 1986 et a donné lieu depuis à trois avenants. 885 MF ont été consacrés au plan routier au cours des trois années 86/87/88. Ce plan se fixe trois objectifs: assurer l'acceuil du lien fixe, créer un corridor sud-ouest et désenclaver les ports de Rouen et du Havre. Ce plan s'insère dans le nouveau Schéma Directeur Routier National qui, après une large consultation régionale, était définitivement approuvé par décret le 18 mars 1988. Parmi les principaux objectifs de ce plan national figurait la meilleure liaison des ports de notre façade ouest aux régions européennes les plus dynamiques et la création de grandes transversales évitant Paris et irriguant l'ouest du territoire. Il y avait donc cohérence avec le plan Transmanche.

Assurer l'acceuil du lien fixe

17. Les principales liaisons appelées à diffuser le trafic empruntant le futur tunnel sont:

-du nord de la France vers la région parisienne, l'est et le sud-est; l'autoroute A 26, Calais Reims, achevé en 1989, l'autoroute A 1, Paris-Lille, dont l'élargissement à deux fois trois voies est terminé, l'autoroute A 16, Paris-Amiens-Boulogne, dont la mise en service sera coordonnée avec celle du lien fixe;

-vers la Belgique et l'Europe du nord, l'axe Calais-Dunkerque-Frontière belge, aménagé en véritable autoroute sans péage (RN 1);

-vers l'ouest et le sud'ouest, la RN 1 et l'A 16 vers Abbeville puis Amiens ou Rouen par la RN 28.

Créer un corridor sud-ouest

18. L'un des objectifs du plan routier de 1986 était d'offrir au trafic généré par le lien fixe et se dirigeant vers la Péninsule Ibérique, un itinéraire qui contournerait la Capitale par l'Ouest. L'aménagement en véritable autoroute, libre de péage, de la RN 28, de Rouen à Abbeville, retenu au Schéma Directeur, permet d'atteindre ce résultat, Rouen étant relié en autoroute à péages à Alençon, Le Mans et Tour.

Désenclaver les Ports de Rouen et du Havre

19. Une étude, effectuée par les deux ports autonomes, évaluait à 2 millions de tonnes par an le trafic provenant des départements du Nord, de la Somme, des Ardennes, de l'Oise et de l'Aisne, capté par les ports belges et hollandais du fait de l'insuffisance de la desserte du Havre. Avec la création du lien fixe, les côtes anglaises seront reliées par la route au continent, ce qui va bouleverser les

données du marché du transport maritime, en facilitant
l'utilisation des ports continentaux par les usagers du
Royaume-Uni et les ports britanniques par les usagers du
continent. Améliorer la desserte des ports de la Manche,
c'est moderniser les axes Le Havre-Rouen-Abbeville-Calais et
Le Havre-Amiens-Saint-Quentin. La A 29 Le
Havre-Neufchatel-Amiens-Saint-Quentin complétera la desserte
terrestre de Rouen et du Havre en les reliant au tunnel et à
l'Est.

LES INFRASTRUCTURES PORTUAIRES

20. La construction du tunnel est, à l'évidence, de nature
à modifier les équilibres économiques de la filière maritime.
Déjà frappés par la quasi-disparition de la construction
navale, les ports du Nord Pas-de-Calais riquent en outre de
voir leurs activités de transit maritime bouleversées. Or le
transport maritime continuera à jouer un rôle essentiel,
d'abord en évitant le monopole naturel que le tunnel pourrait
constituer si la concurrence potentielle du maritime n'en
faisait pas un marché "disputable", ensuite en assurant le
maintien et le développement de la diversification des
activités portuaires.

La concurrence potentielle

21. Les activités à rendement croissant comme celles
engendrées par le tunnel, dont les coûts fixes sont très
importants et les coûts variables relativement faibles, ne
peuvent en principe fonctionner selon les principes de la
concurrence. Les économistes s'accordent cependant sur la
faible probabilité d'une guerre de prix qui justifierait une
réglementation. Il n'en reste pas moins indispensable de
compter sur une "force de dissuasion" à l'égard du monopole.
Il a même été imaginé de faire financer par les usagers du
tunnel, sous forme d'une taxe specifique qui serait levée par
les régions portuaires, le maintien en état de marche des
ports adaptés à des transports par transbordeurs.

22. De 1980 à 2000, la technologie du transport par
transbordeurs évolue profondément, par le passage à des
unités plus grosses (600 UVP) voire plus rapide (5 rotations
par jour en période pleine). Pour les touristes voyageant
avec leurs voitures le gain de temps n'est pas un élément
déterminant du choix modal et les salons et boutiques d'un
car-ferry se comparent avantageusement à l'intérieur d'une
voiture immobilisée dans une navette. Les marchandises
dangereuses et hors-gabarit continueront, de plus, à
emprunter la voie maritime. Il y a donc tout lieu de penser
qu'un trafic par mer, sans doute réduit de moitié par rapport
au trafic actuel (plus pour les passagers, moins pour les
marchandises), se maintiendra sans taxe spécifique en plus
des trafics transversaux provenant des ports normands.

Redéploiement et diversification des activités portuaires

23. Le développement d'activiés portuaires diversifiées et
la valorisation des chances des ports français dans

l'acheminement du trafic transocéanique à destination et en provenance de Grande-Bretagne seront favorisés. Calais a entrepris la réalisation d'un nouveau port en eau profonde à l'est qui, dès l'automne 89, sera capable d'acceuillir simultanément trois cargos de type "Panamax". L'ensemble des travaux du port à l'est représentera un investissement total d'environ 500 millions de francs. A cet investissement sur Calais il faut ajouter le prolongement du quai à conteneurs de Dunkerque selon un calendrier fonction de l'évolution du trafic, la restructuration des installations de réception et de traitement de la pêche sur la zone de Capécure à Boulogne et l'aménagement d'un nouvel avant-port à Dieppe.

LA VALORISATION DES INFRASTRUCTURES DE TRANSPORT
Les effets à court et long terme
24. Les effets de la mise en service d'une nouvelle infrastructure de transport s'analysent en distinguant les effets directs à court terme résultant d'abord des dépenses de contruction, ensuite de l'amélioration du niveau de service et les effets socio-économiques à moyen et long terme conduisant à la re-localisation des activités at au développement de certaines d'entre elles. Une idée trop fréquemment répandue veut qu'une nouvelle offre de transport favorise automatiquement le développement des régions desservies or si cette condition peut apparaître fréquemment comme une condition nécessaire au développement régional, elle n'est pas une condition suffisante d'ou la nécessité de mesures d'accompagnement.

Les infrastructures d'accompagnement
25. Il est primordial de concentrer les énergies sur quelques objectifs précis, les ressources financières étant toujours limitées et d'éviter le saupoudrage. Ces mesures d'accompagnement se rattachent, pour l'essentiel, à des réalisations d'infrastructures puisqu'elles portent sur la création de zones d'activités, notamment touristiques, impliquant le montage d'opérations d'urbanisme; la réalisation de plate-formes multimodales de traitement des marchandises avec les infrastructures de rabattement associées et la modernisation des terminaux de voyageurs dont les gares SNCF. Les aides financières ou fiscales et la mise en oeuvre des outils de planification spatiale ne relèvent pas directement des infrastructures mais les interactions sont évidentes.

L'aménagement du territoire
26. Les outils classiques de l'aménagement du territoire concernent essentiellement la maîtrise foncière par la collectivité publique et la planification spatiale. La maîtrise foncière implique la procédure d'expropriation et notamment la Déclaration d'Utilité Publique et la création d'une Zone d'aménagement différé pour éviter la spéculation foncière. Les outils de planification spatiale incluent le

Schéma Directeur, le Plan d'Occupation des Sols ou P.O.S. et la Zone d'Aménagement Concerté ou Z.A.C. Il faut ajouter à ces outils classiques les procédures exceptionnelles liées au caractère de grand projet.

27. L'inventaire complet des mesures de valorisation serait à la fois fastidieux à développer et incomplet car les études sont loin d'être achevées et de nouvelles initiatives verront le jour relevant de nombreux acteurs publics et privés. Il suffit de constater la prise de conscience de la Région Nord - Pas-de-Calais du parti qu'elle pourra tirer de sa situation au centre d'un vaste réseau de communications internationales reliant Londres, Paris, Bruxelles, Amsterdam, Cologne, Hambourg etc. Il s'agit pour elle, non seulement de minimiser les effets négatifs du tunnel, mais de tirer un maximum de profits de cette opportunité exceptionnelle. En d'autres termes, la Région ne veut pas se borner à "regarder passer les trains". Il reste à être capable de mieux cerner les effets des infrastructures.

LA RECHERCHE D'UN OPTIMUM SOCIO-ECONOMIQUE

28. Un catalogue fait toujours craindre non seulement l'ennui de toute énumération mais encore un effet d'entraînement par crainte d'exclusion. Il ne faut oublier ni la route, ni le fer, ni la voie maritime et, bien entendu ne pas ignorer le transport aérien qui reste un concurrent dominant fort dynamique. Le système de transport est d'envergure internationale, non seulement européen mais même intercontinentale si l'on tient compte du rôle de feeder que peut jouer la redistribution du trafic entre les terminaux de part et d'autre de la Manche. Les infrastructures doivent donc être conçues en conséquences. Mais les Régions et les ports côtiers ne veulent pas être sacrifiées à des ambitions de croissance globale des échanges. Et, si l'aménagement regional doit être pris en compte, il ne doit pas se limiter aux seules régions du Kent et du Nord - Pas-de-Calais. Le Bénélux ne veut pas être oublié et pas davantage la Picardie et notamment Amiens.

29. Dans ces conditions, comment se protéger contre le sur-investissement? Il ressort clairement de cet exposé que l'on se trouve devant une situation paradoxale. D'abord plus d'un siècle de tergiversation avant d'entreprendre la réalisation du tunnel. Pendant cette période une montagne d'études intéressant non seulement le génie civil mais encore l'économie du projet. Econométres, à vos modèles! il y a de quoi se faire plaisir. La décision est prise de lancer la consultation en vue d'une maîtrise d'ouvrage privée et c'est le paroxysme des études de génie civil et économico-financières. La concession est signée et dans les deux ans qui suivent un volume d'investissements, dits d'accompagnement, est décidé, d'un montant supérieur à celui du tunnel, avec des études économétriques qui ont échappés pour l'essentiel, à la perspicacité du présent auteur!

30. Il ne faut pas en déduire que ces investissements sont mal fondés mais que l'image du "maillon manquant" est trompeuse si elle laisse supposer qu'il suffit de réaliser ce maillon pour mettre les infrastructures à niveau. Un système de transport forme un tout et son insertion spatiale est un élément indissociable de son utilité qui ne peut s'évaluer à partir de la seule satisfaction des usagers en dehors de toute considération d'effets externes. Le caractère de "jamais vu" de l'ouvrage de génie civil et le montage financier spécifique a pu masquer auprès de l'opinion l'aspect global du projet. Mais la réalité de sa rentabilité, lié à la performance de bout en bout et la sensibilité des collectivités locales ont rapidement remis les montres à l'heure et fait apparaître le projet comme un élément d'un tout indissociable qui inclut **des** infrastructures et non le seul tunnel, **des** techniques d'exploitation et des opérations de valorisation.

CONCLUSION

31. Un ouvrage exemplaire justifie des moyens exemplaires de valorisation et de suivi. Pour les économistes du transport, la réalisation de cet ouvrage exceptionnel est une chance à saisir pour améliorer nos connaissances sur les effets des infrastructures et les possibilités de les valoriser. Il serait regrettable que cette préoccupation d'amélioration des connaissances ne soient pas prise en compte car il en résultera une meilleure efficacité des investissements futurs.

32. Le montage d'un observatoire régional est en cours, tout au moins en France, mais les moyens, pourtant modestes, d'harmonisation du receuil des données internationales et de constitution d'une base de référence pour évaluer les effets de la mise en service de l'ouvrage, sont difficiles à rassembler. La dynamique de marché peut faire illusion sur la possibilité de se passer de planification. L'ampleur des infrastructures d'accompagnement, financées par les collectivités publiques, devrait faire perdre l'essentiel de ces illusions qui peuvent coûter très cher.

SUMMARY. The accompanying infrastructures to the Channel Tunnel appear in the State/Region contracts concerning railways, roads and portal infrastructures.

1.   To talk about accompanying infrastructures is to refer to three facts:
-the first results from the concept of the transport system itself, which relates the system's interactive components to the objective of improving the quality of service, i.e. infrastructures, vehicles and operating techniques;
-the second results from the effect of a concentration of traffic that any major infrastructure would not be without;
-the third is related to the positive external effects of any structural investment, which first of all need to be amplified and then developed to the profit of the regions concerned.

2.   The Government has decided that the creation of the TGV north and the interconnecting north/south route at Ile de France via Roissy-Charles de Gaulle airport will coincide with the Channel Tunnel (Spring 1993).

3.   The TGV route through Lille will provide the opportunity for an important urban development around the future station. The town of Lille has therefore decided to create an international business centre, to be developped between the existing station and the station that will be built 400m away for the TGV.

4.   The principal road links designated to distribute the future tunnel traffic are:
-from the north of France to the Parisian region, the east and south-east: the A26 Calais-Reims motorway, completed in 1989, the A1 Paris-Lille motorway, where expansion to three lanes in both directions is now finished, the A16 Paris-Amiens-Boulogne motorway, which will be opened to coincide with the Fixed Link;
-to Belgium and northern Europe; the Calais-Dunkerque-Belgian frontier axis, converted into a proper motorway without a toll (RN1);
-to the west and south-west, the RN1 and A16 to Abbeville, then Amiens or Rouen via the RN28.

5.   One of the objectives of the 1986 road plan for traffic generated by the Fixed Link and heading for the Iberian peninsular, was the provision of a route that bypassed the capital to the west. The conversion of the RN28 from Rouen to Abbeville into a proper motorway without a toll, retained in the main outline, would produce this result, Rouen being linked by toll motorways to Alençon, Le Mans and Tour.

6.   An all too common idea is that a new transport project automatically favours the development of the regions it serves. Now, if this condition frequently appears to be a necessary condition for regional development, it is not sufficient in itself, hence the need for accompanying measures. The regions and coastal ports do not want to be sacrificed to the ambitions of an increase in global trade and if regional development must be taken into account, it should not be limited solely to Kent and the Nord-Pas de Calais. Benelux does not want to be forgotten, any more than Picardie, and notably Amiens.

7.   As financial resources are always limited, and to avoid their fragmentation, it is essential that energies are concentrated on a few precise objectives. These accompanying measures focus on the creation of activity zones, notably tourist, and involve setting up urban projects; the creation of multi-modal goods platforms with the associated infrastructures and the modernisation of passenger terminals, including SNCF stations. What emerges from this presentation is that the "missing link" image is misleading, because the tunnel only has any meaning if it is part of a whole, within which it is no longer even dominant.